Excel

Get the Results You Want!

SmartStudy 9

Science

Jim Stamell
& Geoffrey Thickett

PASCAL PRESS

Reprinted 2016, 2018, 2020, 2022, 2023, 2025

ISBN 978 1 74125 422 8

Pascal Press
PO Box 250
Glebe NSW 2037
www.pascalpress.com.au

Publisher: Vivienne Joannou
Project editor: Leanne Poll
Edited by Karen Pearce
Indexed by Mary Coe
Typeset by Precision Typesetting (Barbara Nilsson)
Cover and page design by DiZign Pty Ltd
Printed by Vivar Printing/Green Giant Press
Photos by Dreamstime and iStockphoto. Diagram on page 108 from Bush, S, Dickson, A, Harman, J and Anderson, J 1999, *Australian Energy: Market Developments and Projections to 2014-15*, ABARE Research Report 99.4, Canberra, p. 26. Data for diagrams on page 194 from Australian Bureau of Statistics, cat. no. 3101.0.

Students
All care has been taken in compiling this study guide, but please check with your teacher about the exact requirements of the course as these can change from year to year.

TABLE OF CONTENTS

TAKE THESE REVISION STEPS TO SUCCESS!

Step 1 Quick Revision

- Check that you know the **key points** in each topic that you are studying by filling in the missing words in this section.
- Once you have completed the revision, **mark your work** by looking at the answers at the bottom of the page. This is **instant feedback** for you.

Step 2 Revision Summaries

- If you got any of the answers wrong in the **Quick Revision** section you can revise by reading the corresponding point in the **Revision Summaries** section. The numbered points in the **Quick Revision** section correspond to the numbered points in the **Revision Summaries** section.
- If you got all the answers right in the **Quick Revision** section then just skim over this section to be 100% sure you know the material.
- The **checklist** at the end of this section will help you **double-check** the essential information that you should know from this section.

Step 3 Revision Test

- These are test-style questions that you should be able to answer in order to prepare for the class test or common test. In this section you **recall** and **apply your knowledge**. Make sure you have **fully revised** your work in the **Revision Summaries** section.
- **Hints** for some questions are provided in case you need extra help with them. They are found at the bottom of the page.
- **Marks** are allocated for each question. Always check the marks and use them as a guide for **how much to write** in your answer.
- Time yourself—**check** how much **time** you have got to complete the test. Also look at the **total marks** of the test to calculate approximately how much time you should spend on each question. For example, if there are twenty marks in total and twenty minutes have been allocated for completion of the test, then spend about one minute on a one-mark question and six minutes on a six-mark question. If you cannot complete all the questions within the suggested time, you may need to revise the topic.
- Fill in the **Your Feedback** panel once you have marked your work in order to calculate your percentage mark. Then complete the **Test & Exam Results** page to keep a running total of all your test marks.

Step 4 Check Your Answers

- **Complete answers** to the **Revision Test** and **Sample Exam Paper** questions are also found at the back of the book. When you have finished a test, turn to this section to check the answers. Make a note of any questions that you got wrong.
- The **ticks** that appear in the answers indicate those parts of the answer which receive marks. Therefore, even if your answer is incomplete, you may be entitled to some marks for what you have written. Compare the answer to your own answer to find out whether you are entitled to any marks for the question.
- If you still **cannot understand** how the correct answer was obtained, revise that part of the topic and, if necessary, refer to your class notes or ask your teacher for assistance. It is important to learn from your mistakes. Remember: the questions in this book are very similar to the ones you will get in your examination. Using a pencil, write down the marks you get for each question, then add them up to get a total.
- Remember that there is an ***Excel*** *Year 9 Science Study Guide* available to help you revise further.

Step 5 Test & Exam Results

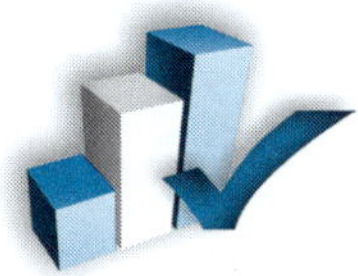

- Go to the **Test & Exam Results** section on page 252 to record your test score as a percentage. When you have completed all topics you will be able to determine your areas of weakness and your areas of strength.
- It is important that you know which **areas need further work**—to 'know what you don't know'. The more you prepare for the topic tests, the more successful you will be and the more you will remember when you sit your end-of-term/semester/year test or exam.

Step 6 Sample Exam Papers

- There are two Sample Exam Papers in this book. Sample Exam Paper 1 is similar to a **half-yearly test** and covers the Biological Sciences and Chemical Sciences strands. Sample Exam Paper 2 is similar to an **end-of-year test** that covers the entire contents of the book.
- Before attempting the **Sample Exam Papers,** make sure that you have completed all of the **Revision Tests** and have worked through the solutions to all the questions you answered incorrectly.
- Set aside the **time allowed** for the paper and complete it under **exam conditions**—no sneaking a look at your notes or textbook! That way you will be better prepared for your final exam.
- **Complete answers** to the Sample Exam Papers are provided at the end of the book.
- **Mark** your test to see how well you have done. Use the **Test & Exam Results** page to determine your areas of weakness and your areas of strength.

HOW TO USE THIS BOOK TO STUDY FOR A CLASS TEST, HALF-YEARLY OR END-OF-YEAR EXAM

Depending on your teacher or school, you will be given a variety of tests and exams each year. There may be a single-topic test, a test that covers a number of topics, a semester test or exam, or even a half-yearly or yearly exam.

Step 1 **Find out** which topics will be covered in the class test.

- To do this, look at your class workbook/textbook, laptop/tablet or online study program.
- Ask your teacher if you are still not sure what topics they are. For example, your class test may be on The body systems.

Step 2 **Match** the topics that your test is on to the topics in this book.

- For example, the first four units in this book cover The body systems.

Step 3 **Use** this book to study the topics being tested.

- For example, the first four units in this book contain the topics you will study for your class test!

Note:

- When you are using the book to study for a **half-yearly** test, follow the same steps as above—the only difference being that will be you will have more topics to revise of course!
- When you are using this book to study for an **end-of-year** test, you will more than likely need to study the whole book!

REVIEW OF MAJOR BODY SYSTEMS

The body systems

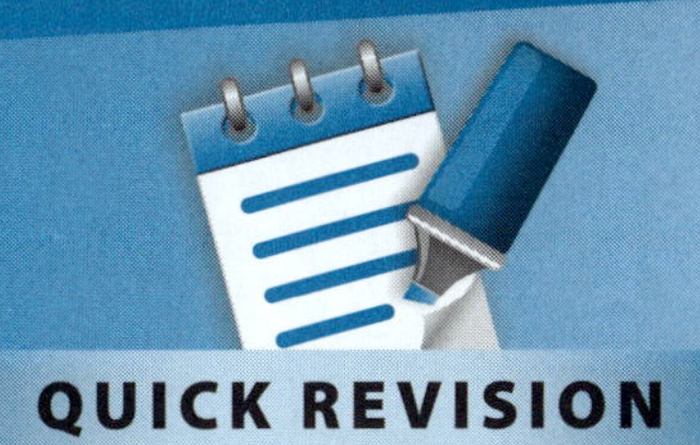

QUICK REVISION

1 Our organs are composed of different types of __________. Each tissue contains specialised cells. Organs can belong to one or more __________ systems. For example, the lungs are organs in the __________ system as they transfer oxygen from inhaled air into the bloodstream. The lungs are also __________ organs because they are used to exhale waste carbon dioxide from the body. The organ systems work __________ to ensure the body remains healthy and active. The skeletal and __________ system provides support for the body and allows it to move from one place to another. The reproductive system produces __________ cells. In males the testes produce __________ and in females the __________ produce eggs. Fertilisation involves the __________ of a sperm and an egg combining. The __________ egg develops into a baby in the womb or uterus.

2 The bloodstream is an important component of the circulatory system. The __________ is pumped around the body by a muscular __________ called the heart. The heart has four chambers. Blood leaves the __________ of the heart through vessels called arteries. Blood returns to the atria of the heart through vessels called __________. The smallest blood vessels are called __________. In the capillaries oxygen from the lungs is absorbed into the bloodstream. The oxygen is __________ through a series of increasingly larger blood vessels and then finally back to the heart via the __________ vein. The oxygenated blood is then pumped through a large artery called the __________ which branches to carry blood to all the body's organs, including the brain. The blood that emerges from these organs is __________ and also contains metabolic wastes, including carbon dioxide. This blood eventually flows into the major veins and the largest vein, called the__________, transports this blood back to the __________. The pulmonary __________ then pumps this blood to the lungs to allow carbon dioxide to be released.

3 Our digestive system contains many organs that work together to ensure nutrients from our food are supplied to our body's __________ to keep them alive. Digestion begins in the __________ where the teeth break and crush the food into small pieces. Saliva from the __________ glands is mixed with the food to __________ it and also to begin carbohydrate digestion. The moist food is swallowed and passes down the __________ into the stomach. The stomach contains acidic digestive juices and __________ that break proteins down into simpler nutrients. The liquidised food passes into the __________ intestine. Juices from the pancreas enter at this point to continue carbohydrate metabolism and __________ enters from the liver and gall bladder to help __________ fats. Once digestion is complete the soluble nutrients __________ through the wall of the small intestine and then into the bloodstream. Water and various other nutrients are removed when the waste materials are in the __________ intestine.

4 The respiratory system involves the absorption of __________ and the removal of carbon dioxide from the body. In order to inhale, your __________ cavity needs to become larger. This is achieved using the diaphragm muscle and the muscles of the __________ cage. Air is drawn into the lungs when the chest cavity __________. Once the air is in the lungs it travels down through smaller and smaller tubes to __________ which are surrounded by blood capillaries. Here the oxygen diffuses into the capillaries. At the same time the carbon dioxide diffuses the __________ way, into the alveoli and into the lungs. As the diaphragm and rib cage muscles relax, the chest becomes __________ and air is exhaled. Not all the oxygen has been removed from the air in this process.

(cont.)

REVIEW OF MAJOR BODY SYSTEMS *(continued)*

The body systems

QUICK REVISION

5 The excretory system is used to __________ water-soluble wastes from the body. As the blood circulates to each body organ, it removes various metabolic __________ products. Some of these are toxic and the liver's role is to detoxify them. When the blood reaches the kidneys, glucose is returned to the bloodstream while the wastes are __________ together with water to form a solution called urine. The urine flows down tubes called __________ into the bladder where it is stored until it is eliminated. This elimination occurs through a tube called the __________. The lungs are also excretory organs as they remove __________ dioxide from the body.

A dialysis machine tries to imitate some of the functions of the human kidney. One of the kidney's main tasks is to remove urea and certain salts from the blood. In a dialysis machine, blood from the patient passes through tubes made of a semi-porous __________. Red and white blood cells and other important blood components are too large to fit through the __________ in the membranes, but urea and salt flow through into a sterile solution outside the membrane and are removed.

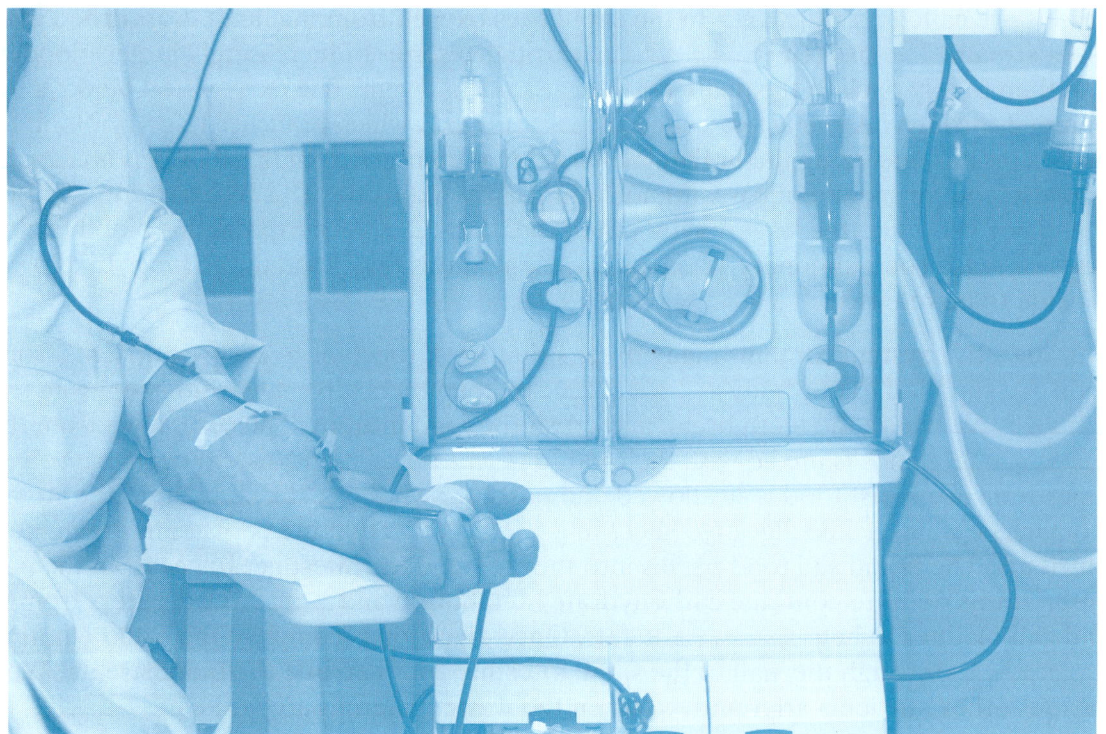

Answers **1** tissues; organ; respiratory; excretory; together; muscular; sex; sperm; ovaries; nuclei; fertilised **2** blood; organ; ventricles; veins; capillaries; transported; pulmonary; aorta; deoxygenated; vena cava; heart; artery **3** cells; mouth; salivary; moisten; oesophagus; enzymes; small; bile; emulsify; diffuse; large **4** oxygen; chest; rib; expands; alveoli; opposite; smaller **5** eliminate; waste; filtered; ureters; urethra; carbon; membrane; pores

REVIEW OF MAJOR BODY SYSTEMS

The body systems

REVISION SUMMARIES

1 The following table summarises the key points about some of the major human body systems.

System	Components	Job
circulatory	heart, veins, arteries	transport food around the body; transport gases around the body; carry wastes to excretory organs
digestive	stomach, liver, pancreas, small intestine, large intestine	change food into forms usable by body cells; eliminate unused food
respiratory	lungs, diaphragm	exchange gases
excretory	kidney, bladder	clean the blood; eliminate nitrogenous wastes
skeletal/muscular	skeleton, muscles, tendons, ligaments	support; movement; protection
reproductive	testes, ovaries, uterus	produce male (sperm) and female (ova/eggs) sex cells; provide an organ to protect the developing baby

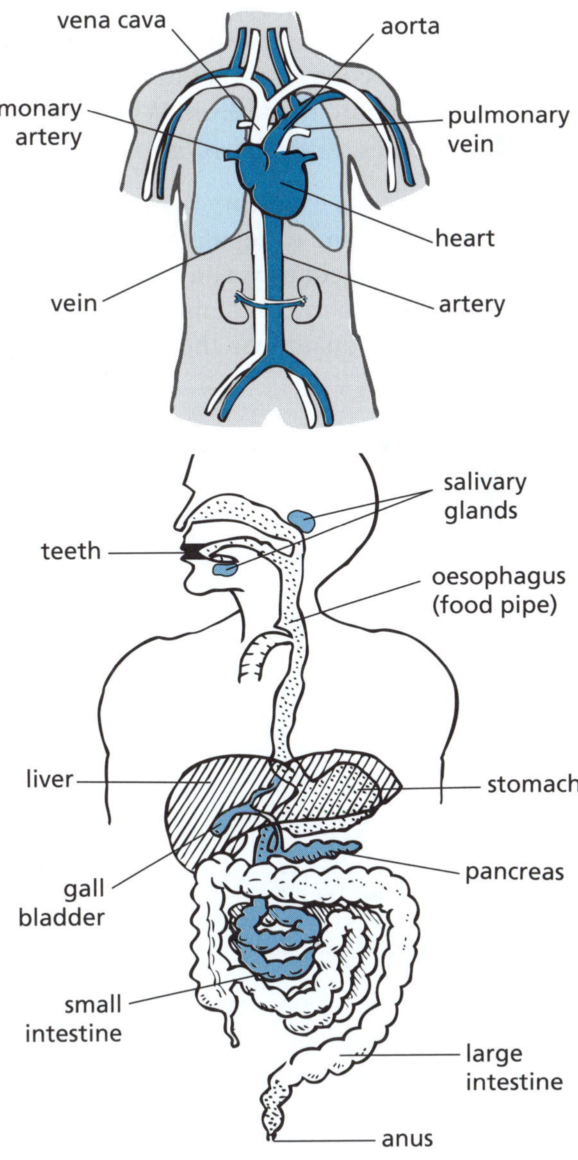

2 The circulatory system is vital to the proper functioning of the body. The bloodstream transports oxygen to the cells of the body and transports wastes back to the kidneys or lungs where they are removed. Arteries are blood vessels that have thick muscular walls and transport **oxygenated blood** away from the heart. Oxygenated blood is bright red in colour. The **aorta** is the largest artery. Veins have thinner walls and transport **deoxygenated blood** (which is bluish-red in colour) back to the heart from the body. The two branches of the vena cava are large veins that return deoxygenated blood from the body to the heart. The pulmonary artery transports blood from the heart to the lungs to be oxygenated. The pulmonary vein transports oxygenated blood back to the **heart**.

3 The digestive system converts food into usable **nutrients** and **energy**. These nutrients are then transported to all cells and organs via the circulatory system. The diagram on the right shows the major organs of the digestive system. The digestive system is composed of the **alimentary canal**, the liver and the pancreas. The alimentary canal is a tube running from the mouth to the anus. The teeth chew food into smaller pieces and mix it with saliva from the salivary glands to start **carbohydrate** digestion. The bolus of food then travels down the oesophagus into the stomach where acids and enzymes start to digest the food. The partly digested food moves into the small intestine

(cont.)

where it is mixed with bile from the gall bladder and liver. The bile emulsifies **fats**. The pancreas releases digestive juices into the small intestine to aid in carbohydrate digestion. The nutrients produced in the digestive process **diffuse** through the intestinal wall into the bloodstream. Waste and water move into the large intestine where water is reabsorbed into the body and waste is eliminated through the anus.

4 The respiratory system allows **gaseous exchange** between the lungs and the air. During **inhalation** the **diaphragm** and rib cage muscles cause the chest cavity to expand. Air inhaled through the nasal cavity moves down the trachea into the lungs. The lungs contain tiny structures called alveoli that increase the surface area of the lungs so oxygen can be more rapidly absorbed by the blood. Oxygen diffuses through the membranes of the alveoli and enters the blood stream via its fine **capillaries**. In the reverse direction, carbon dioxide waste diffuses from the blood back into the alveoli. When the diaphragm and rib cage muscles relax, the air is exhaled.

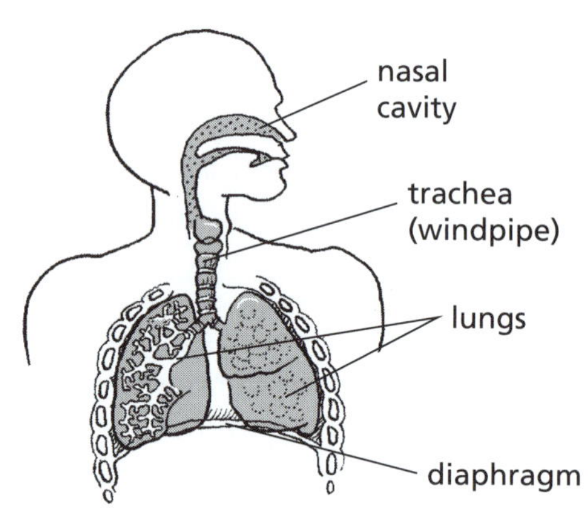

5 The role of the excretory system is to remove water-soluble waste materials from the bloodstream and to regulate water and salts in the body. The kidneys filter the blood and remove **nitrogenous waste** such as urea. These wastes are produced when the liver **detoxifies** the blood of harmful chemicals. Urine is produced by the **kidneys** and is stored in the **bladder** until it is eliminated via the urethra. The lungs are also excretory organs as they rid the body of carbon dioxide.

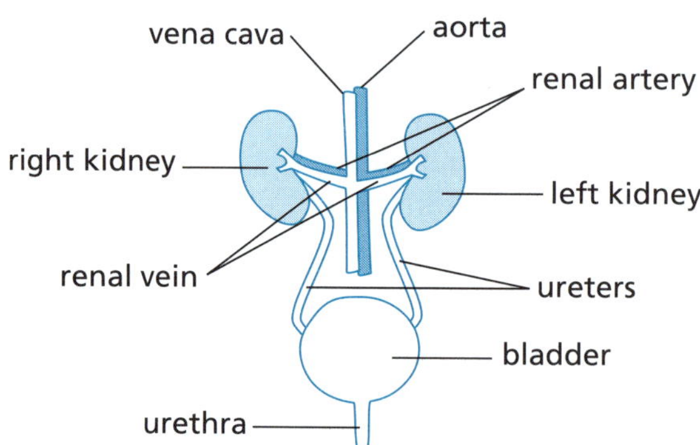

Checklist

Can you:

1. *Name the major organ systems of the human body?* ☐
2. *Name the organs of the circulatory system and describe their roles?* ☐
3. *Name the organs of the digestive system and describe their roles?* ☐
4. *Name the organs of the respiratory system and describe their roles?* ☐
5. *Name the organs of the excretory system and describe their roles?* ☐

REVIEW OF MAJOR BODY SYSTEMS

The body systems

30 MINUTES

REVISION TEST

1 Complete the following sentences.

a Bile is made in the __________. (1 mark)

b The digestive process starts in the __________. (1 mark)

c The heart, blood, veins and arteries are part of the __________ system. (1 mark)

d The bladder is an organ of the __________ system. (1 mark)

e The aorta transports __________ blood from the heart to the rest of the body. (1 mark)

2 The following diagram shows the different stages the chest cavity goes through when a person is breathing in (inhaling) and breathing out (exhaling), but the steps are not in order. Match the following randomly listed events to the correct stages, labelled A to F on the diagram.

a air is forced out from the lungs to the exterior (1 mark)

b chest cavity pressure decreases (1 mark)

c chest cavity volume increases (1 mark)

d chest cavity volume decreases (1 mark)

e chest cavity pressure increases (1 mark)

f air drawn in from outside into the lungs (1 mark)

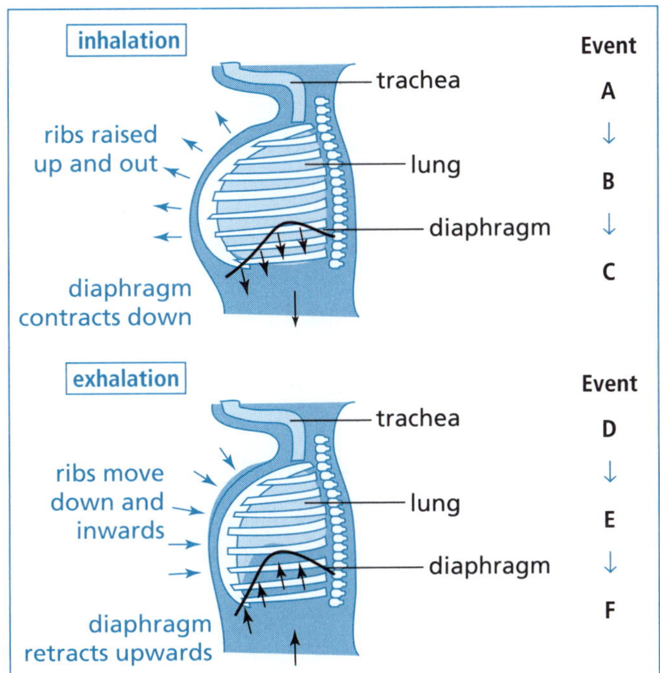

3 Match the key organs of the excretory system (labelled A, B and C in the diagram) with one of the following roles. (3 marks)

D detoxifies harmful chemicals before these wastes are filtered by the kidneys

E removes carbon dioxide from the blood stream

F filters the blood to remove waste substances in the form of urine

excretory organ		
kidney	liver	lung
A	B	C

(cont.)

4 Below is a partly labelled diagram of the human heart.

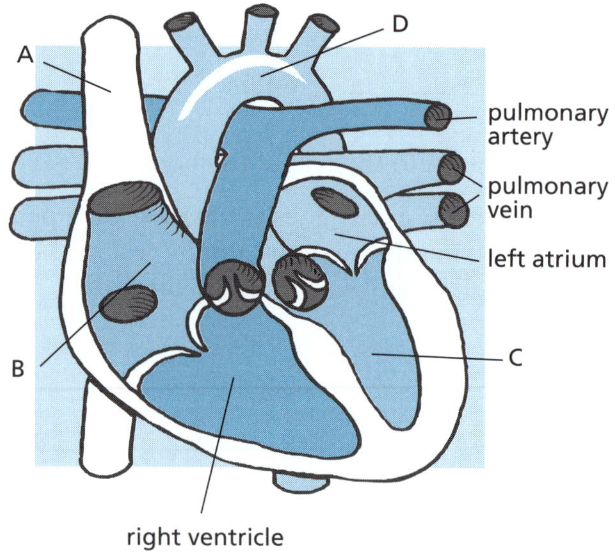

a Name the components A, B, C and D. (4 marks)

b State the function of the:
 i pulmonary vein (1 mark)
 ii right ventricle. (1 mark)

c A baby is born with a hole in the muscular wall of the heart which links the two ventricles. The incomplete wall between the ventricles is known as a ventricular septal defect. Such babies are sometimes called 'blue' babies. Why do they look blue? *Hint 1* (1 mark)

d In diagrams such as these, the right ventricle is shown on the left, while the left atrium is on the right. Suggest a reason for this. (1 mark)

5 True or false?

a The diaphragm is part of the digestive system. (1 mark)
b Saliva contains enzymes that start the digestion of carbohydrates such as sugar. (1 mark)
c Two tubes called urethras transport urine out of each kidney and into the bladder. (1 mark)
d The amount of carbon dioxide in exhaled air is greater than in inhaled air. (1 mark)
e The major component of urine is water. (1 mark)

6 Enzo is a school student who has just recovered from pneumonia. He spent 10 days in hospital under intensive treatment with antibiotics and oxygen therapy. Many of the alveoli in his lungs had become filled with fluid during his infection.

a What is the function of alveoli? (1 mark)
b Explain why Enzo had trouble breathing. *Hint 2* (1 mark)

7 Food spends different amounts of time in the various organs of the digestive system. The following table shows how long food spends in each of the organs.

Organ	Time
mouth	2.5 minutes
stomach	4 hours
small intestine	6 hours
large intestine	20 hours

Draw a sector (pie) chart of this information. *Hint 3* (4 marks)

Hint 1: What colour is deoxygenated blood?
Hint 2: How will fluid alter the lung surface area?
Hint 3: First work out the fraction of 360° that each time represents.

Your Feedback

$\frac{\square}{33} \times 100\% = \square\%$

PAGE 219
PAGE 252

NERVOUS AND ENDOCRINE SYSTEMS

The body systems

QUICK REVISION

1 The organ systems, such as the respiratory, circulatory, digestive and excretory systems, must be __________ so that our bodies work correctly. A failure in coordination leads to disability or disease. It is the role of the __________ system and the endocrine system to carry out this coordination. The nervous system is an __________ system that produces rapid responses. The endocrine system is a chemical system that acts much more __________ and over much longer time periods.

2 The nervous system is divided into two subsystems: the __________ nervous system and the peripheral nervous system. The central nervous system consists of the __________ and the spinal cord which are both __________ by bones. The peripheral nervous system is made up of nerves that travel __________ from the spinal cord to the rest of the body. Nerves are fibrous bodies composed of specialised cells called __________. Electrical signals are relayed along nerve fibres when __________ neurons are stimulated. The signals reach the central nervous system and are then relayed back to __________ or organs via nerve fibres containing __________ neurons. The message leads to an action in the muscle or gland. For example, the signal may cause the muscle to __________ or the gland to release hormones into the blood. The neurons in the nerve do not touch each other but are separated by a small gap known as a __________. When a signal reaches one end of a neuron, it releases a chemical that __________ across the synapse to stimulate the next neuron to fire its electrical signal.

3 Your body is under both conscious and automatic control. You can __________ to move your foot forwards but you cannot control the closing of your __________ in your eye when you walk into bright sunlight. Reflex __________ are another example of responses that are automatic. If you touch a hot kettle your hand is __________ removed. This occurs without your brain directing the action. The brain will only register the __________. The reflex action is called a reflex __________. It involves sensory, __________ and motor neurons only. The sensory receptors in the skin detect the hot object and rapid electrical signals are sent along __________ nerves to the spinal cord. The messages are then relayed by connector neurons in the spinal cord to __________ neurons which relay the message rapidly back to muscles in your arm. These muscles __________ and move your hand away from the hot object.

4 Our brain is __________ by our bony skull. The brain is composed of many parts and each part has specific functions. Humans have a very large __________ which is divided into two parts. About 85% of the mass of the brain is the cerebrum. The cerebrum controls __________ movements as well as our memory and intelligence. Our senses of touch, smell, sight and hearing are controlled here. At the back of the brain is a region known as the __________. This region controls __________ movements and balance. The fine motor skills that allow you to undertake delicate work are controlled here. At the base of the brain is the brain __________. Involuntary actions, like breathing and your heartbeat, are controlled here. Your body temperature is controlled by a region known as the __________, which is found just above the brain stem.

5 Apart from the nervous system, the body has a second control and coordination system called the __________ system. This system is composed of __________ that secrete chemicals called __________ directly into the bloodstream. For example, the ovaries and __________ secrete sex hormones that are essential for our reproductive systems. The pancreas secretes a hormone called __________ which is involved in the __________ of glucose sugar levels in the blood. If this system is not working correctly, then diabetes can result. If glucose levels are too high, the

(cont.)

insulin acts and converts the glucose to a polymer called ____________ which is stored in the liver until more glucose is needed. Glucagon is another pancreatic hormone which converts glycogen back to ____________ when required. The thyroid gland is located in your throat. It produces a hormone called thyroxine that increases our ____________ rate as well as stimulating our general metabolism. The following diagram shows the thyroid gland located in front of and slightly below the larynx.

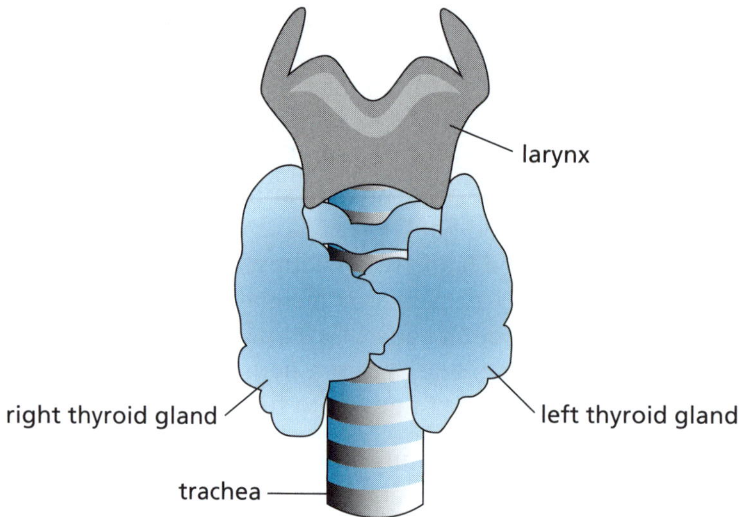

Adrenaline, which is produced by the adrenal glands, also stimulates heart rate and raises our blood pressure. The release of hormones is controlled by a master gland called the ____________ gland which is located at the base of the brain.

Answers **1** coordinated; nervous; electrical; slowly **2** central; brain; protected; out; neurons; sensory; muscles; motor; contract; synapse; diffuses **3** decide; iris; actions; rapidly (quickly); pain; arc; connector; sensory; motor; contract **4** protected; cerebrum; voluntary; cerebellum; involuntary; stem; hypothalamus **5** endocrine; glands; hormones; testes; insulin; control; glycogen; glucose; heart; pituitary

NERVOUS AND ENDOCRINE SYSTEMS

The body systems

REVISION SUMMARIES

1 So that your body can act in an orderly and efficient manner, your various organ systems must be **coordinated**. Two systems that work closely together to do this are the **nervous system** and the **endocrine system**. The nervous system provides very quick controls, and the endocrine system provides relatively slow and long-lasting controls.

In vertebrates such as humans, the nervous system consists of the **brain**, spinal cord and all the nerves that run to various parts of your body. Your brain and spinal cord together are known as the **central nervous system** (CNS). In addition to the CNS is the **peripheral nervous system**, which consists of nerves that connect the CNS to the rest of the body.

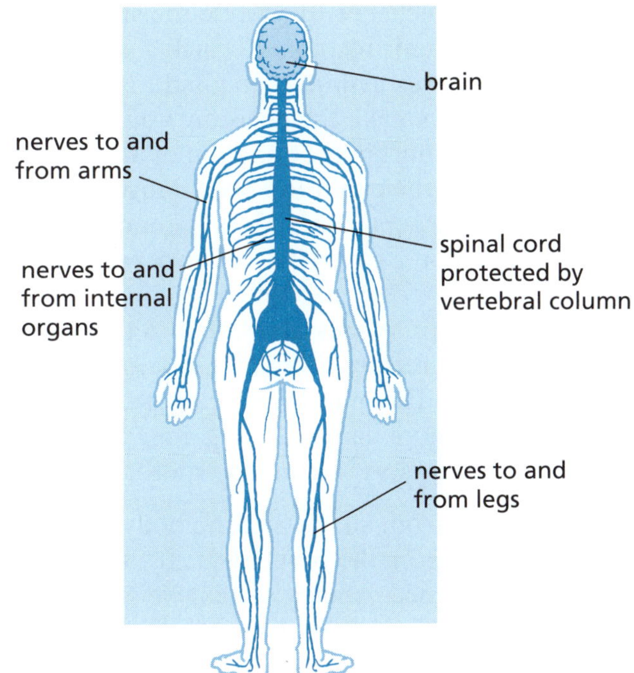

Nerves carry **impulses** (messages) to and from the CNS. All organs in your body have nerve **receptors** called sensory receptors. These receptors constantly send messages to the CNS about what is happening inside and outside your body. Nerves are bundles of nerve cells called neurons. Neurons that carry impulses to the CNS are called sensory neurons. When the sensory receptors on these neurons are stimulated, they send brief electrical pulses along the **sensory neurons** to the CNS. Impulses received by the brain allow you to make decisions about what to do.

Neurons that carry impulses away from the CNS are called **motor neurons**. They are connected to either muscles or glands. Muscles respond by contracting, then relaxing. Glands respond by releasing chemicals called hormones. Motor neurons do not directly connect onto sensory neurons. Between them is another type of neuron called a **connector neuron**. Connector neurons are only found in the CNS. Neurons do not touch each other. There is a small gap between two neurons called a **synapse**. The message is transmitted from one neuron to the other across a synapse by chemicals.

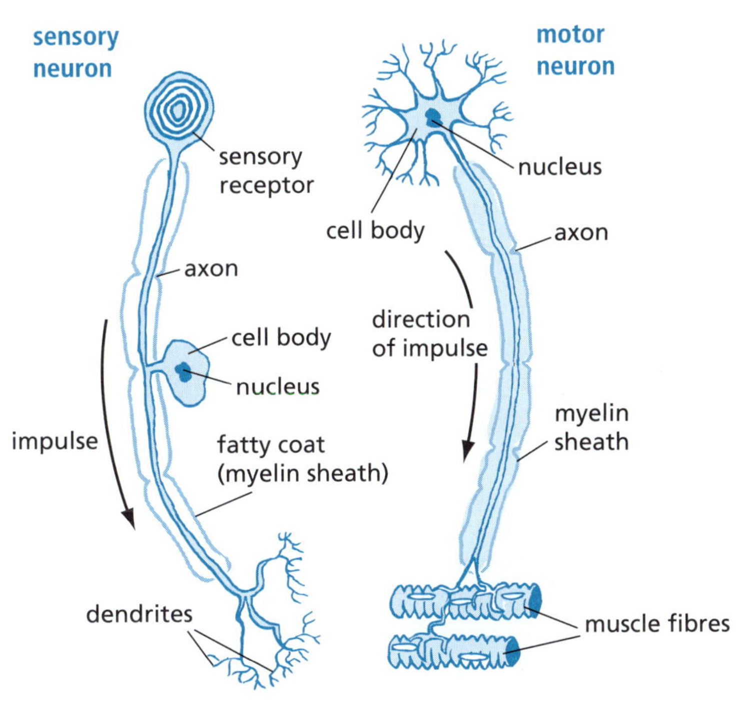

(cont.)

2 Sometimes your body needs to respond very quickly, especially if there is danger. Have you ever suddenly jumped when there was a loud bang behind you? You responded before your brain had a chance to analyse the situation. This automatic response to such a stimulus is called a **reflex action.** If you had to wait for the brain to make the decision, you might have been hurt in that fraction of a second delay, had the explosion been dangerous. Consider a person accidentally touching a hot saucepan on a stove. The **sensory receptor** in the skin detects a sudden change in temperature. A message is sent to the spinal cord and a return message is sent immediately to the muscle in the arm without going to the brain first. The muscle in the arm removes the hand from harm's way. Such a quick action prevents more damage being done to the person's hand. The arrangement of this series of neurons, sensory–connector–motor, is a **reflex arc**.

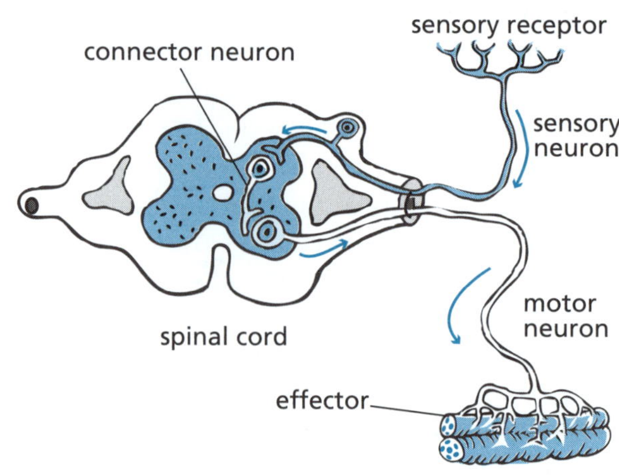

3 Messages from all parts of the body are constantly being sent to the brain and, in return, messages from the brain are sent to all parts of the body. The size of the brain is not an indicator of intelligence. A baby's brain weighs less than half a kilogram. An adult brain weighs about 1.3 to 1.4 kg. (The largest brains are those of sperm whales, about 8 kg, while an elephant's brain is just over 5 kg.)

The brain is not a single organ. It consists of a number of parts.

- The **cerebrum** is the largest part (85% of the brain's mass) and controls such things as voluntary movements, speech, memory, intelligence, vision, hearing, smell and touch.
- The **cerebellum** controls involuntary actions. It plays a crucial role in posture, balance and coordination. When you are doing complex activities that require a number of muscles to work together, the cerebellum makes sure that they work in harmony.
- Messages to and from the brain all pass through the **brain stem**. It contains systems that control important involuntary actions such as breathing, heartbeat and digestion. The hypothalamus on top of the brain stem controls thirst and temperature.

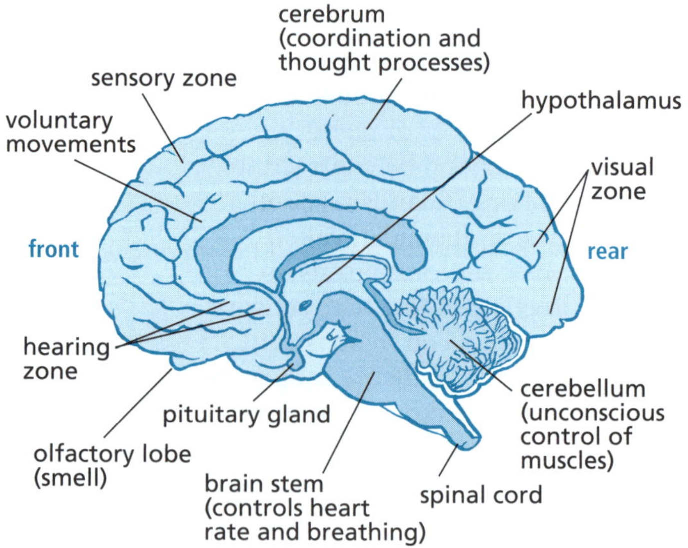

4 Endocrine **glands** control the release of chemicals called **hormones**. Hormones are secreted (released) by these glands in response to specific **stimuli**. The chemicals either diffuse or are carried by the bloodstream to the site of action. Here they modify the activity of specific cells to provide a response to that stimulus. Nerves only act on muscles or glands, but hormones can act on the whole body, parts of the body or on individual organs.

The master gland that seems to control and turn on the other glands is the **pituitary gland** located at the base of the brain. Hormones released from the pituitary stimulate other glands to release their hormones. Sometimes the control is in reverse. A hormone can inhibit some action rather than stimulating it.

The control of the amount of glucose sugar in your bloodstream is due to two hormones. One of these is insulin, which is produced in the pancreas. Insulin is necessary to help convert glucose to glycogen. Glycogen provides a large, inactive molecule in which the energy from many glucose molecules can be stored. Glucagon has the opposite effect to insulin. As blood glucose levels decrease, more glucagon is released to break down the glycogen into glucose. Between the two hormones, the level of blood sugar is maintained between acceptable limits.

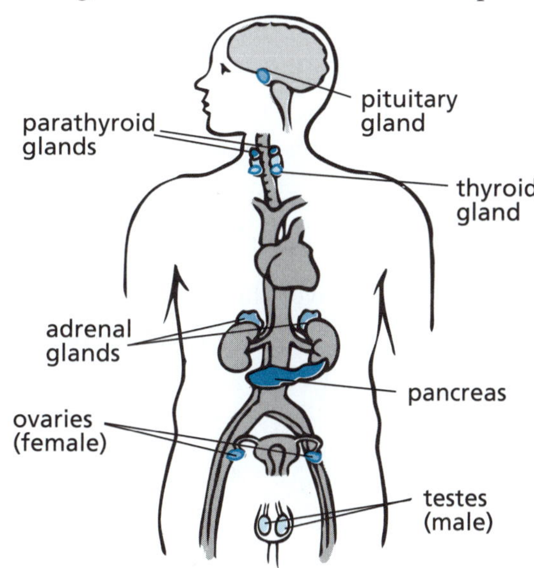

Checklist

Can you:

1. *Describe the functions of the central and peripheral nervous systems?* ☐
2. *Explain how a reflex action is produced?* ☐
3. *Explain the functions of the major parts of the brain?* ☐
4. *Explain the role of hormones in the endocrine system?* ☐

NERVOUS AND ENDOCRINE SYSTEMS

The body systems

30 MINUTES

REVISION TEST

1 The following diagram shows a person responding to burning their finger on hot steam from a boiling pot of water.

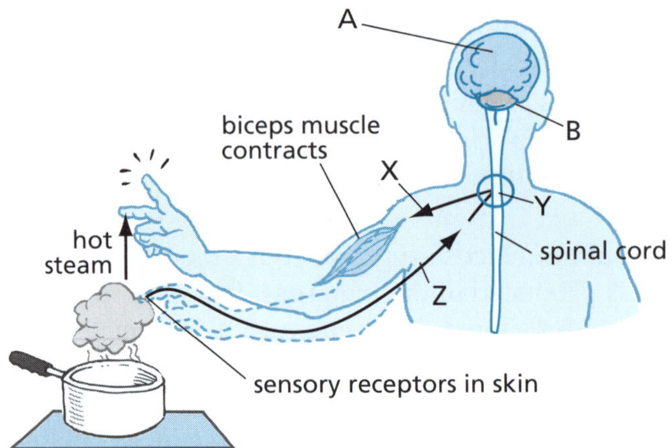

- **a** Name the type of response shown in the diagram. (1 mark)
- **b** Identify the two parts of the brain labelled A and B. (2 marks)
- **c** Identify the three types of neurons labelled X, Y and Z. (3 marks)
- **d** Explain how the brain is involved in this response. *Hint 1* (1 mark)

2 **a** What name is given to the gap between adjacent neurons? (1 mark)

b How are messages transmitted from one neuron to another across this gap? (1 mark)

3 Read the following information and answer the questions below.

> Receptors in your skin and other sense organs detect stimuli. A signal is sent along sensory neurons to the CNS as part of a response to this stimulation. The degree of stimulation does not depend on the size of the impulse. Instead, it depends on the frequency with which pulses are generated. The stronger the stimulus, the higher the frequency with which these pulses move along the neurons.

a True or false?

- **i** Stimuli are detected only by sense organs in your skin. (1 mark)
- **ii** The CNS detects the response of the sensory organ. (1 mark)
- **iii** Strong impulses produce a greater frequency of pulses along the neurons. (1 mark)

b A receptor on your left hand detects a temperature of 20 °C. A receptor on your right hand detects a temperature of 40 °C. Explain how the impulses sent from these two receptors are different. *Hint 2* (2 marks)

4 The following text gives a description of how hormones control part of human digestion. Read the information and answer the questions that follow.

> Digestive enzymes are chemical substances that help break down food as it passes through the digestive system. Secretion of these enzymes is controlled by hormones. Certain cells in the stomach lining secrete the hormone gastrin. Their release is stimulated by food entering the stomach. Gastrin circulates in the bloodstream and causes hydrochloric acid and pepsinogen

to be released from the stomach wall and into the stomach. Pepsinogen is immediately converted to pepsin inside the stomach and begins the breakdown of proteins.

When the mixed food contents from the stomach reach the duodenum, the beginning of the small intestine, it stimulates another hormone to be released. Secretin is a hormone released from the walls of the duodenum. It causes the pancreas to release hydrogen carbonate ions to neutralise the acidity from the stomach contents. It also causes the gall bladder to release bile. Bile is necessary to break up fats into small droplets so that other enzymes can break them down.

Food also stimulates the duodenum to release cholecystokinin. This hormone acts on the pancreas to release trypsinogen which is converted in the duodenum to the enzyme trypsin. This enzyme also breaks down protein. As food passes to the next section of the small intestine, another hormone is released from the small intestine which has a negative effect on stomach acids and enzymes. This hormone, enterogastrone, acts as a 'tap' to turn off stomach acids and enzymes so that they are only present in the stomach when required.

The diagram below shows how the four hormones mentioned in this text act in the digestive system.

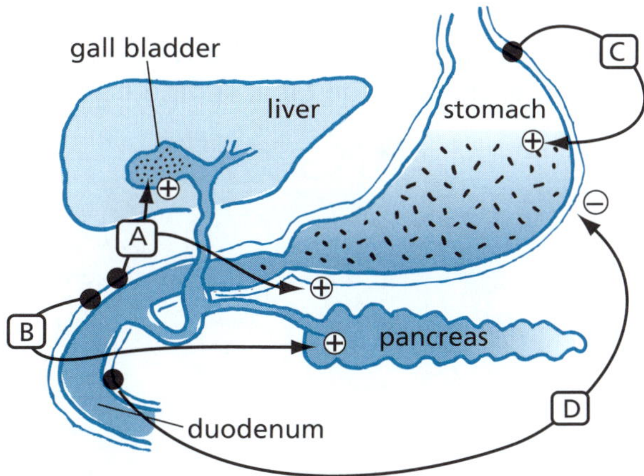

a Complete the table below by answering the following.

  i List the hormones named in the text above next to the letter they correspond to in the diagram. (4 marks)

  ii Write a brief description of what each hormone does. (4 marks)

	Hormone	Function
A		
B		
C		
D		

b Each hormone in the diagram above is shown with a large dot (•), an arrow (⟶) and a plus or minus sign. Explain the meaning of each of these symbols. (4 marks)

(cont.)

5 The table below gives the masses of some animals and their brains.

Animal	Mass of animal (kg)	Mass of brain (kg)	Percentage mass of brain compared to body
elephant	2500	5	
dolphin	136	1.6	
human	60	1.4	
horse	250	0.6	
budgerigar	0.02	0.001	

a Complete the last column of the table by determining what percentage of the mass of the body is made up by the brain. *Hint 3* (5 marks)

b Comment on the size of the animal compared to the size of its brain. (1 mark)

c True or false? The larger the animal, the larger its brain compared to its body. Give evidence to support your answer. (2 marks)

d Humans are the most intelligent of all creatures, therefore their brain, in comparison to their body size, must be large.
Do you agree with this statement? Explain why or why not. (1 mark)

e The following photo shows a healthy human brain. If brain mass was the determining factor to intelligence, which of the animals in the table should be the most intelligent? (1 mark)

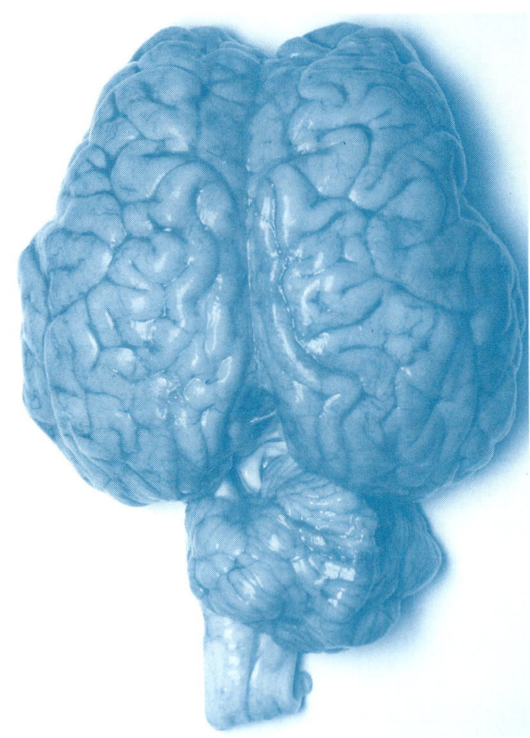

Hint 1: Is this process under conscious control?
Hint 2: The term 'frequency' refers to the number of times per second that the stimulation occurs.
Hint 3: Use the formula: [(mass of brain)/(mass of body)] × 100/1

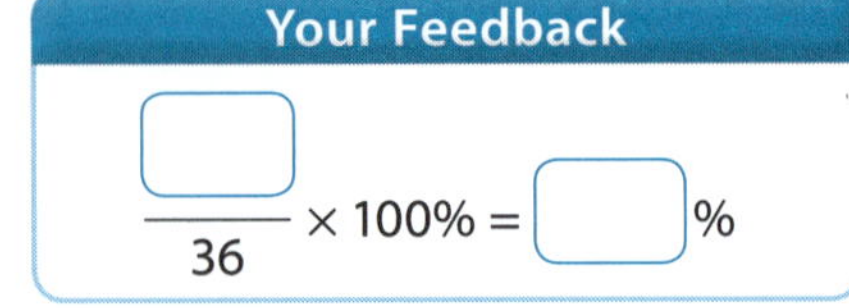

PAGES 219–220
PAGE 252

MICROBES AND DISEASE

The body systems

QUICK REVISION

1 Before the __________ of infectious diseases were discovered, many people invented reasons to account for them. Some indigenous __________ thought enemies had sent __________ magic to destroy them, or that the diseases were a punishment from the gods. Various __________ and other treatments were tried, but with limited success. Some peoples tried __________, dancing and even making sacrifices to the spirits to __________ them from such retribution.

2 Microorganisms (sometimes called __________) are the smallest living things. They are so small that they cannot be seen without the help of a __________. There are four types of microorganisms that cause infectious disease.

- Bacteria. These are simple organisms that are made up of only one __________. While most bacteria cause no harm, those that do are often known as __________. Harmful __________ can cause food poisoning or very serious diseases like tuberculosis, tetanus or meningitis.
- Viruses. These do not have any of the __________ found in living cells. They are smaller than bacteria and consist of a protein capsule that contains a strand of __________ (the code that allows them to duplicate).
- Fungi. These are plant-like organisms that lack the green pigment __________ and include moulds and yeasts. They can occur on the skin and inside the body. All fungi need existing organic matter for their __________. They cause infections such as athlete's foot, jock itch and ringworm.
- Protozoa. These comprise the __________-celled microscopic animals. Malaria is one of the top five 'killer' infectious diseases in the world. It is caused by protozoans from the genus *Plasmodium*. Nearly 800 000 people die from malaria every year, which is transmitted by the female *Anopheles* __________.

3 Antibiotics are medicines prescribed by __________ when harmful microbes have made you ill. These chemicals are substances that harm and kill __________. Some antibiotics stop bacteria reproducing and others kill bacteria directly. They can treat diseases caused by bacteria, such as tuberculosis, but they do not harm __________. Antibiotics cannot treat diseases such as colds and flu, which are caused by viruses. Bacteriostatic agents, slow __________ or stall bacterial growth.

Antiseptics are substances which __________ the growth and development of microorganisms. These are usually applied to the __________ or mucous membranes. Their uses include cleaning the skin after injury, and preparing the skin surfaces before giving __________ or performing surgery. Antiseptics are also used for disinfecting non-living objects, such as instruments and furniture surfaces. Aqueous and alcoholic __________ of iodine and of hydrogen peroxide have been used as antiseptic __________.

4 Non-infectious __________ can't be passed from one person to another. Instead, these types of diseases are caused by factors such as the __________, genetics and lifestyle. Many lifestyle diseases have their start later in an individual's life and need a longer __________ to cause death. As medicines improve and we conquer many infectious diseases and our life spans become longer, non-__________ diseases will become more prominent. According to estimates from the United Nations, nearly two-thirds of deaths in the world are caused by non-infectious diseases such as cancer, diabetes, __________ and lung disease.

Answers **1** causes (reasons); peoples; bad (harmful); herbal; chanting (singing); spare (save) **2** microbes; microscope; cell; germs; bacteria; structures (organelles); DNA; chlorophyll; food (nourishment); single; mosquito **3** doctors; bacteria; viruses; down; prevent (delay); skin; injections; solutions; agents **4** diseases; environment; time; infectious; heart disease (or stroke)

MICROBES AND DISEASE

REVISION SUMMARIES

1 From ancient times up until the late 19th century **bloodletting** was a common medical practice used to 'treat' a wide assortment of diseases. Bloodletting is withdrawing small quantities of blood from a patient to cure or prevent illness and disease, and it became a standard treatment for almost every complaint. Besides cutting into a vein, **leeches** were also used for bloodletting. Attached to the skin, this type of worm can suck several times its original body weight in blood. This was based on an ancient system of medicine where blood and other bodily fluid were considered to be 'humours' and the body needed a proper balance to maintain health. Since curing diseases remained elusive, bloodletting remained popular as many thought it was better to give any treatment than nothing at all. The psychological benefit of bloodletting to the patient (a **placebo** effect) sometimes overshadowed the problems it caused to the normal functioning of the body. Today it is recognised as a harmful practice as it can weaken the patient and aid infections.

Before traders or Europeans arrived in Australia, the Indigenous peoples that lived here were not afflicted by many of the **diseases** which plagued Europeans. Spiritual doctors were able to cure a number of sicknesses and injuries. Plant remedies could treat stomach troubles or snake bites, and heat was used to take care of aches and pains. Europeans unwittingly introduced diseases such as measles, tuberculosis, venereal disease, cholera, whooping cough, influenza and even the common cold. However, the biggest killer of Aboriginal people was smallpox. This disease is caused by either of two virus variants: *Variola major* and *Variola minor*. It is estimated that within the first few years of British settlement, most of the Indigenous people living around Sydney were killed by smallpox. As European settlers moved to other parts of the country these incidences of smallpox spread. With the death of many family and clan members, social system associations between generations were damaged and surviving Aboriginal groups could not live as they had previously.

2 **Contagious** diseases are caused by **microorganisms** (also called microbes). Organisms that cause disease are called **pathogens**. Many pathogens are also **parasites**, organisms that live off another organism (their host), but provide no benefit to the host. There are four types of disease-causing microbes.

- **Viruses.** These microbes typically consist of a nucleic acid molecule in a protein coat. It is a point of debate whether viruses are actually alive, since they don't grow, respire, excrete or carry out many functions of living things. The only way they can grow or reproduce is to invade living cells and use their chemical machinery to survive and to make multiple copies of themselves. Diseases caused by viruses include flu (influenza), common cold, measles, mumps, German measles (rubella), smallpox, chicken pox and rabies.
- **Bacteria.** These are unicellular (single-celled) spherical, spiral or rod-shaped microscopic organisms that occur singly or in chains. They lack organelles and an organised nucleus. They can exist either as independent (free-living) organisms or as parasites. They need a food supply to allow them to grow and multiply. Diseases caused by bacteria include cholera, tuberculosis (TB) and anthrax.
- **Fungi.** These spore-producing organisms lack chlorophyll and vascular tissue and include moulds and yeast. They may be unicellular or multicellular, and they all feed on organic matter. Not all fungi are microscopic. Mushrooms, for example, are fungi that can be easily seen. Diseases caused by fungi include tinea, ringworm and thrush.

- **Protozoa.** This is an assorted group of unicellular organisms. They have true nuclei and a cell membrane, so they are similar to our body cells. Many protozoans move about using appendages known as cilia or flagella. They include the amoebas, flagellates and ciliates. Diseases caused by protozoans include malaria, sleeping sickness and dysentery. Some of these diseases are spread to humans by insects. For example, malaria is spread by a certain type of mosquito, while trypanosomiasis (sleeping sickness) is transmitted by the tsetse fly.

Most microorganisms do not cause disease. Many bacteria, for example, are harmless to humans; the human skin is covered in them and the digestive system contains many bacteria that cause no problems at all. The characteristic flavour of blue cheeses is due both to the mould and to types of bacteria encouraged to grow on cheese.

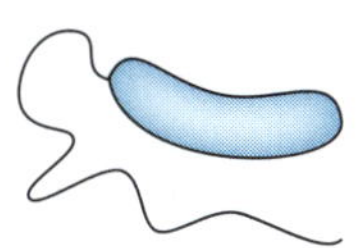

cholera bacteria

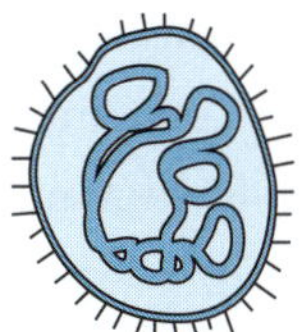

mumps virus

Cold and flu viruses are spread more easily during winter months as people are more likely to huddle together in close quarters due to the colder temperatures. While many of these viruses spread through the air, others spread through contact. This is why people are asked to wash their hands after going to the toilet, or before eating a meal.

3 When pathogenic bacteria infect the body, they can reproduce quite fast. Often they produce chemicals called toxins that cause an illness. Bacteria can usually be killed by **antibiotics** such as penicillin. Viruses cannot be killed by antibiotics.

Before the early 20th century, treatments for infections were based mainly on tradition and experience. Certain herbs and treatments worked, but people didn't know why. Mixtures with **antimicrobial** properties were used in treating infections over 2000 years ago. Many ancient cultures, such as the ancient Egyptians and ancient Greeks, describe the healing powers of using specially selected moulds, plant materials and other extracts.

Modern antibiotics include a broader range of antimicrobial compounds, such as antifungal and other compounds. Most modern antibacterial compounds are semi-synthetic modifications of various natural compounds. These include the penicillins (produced by fungi in the genus *Penicillium*). **Penicillin** was the first natural antibiotic to be identified. It was discovered by Alexander Fleming in 1928. Howard Florey and Ernst Chain succeeded in purifying the first penicillin in 1942, but it was not widely available before 1945. Purified penicillin exhibited effective antibacterial activity against a wide range of bacteria and was harmless to humans. The development of penicillin led to renewed curiosity in the exploration for other antibiotic compounds. Chain, Florey and Fleming shared the 1945 Nobel Prize in Medicine for their discovery and development of penicillin as a beneficial drug.

Antibiotics affect how bacteria function or grow. Some target the bacterial cell wall or the cell membrane while some interfere with essential bacterial chemical activities. Narrow-spectrum antibiotics damage specific types of bacteria, whereas broad-spectrum antibiotics affect a wide range of bacteria.

(cont.)

Bacteria can frequently **mutate** (change in form), increasing their resistance to certain antibiotics. As their resistance becomes more common, a greater need for alternative treatments arises. Viruses, too, can frequently mutate. This is why annual influenza vaccines (flu shots) need to be regularly modified to protect against the highly variable influenza virus.

4 Non-infectious diseases are not caused by specific organisms and cannot be passed from person to person. There are three main types of non-infectious conditions.

- Conditions that are **inherited** and arise due to abnormalities in genes or chromosomes. These include cystic fibrosis, Down syndrome, sickle cell anaemia and haemophilia.
- Conditions caused by **environmental factors** such as stress, physical and mental abuse; diet; and exposure to pollutants, contaminants, radiation and chemicals. For example, asbestos in the environment causes asbestosis and a rare form of cancer called mesothelioma. People have also been poisoned by substances containing lead or mercury.
- Conditions caused by **lifestyle factors** such as heart disease (which can be caused by being overweight and not exercising), skin cancer (resulting from too much exposure to sunlight) and diseases from drinking too much alcohol or smoking. Chemicals in cigarette smoke, as well as alcohol in wines and beer, are metabolic poisons.

Most chronic non-infectious diseases can be prevented, especially the ones causing the most deaths in the developed world: heart disease, stroke, diabetes and cancer. There is now evidence to suggest that some microbes may contribute to several non-infectious persistent diseases such as some forms of cancer and heart disease. Signs banning smoking (see below) are common, not just because of annoyance but also because second-hand smoke is a health hazard.

Checklist
Can you:

1 *Outline some beliefs about infectious diseases before modern medicine?* ☐
2 *Describe the four types of microbes that cause infectious diseases?* ☐
3 *Describe what antiseptics and antibiotics are?* ☐
4 *Name different types and causes of non-infectious diseases?* ☐

MICROBES AND DISEASE

The body systems

30 MINUTES

REVISION TEST

1 True or false?

a The common cold is the most common illness in the world. *Hint 1* (1 mark)

b The common cold and influenza are caused by bacterial infections. (1 mark)

c Cold weather carries germs more easily. (1 mark)

d Antibiotics are suggested for the common cold and influenza. (1 mark)

e New cold and flu viruses are always emerging. (1 mark)

f Washing your hands with soap and water is an easy and effective way of preventing colds or giving them to others. (1 mark)

g Antibiotics are used to treat bacterial infections and they work by killing bacteria or stopping them from growing. (1 mark)

h Some mosquitoes are capable of carrying diseases and infecting people or animals. (1 mark)

i Non-infectious diseases can be shared from one person to another. (1 mark)

j Causes of non-infectious disease include genetic disorders and the environment. (1 mark)

2 Smallpox was a highly contagious viral disease exclusive to humans. Around 30% of all smallpox cases resulted in death. Smallpox epidemics killed millions of people around the world. For instance, Inca populations in South America were reduced by 60 to 90% when European colonists arrived. Thousands of native Hawaiians died from the many new diseases brought by Westerners, including smallpox, and by the mid-1800s whole villages were wiped out by this disease.

a What type of microbe causes smallpox? (1 mark)

b Suggest why smallpox was such a deadly disease to native peoples. *Hint 2* (2 marks)

c Smallpox was a major health problem around the world for thousands of years. It was eradicated in the late 1970s after a successful worldwide vaccination program. How did vaccination help to rid the world from smallpox? (1 mark)

d Smallpox is not spread by insects or animals. People caught smallpox by being in close contact with an infected person or their personal items. In Europe smallpox was more common in cities than in the country. Suggest a reason for this. (2 marks)

3 Bloodletting was popular in the United States, as it was in Europe, up until the late 19th century. George Washington was bled heavily after he developed a throat infection from weather exposure. He died soon afterwards from the throat infection in 1799.

a What is bloodletting? (1 mark)

b Why was it in common use by many cultures for over 2000 years? (1 mark)

c Leeches became especially popular for bloodletting in early 19th-century Europe. How were they used? (1 mark)

d The exact cause of Washington's death has been the focus of much discussion. Most people cannot help but think that the bloodletting was a major contributing factor. Suggest a reason for this. (2 marks)

e The exact quantity of blood removed from the ailing Washington is not known, but one report gave this information.

> Albin Rawlins drew 13 ounces; Dr. James Craik drew 20 ounces, then he drew another 20 ounces, and finally he drew another 40 ounces. Dr. Elisha Dick drew 32 ounces. This occurred over a period of nine to ten hours on Saturday, December 14th, 1799.

Given one ounce is approximately 30 mL, how much blood was removed from Washington? (2 marks)

(cont.)

f Washington's weight towards the end of his life was roughly 90 kg. Given a healthy adult male has around 75 mL of blood for each kilogram of mass, approximately how much blood did he have before being bled? (1 mark)

g Calculate the percentage of blood removed from Washington by bloodletting. (1 mark)

4 Read the following information and then answer the question below.

> Living things are characterised by the ability to metabolise (use chemical processes to break down energy-rich substances into smaller units to be used as building blocks or as energy sources), grow, react to stimuli and reproduce. Viruses are not living things. They are small (about one-hundredth the size of the average bacterium) infectious agents that can reproduce only inside the living cells of an organism. They can infect all types of living things, such as animals, plants and bacteria. Viruses are complex collections of molecules including proteins, nucleic acids, lipids and carbohydrates. On their own they can do nothing until they enter a living cell. Without cells, viruses are unable to multiply.
>
> Viruses do not consist of cells and do not have cell membranes or other components of living cells. Hence living host cells are needed for their reproduction. Outside of the host, they are non-living chemicals. They neither metabolise nor respond to stimuli, but they do have genetic material and can therefore mutate and evolve. They form parasitic relationships with living hosts; the virus benefits at the cost of the living organism.

Viruses are not usually considered as living organisms. Give reasons for this. *Hint 3* (3 marks)

5 The following diagram shows some of the natural barriers the body has to stop harmful microbes getting inside it. Explain how four of these barriers work. (4 marks)

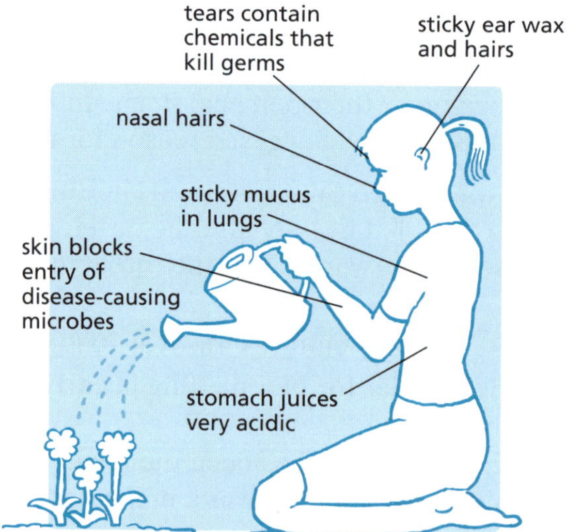

6 Diseases affect not only humans but also plants and animals. Potato blight was the cause of the Irish Potato Famine. Potato blight is caused by the fungus *Phytophthora infestans*. It had a profound impact because it devastated a staple food that fed much of Ireland in the mid-1800s.

The following table gives the Irish population (based on census data at the time).

Year	1841	1851	1861	1871	1881	1891	1901	1911	1921
Population (millions)	8.2	7.6	6.0	5.5	5.3	5.0	4.8	4.8	4.9

a Draw a line graph showing this information. *Hint 4* (4 marks)

b The primary food supply for Ireland at the time was potatoes and about half of the farming population relied on this crop. Can you explain the shape of your graph using this information? (2 marks)

7 Malaria is a disease of tropical and subtropical regions caused by a single-celled parasite of the genus *Plasmodium*. There is currently no vaccine offering a high level of protection against this disease, although efforts to develop one are continuing. Several medications are available to prevent malaria for travellers to malaria-endemic countries, and a range of anti-malarial medications are accessible. The World Health Organization estimates there are over 216 million documented cases of malaria in the world. The following diagram shows the life cycle of the malaria parasite.

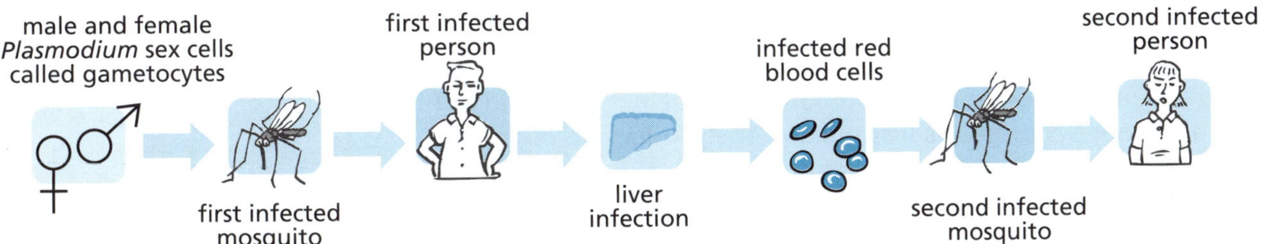

a Arrange the following sentences in the correct sequence. *Hint 5* (4 marks)

i Some of the parasite cells develop into sexual forms, called male and female gametocytes, that circulate in the bloodstream.

ii The parasites travel to the liver and invade liver cells.

iii In a week or two, forms of the *Plasmodium* parasite travel to and invade the mosquito's salivary glands.

iv The cycle of human infection begins again when the mosquito takes a blood meal.

v A female *Anopheles* mosquito carrying malaria-causing parasites injects the parasites into the bloodstream as it feeds on its human victim.

vi When another mosquito bites an infected human, and drinks blood, it ingests the gametocytes.

vii Over a week or two, the parasites grow, divide and produce many thousands of copies in each liver cell.

viii The parasites exit the liver cells and, in the bloodstream, invade red blood cells.

ix In the mosquito gut the infected human blood cells burst, releasing the gametocytes.

b In many poor countries where malaria occurs (especially Africa), medicines are not readily available. Suggest two methods that can be used to reduce the number of malaria cases. (2 marks)

Hint 1: You might need to use your experience to determine whether this is true or not.

Hint 2: Consider what medicines were available at those times to fight the disease.

Hint 3: Consider what features of living things are. Do viruses have them?

Hint 4: Don't forget to label your graph properly.

Hint 5: Use the diagram to help you correctly sequence these sentences.

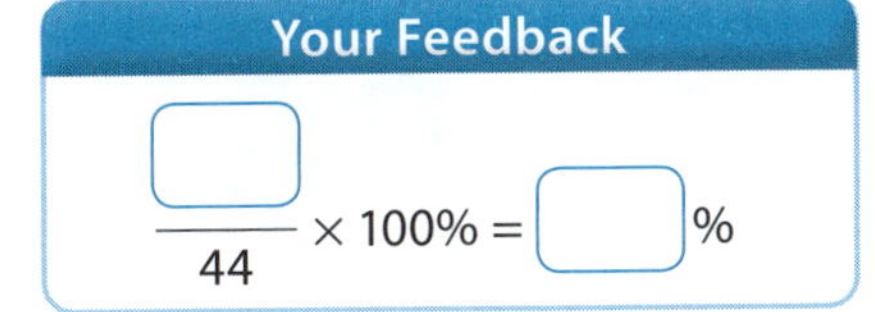

HUMAN HEALTH

The body systems

QUICK REVISION

1 Microbes invade the bodies of plants and __________ in order to reproduce and complete their life __________. They are not there because they want to deliberately cause us harm. Over a very long period, organisms have developed various means to __________ these infections. The human body has defences to attack and destroy the __________ pathogen. In the process, your body acquires __________ to that particular disease. This means that should this particular pathogen attack again, the body can deal with it using special __________ called antibodies. Antibodies are __________ found in blood and other bodily fluids that are used by the __________ system to identify and neutralise foreign objects, such as __________ and viruses. There are a number of different antibodies performing different roles. This vast diversity of antibodies allows the immune __________ to recognise an equally broad selection of antigens. An __________ is a toxin (__________) or other foreign substance (such as bacteria, foreign blood cells or the cells of transplanted organs) that induces the production of one or more __________. The antibody binds to a specific antigen somewhat like the fit between a __________ and a key. The immune system will try to wipe out or neutralise any antigen that is identified as a foreign and potentially damaging invader. (This term originally came from *anti*body *gen*erator as it stimulates the production of an __________.)
The following photo shows antibodies attacking a virus.

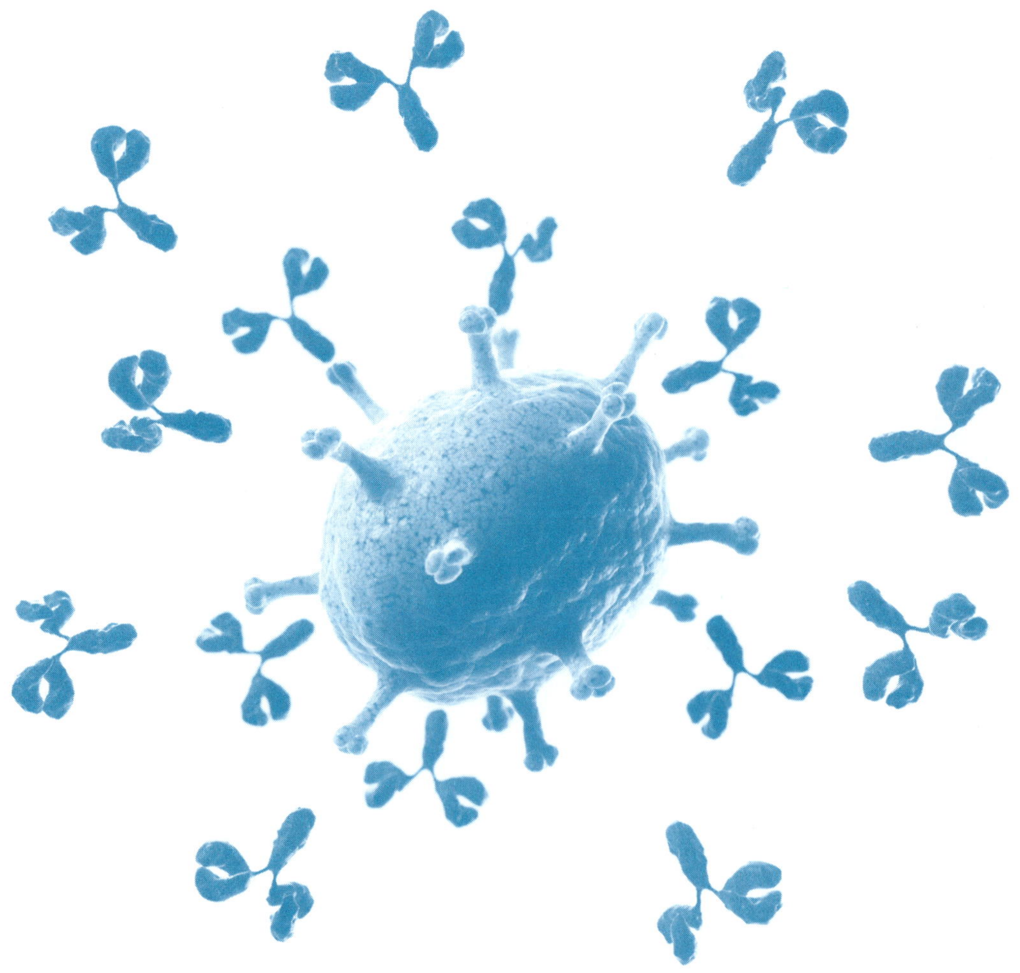

HUMAN HEALTH

The body systems

QUICK REVISION

2 A vaccine is a substance used to encourage the production of antibodies and provide ________ against one or more diseases. Scientists have developed vaccines from killed or weakened bacteria and ________ to stimulate antibody production. This means that the antibodies are ready in your system to ________ any infection from that particular bacterium or virus.

Immunisation is the process of ________ people or animals with dead or inactive versions of microorganisms. Once a person has been vaccinated, their body starts to make ________ that destroy the microorganism. Because the microorganisms in the vaccination are not dangerous, the person does not get ________. However, if they later come into contact with a live (dangerous) ________, then they already have the antibodies they need to ________ it before it multiplies and makes them ill.

The terms *vaccination* and *immunisation* are often used interchangeably but mean slightly different things. Vaccination means receiving a ________, usually by an injection; immunisation means both getting a vaccine and becoming ________ to a disease as a result of it.

3 Technology plays an important part in modern medicine. A common, well-known technology is the use of X-ray images, which have been available for over a ________. High-tech imaging can be used to obtain detailed ________ of body parts and to accurately detect and highlight injected dyes. Use of fibre ________ in medicine and brain scans are other useful innovations. Robot-assisted surgery gives surgeons precise ________ over surgical tools, allowing operations a greater chance of success.

Technology has also produced artificial devices to assist ________. Cochlear implants are probes with several electrodes implanted into the ________ allowing people with severe ________ loss to recognise some sounds. Plastic ________ replace those in the eye which become dulled due to cataracts. New lenses are being developed all the time to make the surgery less complicated for ________ and the lenses more helpful to patients. Another new type of lens blocks both ultraviolet and blue light ________, which research indicates may damage the retina. Mechanical heart valves are made from materials such as metal (stainless ________ or titanium) or ceramics, and are used to replace diseased heart ________.

Answers **1** animals; cycles; fight (attack); invading; immunity; molecules; proteins (chemicals); immune; bacteria; system; antigen; poison; antibodies; lock; antibody **2** immunity; viruses; fight (attack); inoculating; antibodies; ill (sick); microorganism; destroy (kill); vaccine; immune **3** century; pictures (images); optics; control; patients; ear; hearing; lenses; surgeons (doctors); rays; steel; valves

1 A **pathogen** or germ is a microorganism (such as a virus, bacterium or fungus) that causes its host to become ill. The term is most often used for biological agents that disrupt the normal functioning of a multicellular animal or plant. However, pathogens can infect any organism from all of the biological kingdoms.

Transmission of pathogens to humans occurs through several different routes:

- infection through **droplets** or particles in the air
- direct **contact** with either the sick person or materials they have been in contact with
- infection through **blood** and tissue
- infection through **contaminated** water and food.

In most cases an organism's **immune system** keeps it healthy and prevents infections. Plants and animals have a non-specific immune system as the first line of defence. In vertebrates, **white blood cells** (leucocytes) are not tightly linked with a particular organ or tissue, so they can function like independent, single-celled organisms. Leucocytes can move freely around the body to interact with and capture cellular debris, foreign particles or invading microorganisms.

A foreign substance that invades the body is called an **antigen**. When an antigen is detected, several types of cells work together and respond to it by producing antibodies. Antibodies are produced by a type of white blood cell, and they act by:

- combining with some antigens, such as bacterial toxins, to neutralise their effect
- removing other substances from circulation in body fluids
- binding certain bacteria or foreign cells together.

Antibody molecules are specific in that they react with only one kind of antigen. This means that different antibodies are needed for different antigens. Antibodies are not capable of destroying the antigen without help. This is where other cells, such as T cells, come in. Once the danger is over, antibodies remain in your body, so you acquire immunity to that particular disease. Should that microbe attack again, the antibodies are there ready to deal with it.

This principle forms the foundation of **immunisation**. Immunisation introduces an antigen to the body in a manner that does not make a person ill, but allows the body to make antibodies that will guard that person from future assault.

2 Immunisation protects people against harmful infections before they come into contact with them. It uses the body's natural defence mechanism, called the **immune response**, to build resistance to particular infections and allow people to remain healthy. When an immunised person comes into contact with that disease in the future, their **immune system** can respond promptly to prevent the person developing the disease. A newborn child doesn't have any opportunity to develop protective antibodies on its own. Its immune system is compensated by having some of its mother's antibodies in its system. Many of these are transferred across the placenta before birth and through the mother's milk after birth.

Vaccines contain a minute dose of either:

- a live, but weakened, form of a virus
- a killed bacteria or virus, or some small parts of bacteria
- an altered toxin produced by bacteria.

In general, the normal immune response takes around a fortnight to work. That is, protection will not occur straight away after immunisation. Many immunisations need to be repeated several times to build long-lasting protection. However, there are some vaccines which give protection after only one dose. The protective effect of some vaccines can be several decades, while others last for shorter periods. **Booster doses** are required because immunity lessens over time. Due to repeated changes (called mutations) to the influenza virus, annual influenza vaccination is necessary to safeguard against the latest virus.

The protection levels afforded by vaccines differ. For example, the measles, mumps and rubella (MMR) vaccine protects over 95% of children who have completed the course. While a small percentage of those vaccinated with MMR might still catch one of these diseases, it is often milder than it would be otherwise. However, most children who are not vaccinated and exposed to measles will catch the disease, which carries with it a high risk of complications like lung infection (pneumonia) or brain inflammation (encephalitis). Vaccines are also more economical in preventing the spread of diseases than trying to treat outbreaks afterwards. Whole communities can be protected because contagion is lessened or prevented altogether.

There are strict requirements in Australia that all vaccines are rigorously tested for effectiveness and monitored for safety before being released for general use. Laws are stringent so many vaccines are trialled with large numbers of volunteers and checked for safety. It can take many years before they are ready to be released. While some children experience minor **side effects** following immunisation, these are not long-lasting and the child recovers without any problems. Very rarely (one in every million), a vaccine causes a severe allergic reaction but, given some diseases can cause permanent damage or even death, the very slight risk is worth it.

3 Technological advances often have important applications in medicine. One of the first uses of electromagnetic radiation, of which visible light is a small part, was X-ray imaging. Wilhelm Röntgen discovered **X-rays** in 1895, and soon afterwards they were being used for diagnosis in medicine. (Röntgen received the first Nobel Prize in Physics in 1901.) Medical X-rays are generated when a stream of fast electrons comes to a sudden halt on a metal plate. The images formed by X-rays result from the different **absorption rates** by different tissues. Calcium in bones absorbs X-rays the most, so they look white on a film recording of the image, called a radiograph. As fat and other soft tissues absorb fewer X-rays, they look grey. Air absorbs the least, so lungs look black on a radiograph.

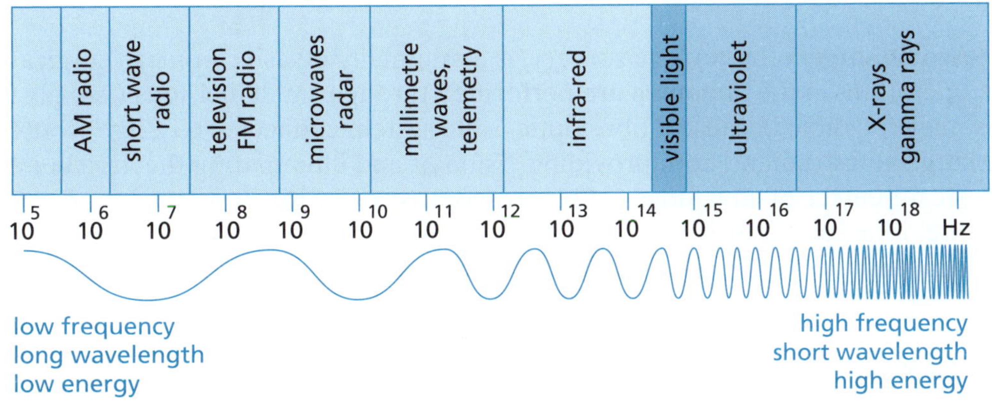

Other bands in the electromagnetic spectrum are also used in medicine.

- **Microwaves** can be used to detect cancerous breast tumours.
- **Infra-red** (or heat) lamps can be used to treat sore muscles and stiff joints, and infra-red cameras can detect cool and hot regions on the skin. Hot regions may indicate infection.

(cont.)

- **Ultraviolet (UV)** radiation is used to sterilise equipment in hospitals as it kills bacteria and viruses. UV lamps can diagnose fungal skin infections and detect foreign objects in the eye. UV lamps can also be used in dentistry to speed up the polymerisation process in tooth filling as shown in the photo below.
- **Gamma rays** can be used for sterilisation, and to treat certain types of cancers and tumours.

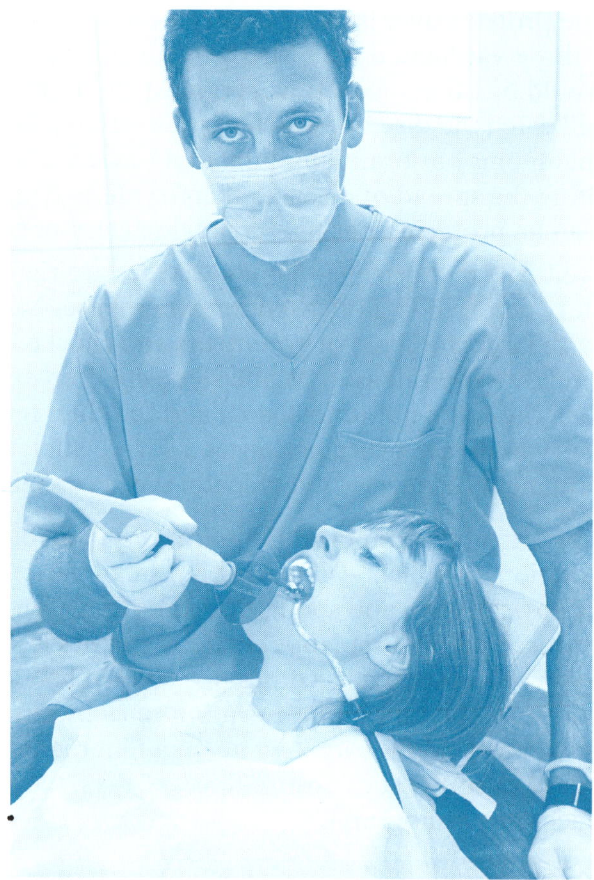

Laparoscopic surgery, or keyhole surgery, is a minimally invasive modern surgical technique where operations in the abdomen are performed through small incisions (usually 0.5 to 1.5 cm). This is made possible because a fibre optic cable system connected to a light source is attached to the surgical instrument, both providing vision of and illuminating the area being operated on. There are a number of advantages of laparoscopic surgery for a patient, compared to an open procedure. The smaller incisions mean the patient experiences reduced pain and haemorrhaging (bleeding), and recovery time is shorter.

Checklist

Can you:

1. *Describe how the immune response works against pathogens?* ☐
2. *Describe the benefits of vaccination and immunisation?* ☐
3. *Outline how electromagnetic radiation is used to diagnose and treat disease?* ☐

REVISION TEST

1 a Give three reasons children should be immunised. *Hint 1* (3 marks)
b Why do children need to get so many vaccinations? (3 marks)

2 There are many ways that harmful microorganisms can spread. Suggest five methods. (5 marks)

3 The following concerns are reasons that some parents give for refusing to immunise their children. Discuss why these concerns should not prevent vaccination.
a the side-effects of immunisation (2 marks)
b a body's natural immunity will protect against infection (2 marks)
c the safety of vaccines (2 marks)

4 a What is the purpose of the body's immune system? (1 mark)
b Discuss the role of white blood cells when pathogenic microbes attack the body. In your answer use the words *antibodies* and *antigens* to show you know what they are. (7 marks)

5 When the body detects a pathogen it begins to make antibodies to fight the infection. The number of antibodies will continue to grow until the infection is under control. Once the infection has passed many of these are no longer needed, but a few will remain to provide 'memory' for your body to react quickly and efficiently to future exposures.

The following graph shows the response of the body after infection by a pathogen. Use this graph to describe how the number of antibodies would vary between a person contracting a disease and getting over it. *Hint 2* (6 marks)

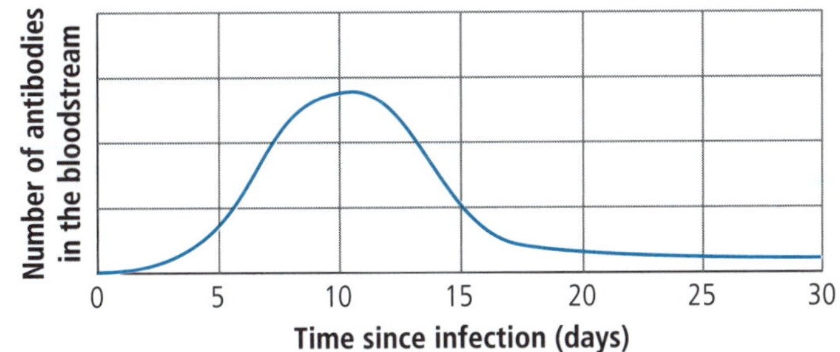

6 When there is an influenza outbreak, it takes time for a vaccine to be prepared and then distributed.

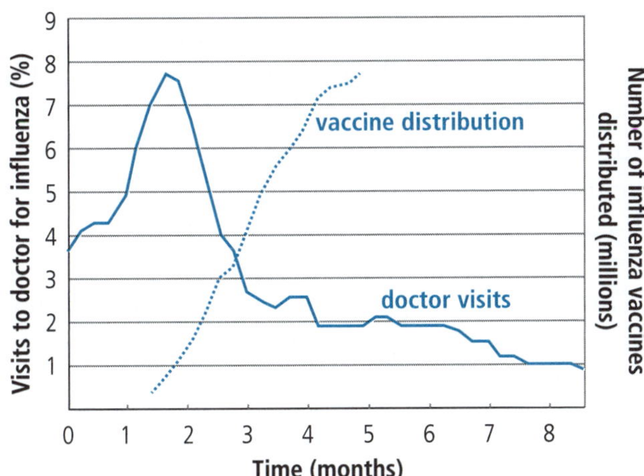

(cont.)

a Why does it take some time for the influenza vaccine to be prepared? (2 marks)

b Does the graph show whether or not the vaccine is effective against influenza? Explain. (2 marks)

c Most people will recover from contracting influenza. Why, then, is a vaccine necessary? (3 marks)

7 Before vaccines, the only means of becoming immune to a disease was to get it and, by fortune, live through it. This is termed naturally acquired immunity. With this immunity, you put up with the symptoms of the disease and a possibility of complications, which can be quite serious or even fatal. In addition, during certain phases of the sickness, the individual may be contagious and spread the disease to others they come into contact with. List five advantages of being vaccinated. (5 marks)

8 Measles is a severe disease caused by an extremely contagious virus, which spreads when people touch or breathe in infectious droplets passed by coughing and sneezing.

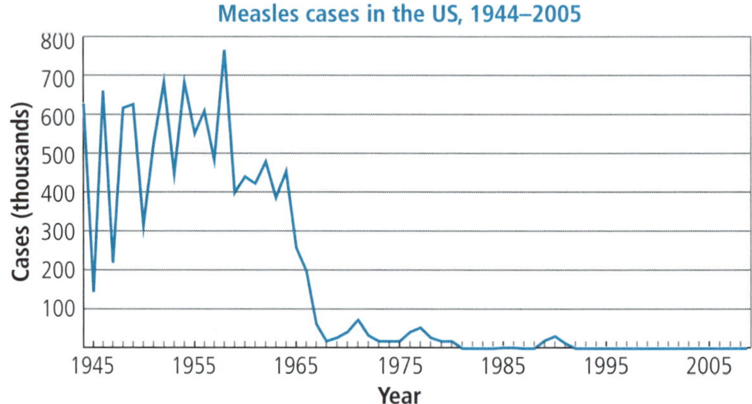

a Approximately how many cases of measles were reported in the United States each year in the late 1950s to early 1960s? (1 mark)

b In which year did the number of reported cases fall below 100 000? (1 mark)

c Use the graph to estimate when a measles vaccine was licensed for public use. (1 mark)

d What evidence is there to support your answer? (1 mark)

9 Most modern composite resins used in tooth fillings are light-cured photopolymers, meaning that they harden with light exposure. The dental LED (light-emitting diode) curing lights use a narrow spectrum of blue light in the range 400 to 500 nm. This takes about 20 seconds.

a What is the function of the curing lights? *Hint 3* (1 mark)

b Before the dental curing light, different materials were used in tooth fillings. Generally this material was a self-curing resin. It was made up of two parts, mixed separately and then placed into the tooth. On mixing, the process started and it self-cured (hardened) fully after 30 to 60 seconds. This presented a problem for the dentist. What was it? (2 marks)

c How did the invention of the photopolymer and the curing light give dentists an advantage? (2 marks)

10 Fibre optics is the technique of transmitting light through transparent, flexible fibres of glass or plastic. Optical fibres can now be used to explore the insides of the body. In medicine, they allow doctors to look and work inside the body through tiny incisions without having to perform major surgery.

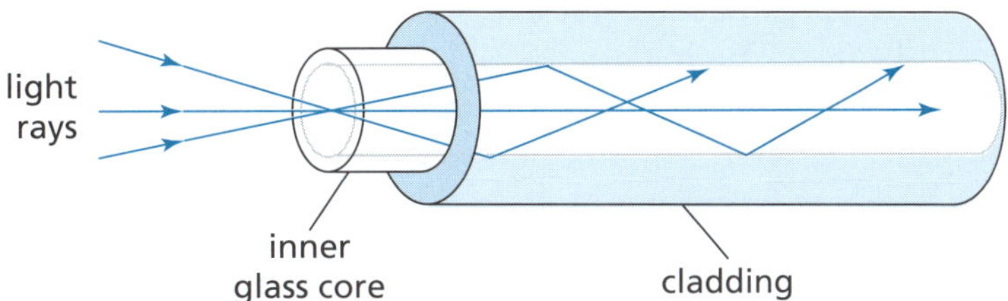

a An endoscope is an instrument for viewing the interior of hollow organs in the body, such as the stomach or intestine. How are optical fibres useful in endoscopy? (1 mark)

b A doctor can use an arthroscope to examine knees, shoulders and other joints. The following photos show arthroscopic images of the cartilage and tendons in a knee. An arthroscope is a tubelike instrument making use of optical fibres to examine and treat the inside of a joint.

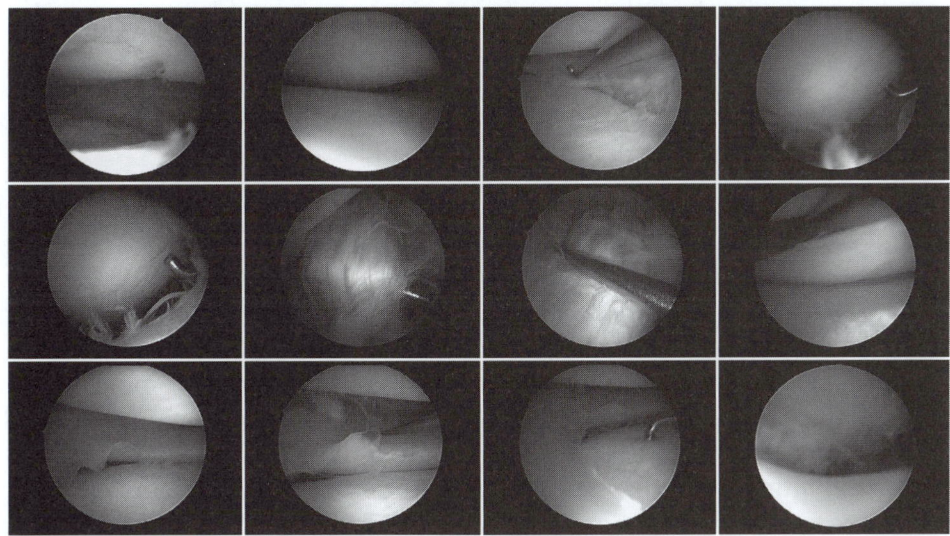

Explain how optical fibres are useful in arthroscopy. (1 mark)

c How might these examinations been done before the invention of optical fibres? (2 marks)

Hint 1: Do young children have as much resistance to diseases as adults do?
Hint 2: Think about why the number of antibodies should rise and fall like this.
Hint 3: A photopolymer is a polymer or plastic that undergoes a transformation in physical or chemical properties when exposed to light.

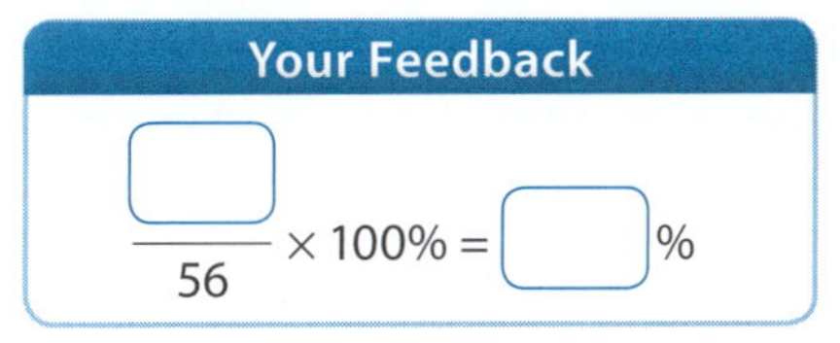

STUDYING AN ECOSYSTEM

Ecology

QUICK REVISION

1 In nature various organisms live together and interact with each other as well as the __________ components of their environment. The water, soil and temperature of the air influence the types of organisms that live in a community. Only organisms that are __________ to cold can live in cold climates. On a rocky headland, only organisms that can withstand strong __________ and sprays of sea salt can survive. In the leaf litter on a forest floor live a wide range of invertebrates. The higher relative __________ levels in the leaf litter prevent organisms from drying out. Soil and leaf litter have __________ acidity levels which affect the types of living things found there. In an open woodland the sunlight penetrates to __________ level, allowing a wide range of green plants to live there. Most grasses, for example, need __________ light levels to grow. However, in a dense forest __________ light reaches the ground, so grass will not readily grow there. These non-living factors are called __________ factors.

The biological or __________ components of the environment also influence the __________ of each organism. Competition for food between members of the same species or with other species is a major __________ issue. There is also competition to __________ and produce offspring. In a lion population there are smaller groups called prides. In a pride there is only __________ dominant male. Other males either leave and establish a new pride or fight the dominant male for his position. Living things need to shelter from rain, wind and other harsh __________ conditions. They require safe shelter for sleeping. Predators are always on the hunt for __________ to eat. Predators often select the weakest members of a prey population to attack and kill, or the one they can easily isolate from a group.

2 Field studies are the best way to investigate ecosystems. Your school may have a large garden in which various __________ are available for study. Other interesting ecosystems are freshwater creeks, woodlands and rocky shores. Let us consider a freshwater creek. The first step in an investigation is to make a __________ of the area of study. On the map you provide a scale and some of the main features, such as the creek and large trees. The next step is to measure __________ in abiotic factors in different zones of the study area. Temperature is easy to measure using a __________. Measure the temperature of the air at ground level as well as the soil temperature at various __________. The temperature of the creek water will also vary. The __________ humidity of the air in exposed and sheltered sites can be measured with a __________ and dry bulb thermometer. The wet bulb thermometer has some damp gauze cloth around the bulb and as the water __________ it cools the bulb. The difference in temperatures between the dry and wet bulb allows the relative humidity to be calculated. The amount of light in the area of study affects the types of organisms living there. In __________ light areas fewer green plants will grow unless they are adapted to low light. Similarly in the creek, the clearer the water the more light will penetrate and the population of submerged water plants will be greater. If the water is __________, however, then light penetration is reduced and the water plant population is reduced. During your field study you need to record the __________ and types of living things. Make a note of their location within the ecosystem. Take photos of these organisms so you can include them in your report.

3 Competition and predation are two of the most important biotic factors in a community of living things. It is important for organisms to __________ long enough to reproduce, otherwise a species may become extinct. Extinction of plant and animal species is becoming increasingly common around the world due to the activities of __________. In a natural population, organisms must find food and water to stay alive. The __________ individuals within a population often do not survive in the struggle to find food and water. They quickly become the ____________ of various predator species. Changes in weather can also lead to population decline. In a drought, grass-eating animals may find that there is little ____________ to eat. Some individuals become weaker and predators prey on them, further ____________ their population. Long-term droughts can lead to local ____________ of a species. In turn, the population of predators declines as their food sources disappear. Plants also compete for water, ____________, minerals and sunlight. Some species of plants are more competitive than others. Introduced plants may out-compete ____________ plants and gradually drive their population numbers ____________ to very low levels. Overgrowth of some surface algal species in rivers can lead to a decline in water plants that live on the river bed. The algae growing at the surface ____________ sunlight from penetrating into the water. The algae also stop oxygen diffusing from the ____________ into the water. The following photo shows algae growing on the surface of a pond.

Answers **1** non-living; resistant (adapted); winds; humidity (moisture); variable (different; higher); ground; high (strong); less; abiotic; biotic (living); populations; survival; mate; one; environmental; prey (animals, food) **2** habitats; map (diagram); variations (differences); thermometer; depths; relative; wet; evaporates; low; turbid; number **3** survive; humans (people); weakest; prey (food); grass; reducing; extinction (disappearance); soil; native; down; block (prevent); air

STUDYING AN ECOSYSTEM

Ecology

REVISION SUMMARIES

1 An **ecosystem** is a community of living organisms interacting with the non-living components of their environment. The non-living (**abiotic**) environment includes both physical and chemical components. These components can be measured and include:

- air and water temperature, measured using a thermometer
- wind speed and direction, measured using an anemometer
- relative humidity of the air, measured using wet and dry bulb thermometers
- light intensity, measured using a light meter
- soil and water acidity, measured using indicators or a pH meter.

Some of these components can vary daily due to weather changes and others change over a longer period of time.

The biological (**biotic**) components of the environment include the following.

- Competitors: living things that compete with each other for food, mating partners, shelter or food.
- Predators: living things that kill and feed off other living things.
- Shelter: places to rest and remain safe.
- Food: source of nourishment that provides nutrients for growth, reproduction and maintaining life. Different populations of organisms have different food requirements.

2 In ecosystems there are many habitats or places where different populations of organisms live. Consider a freshwater creek and the surrounding bushland for example. The communities that live on the surface of the creek are different to those that live on the bottom. Even the edge dwellers of the creek are quite a distinct **population** from communities in other areas of the ecosystem. When you study a freshwater creek ecosystem, you need to map out the area and take a variety of measurements of the abiotic components. Let us examine some of these.

- **Temperature.** If you are investigating the creek, find out how the water temperature changes with depth. Do this by placing a sample bottle into a weighted tin and lowering it by a string to the depth you want. Measure the temperature straight away. Do this for different depths. If you are investigating the bushland leaf litter, measure the temperature on the top and at different depths in the litter.

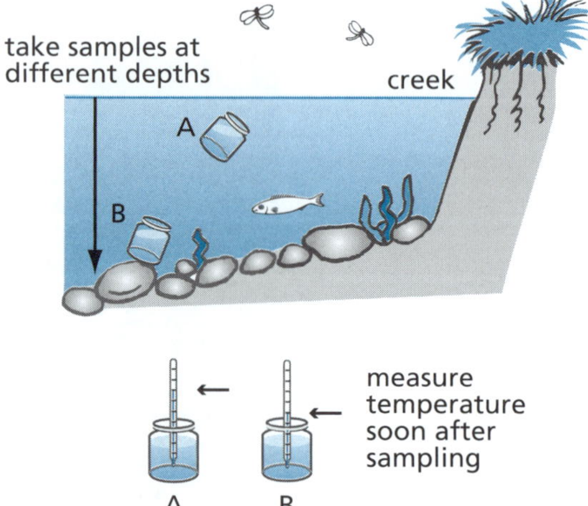

- **Water turbidity.** The presence of unsettled sediment in the water cuts down how much light penetrates the water. Turbidity is the **cloudiness** or haziness of a liquid due to individual particles (suspended solids) that are generally unseen by the naked eye. Measuring turbidity is an important test of water quality. In order to measure turbidity you can use a turbidity disc. The turbidity disc is a metal or wooden disc that is painted black and white and is weighted at the centre. Supporting strings attached at the edge of the disc are attached to a long (2.5 m) central string. Lower the disc into the water until the white surface just disappears. Mark this depth on the string. Use a tape measure to measure the length of the string that was underwater. Record this depth and compare it to the real depth of the creek at this point.

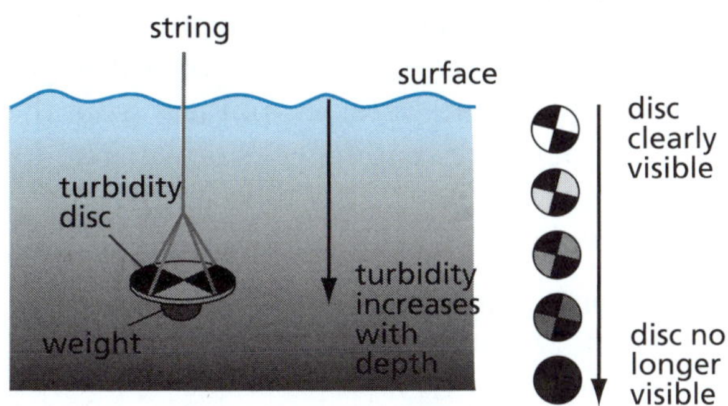

- **Wind speed and direction.** You can classify the wind speed qualitatively using terms such as strong wind, light wind and calm, or you can use a wind meter (anemometer) to record an actual speed (this is a quantitative measure). Estimate the approximate direction of the wind by observing the motion of tree branches or by throwing a few small leaves into the air and noting which direction they move in. Use your magnetic **compass** to find the direction the wind is coming from.

3 **Competitors** and **predators** are important examples of biotic components of an ecosystem. Competition occurs between organisms within a species and between species. This usually occurs where there is a limited supply of food, water or space to live. The following are some examples of competition.

- Trees can compete with other trees for available light, water and minerals in the soil. In some forests, the light reaching the ground is reduced due to the dense canopy of trees. Grasses may find it difficult to grow in such low light conditions.
- Communities of birds may compete for the same food sources in an ecosystem. The more aggressive species may out-compete the less aggressive species and so drive them out of the area. This can also happen with bird populations that are migratory.
- Some plants and fungi produce toxic chemicals to assist them to compete for available habitats. *Penicillium* moulds secrete a chemical (penicillin) that kills or inhibits the growth of bacteria.

Predators kill and eat other organisms. The **prey** is the organism that is eaten. For example, a snake is a predator of some frog species. The snake, however, may be the prey of a kookaburra. In any ecosystem the population of the prey is greater than the predator population. Predators compete amongst themselves for prey. The population of more successful predators will rise and the population of less successful predators will fall. In some cases the population of the least successful predator may drop to zero and it becomes **extinct** (disappears altogether) in that ecosystem. (Scientists use the term **extirpated** when a species disappears from an area.)

Checklist

Can you:

1 *Distinguish between the abiotic and biotic components of an ecosystem?* ☐
2 *Describe ways of measuring some abiotic features of an ecosystem?* ☐
3 *Distinguish between predators and prey in an ecosystem?* ☐

STUDYING AN ECOSYSTEM

Ecology

REVISION TEST

1 The relative humidity of the air can be measured using wet and dry bulb thermometers (called a hygrometer). The wet bulb thermometer has a piece of cotton gauze wrapped around the bulb and this gauze is kept damp. As the water evaporates from the gauze surface it removes heat from the bulb and the temperature reading goes down. The following table shows a sample calculation of the relative humidity when the dry bulb temperature is 24 °C and the wet bulb temperature is 20 °C. The answer is 67% relative humidity.

Wet-bulb readings (°C) \ Dry-bulb readings (°C)	15	16	17	18	19	20	21	22	23	24	25	26	27	28	29	30
15	100	89	80	71	63	56	49	43	38	33	28	24	21	17	14	12
16		100	90	80	72	64	58	50	45	39	34	30	26	22	19	16
17			100	90	81	73	65	59	52	46	41	36	32	28	24	21
18				100	90	82	74	67	60	53	47	42	37	33	29	26
19					100	91	82	74	67	61	54	48	43	39	36	31
20						100	91	82	75	67	61	55	50	45	40	36
21							100	91	83	75	68	62	56	51	46	41
22								100	91	83	76	69	63	57	52	47
23									100	91	84	75	70	64	58	53
24										100	92	84	77	70	64	59
25											100	92	84	77	71	65

temperature change

dry bulb

cotton gauze sack around bulb keeps it moist

wet bulb

water

a Is relative humidity a biotic or abiotic factor in the environment? (1 mark)

b The relative humidity of air near the ground in a deep gully was measured. The dry bulb temperature was 28 °C and the wet bulb temperature was 25 °C. Determine the relative humidity. (1 mark)

c On a day when the relative humidity was 50%, the dry bulb temperature was 22 °C. What was the temperature on the wet bulb? (1 mark)

d A student noticed that the relative humidity of the air in deep leaf litter on a forest floor was much higher than the air 50 cm above the ground. This leaf litter supported large invertebrate populations. Suggest a reason for this. (1 mark)

2 A group of students was given a sample of creek water which was slightly turbid. The aim of their investigation was to measure the mass of suspended material in a known volume of the water. The equipment they used to determine the mass of suspended solids is shown in the diagram to the right.

a Are the students measuring a biotic or abiotic component of the creek ecosystem? (1 mark)

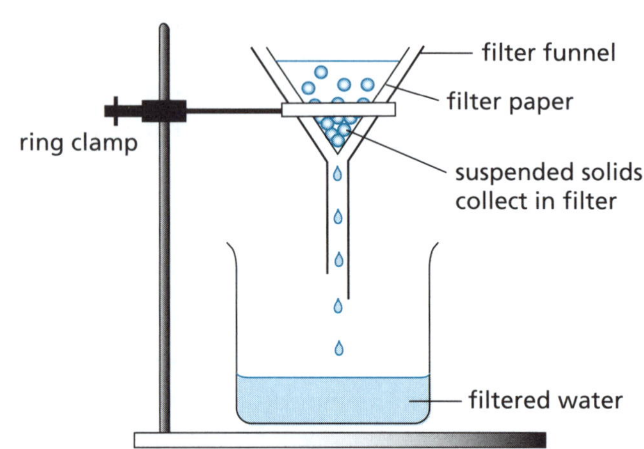

b Discuss how the amount of suspended solids in the creek water can affect the populations of living things in the water. *Hint 1* (3 marks)

c Write the method for the students' experiment. *Hint 2* (6 marks)

3 Some water was collected from a creek during a field study. The water looked clear to the naked eye but when examined with a hand lens with a magnification of 10 times (10×) very small dots could be seen. When viewed under a microscope at two different magnifications the water was found to contain the microscopic invertebrates shown in the drawing below.

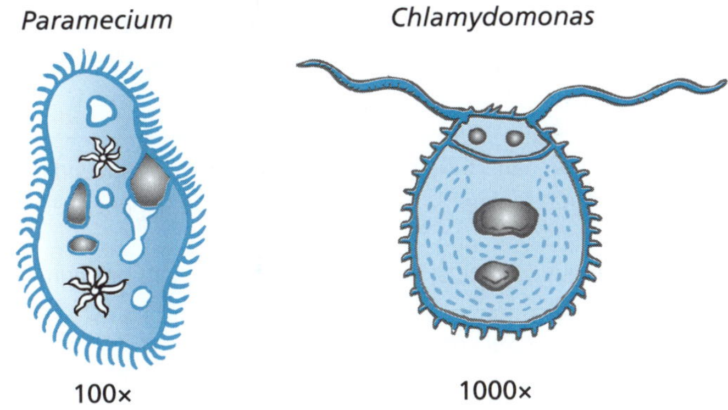

a What do the numbers 100× and 1000× mean in this drawing? (1 mark)

b Which invertebrate is the larger of the two? (1 mark)

4 A group of students mapped an area in a grassy ecosystem in which a purple creeper had started to grow. They drew the grid below to show their results.

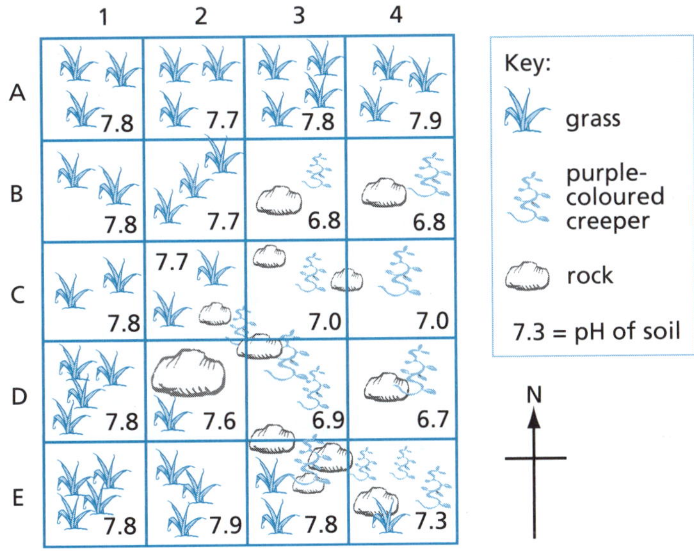

True or false?

a Grasses do not grow near rocks. (1 mark)

b Purple creepers grow on the eastern side of the rocks. (1 mark)

c Grasses and creepers do not occur in the same grid square. (1 mark)

d Grasses are more prevalent than creepers. (1 mark)

(cont.)

e Creepers seem to lower the pH of the soil. (1 mark)

f The largest rock is found in grid reference 2D. (1 mark)

5 Explain the difference between the environment of an organism and the habitat of an organism. (2 marks)

6 The following key can be used to distinguish between and classify segmented worms and insect larvae. Some of these groups can be found in leaf litter communities and freshwater creek ecosystems. In order to use the key, start at question 1 which has alternative answers (1a and 1b). Examine each organism shown below the key and choose the most appropriate alternative. The answer you choose will tell you to go to the next appropriate clue. Repeat this process until you have named the organism. (4 marks)

1a	body soft, wormlike; more than 15 segments; no obvious legs or appendages	go to 2
1b	body wormlike; 15 or more segments; fleshy prolegs (leg-like appendages) may be present on some segments	go to 3
2a	elongated body; numerous segments; bristles sometimes present on segments; 50 mm long or more	Oligochaeta (freshwater worm)
2b	rounder segmented body; suckers on front and rear; up to 100 mm long	Hirudinea (leeches)
3a	larvae with prolegs on the thorax	go to 4
3b	thorax without prolegs	go to 5
4a	one stout leg on the thorax; head with prominent brushes; up to 20 mm long	Simuliidae (sand or black fly larvae)
4b	air of prolegs on first thorax segment; often bright red; up to 18 mm long	Chironomidae (gnat or midge larvae)
5a	cylindrical body; pointed at both ends; abdominal segments with a girdle of fleshy prolegs; 10 to 12 mm long	Tabanidae (horse or marsh fly larvae)
5b	one pair of prolegs on first two abdominal segments; bristles at end of abdomen; U-shaped when resting; 12 to 14 mm long	Dixidae (midge larvae)

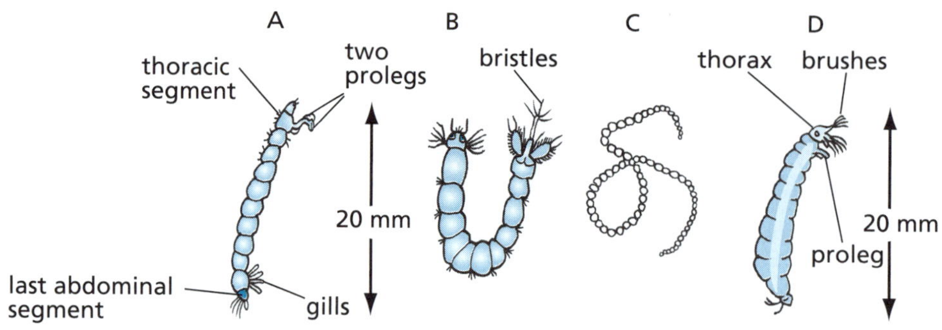

Hint 1: Suspended solids such as mud take a long time to settle and this can influence the depth to which sunlight penetrates into the water.

Hint 2: A method is a series of numbered steps stating what needs to be done.

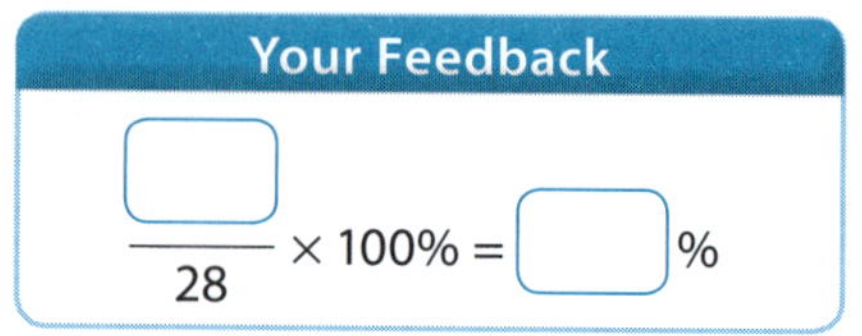

PAGE 224
PAGE 252

FOOD CHAINS AND FOOD WEBS

Ecology

QUICK REVISION

1 Living things depend on the radiant ________ from the Sun. This sunlight is absorbed by organisms called ________. Green ________ and algae are examples of producers. In the upper sunlit layers of the oceans, however, microscopic organisms called ________ are the major producers. Producers require carbon dioxide, ________ and sunlight to produce organic compounds such as ________ and oxygen. This process is called ________. The oxygen that is generated becomes available to all life forms. Consumers use this oxygen to ________ nutrients in their cells in a process called respiration. This process provides usable energy to the consumer. Carbon ________ is then released back into the environment. Consumers can be classified into ________ levels. Herbivores are animals that ________ producers. Herbivores are called ________-order consumers. Second-order consumers feed off ________-order consumers. Higher order consumers (e.g. third, fourth) eat the consumers in the lower feeding levels. All the ________ and higher order consumers are called carnivores. Some organisms are called ________. They eat producers and other carnivores. Humans are omnivores. Organisms called ________ are ones that feed off the remains of animals that have not been completely eaten by other carnivores. Hyenas are scavengers that feed off the remains of herbivores that have been ________ and eaten by lions. Eventually all living things die and their remains are consumed by organisms called ________. Decomposers include ________ and moulds. Minerals and other nutrients are returned to the environment due to the action of these decomposers.

2 Food is essential to ________ life. In the animal kingdom food is obtained by eating plants or other animals. A food chain summarises the feeding order of a selected group of organisms. A food chain is therefore a flow chart. The chain starts with a ________ that absorbs energy from the Sun. The producer makes ________ using the Sun's energy. Organisms that eat these plants are called ________ and they are the second link in a food ________ flow chart. In turn the herbivores are eaten by a ________-order consumer or carnivore. The food chain then continues with increasing orders of consumers. Each box of the flow chart is separated by an arrow. This arrow stands for the words 'is ________ by'. In a food chain that starts with 'lettuce → caterpillar', the lettuce is eaten by the caterpillar. The arrow indicates the flow of energy. In the ocean, the producer is likely to be ________ or phytoplankton. On land the producer is more likely to be a ________ plant such as grass or wheat. On the ________ common herbivores include cows, sheep and horses. In the marine environment, common herbivores include molluscs, some fish and manatees. Further up the food chain are the higher order carnivores. On land these include snakes, tigers and eagles. In the ________ environment these include sharks, rays and sperm whales.

3 Herbivores (or first-________ consumers) eat a variety of different plants. Second-order carnivores eat a variety of different ________. In similar fashion, higher order carnivores eat a wide variety of prey. Consequently there are many different food chains in an ecosystem. Food chains can be combined to form a food ________ which is a more accurate representation of feeding relationships in an ecosystem. In a ________ web, ________ animals can be both first- and higher order consumers as they eat a great variety of different foods. Goats, like humans, are omnivorous.

Answers **1** energy; producers; plants; phytoplankton; water; glucose; photosynthesis; respire (break down); dioxide; feeding (trophic); eat; first; first; second; omnivores; scavengers; killed; decomposers; bacteria **2** sustain (support); producer; food; herbivores; chain; second; eaten; algae; green; land; marine **3** order; herbivores; web; food; omnivorous

FOOD CHAINS AND FOOD WEBS

Ecology

REVISION SUMMARIES

1 **Producers** are plants, algae or phytoplankton that absorb sunlight and use that energy to make food from raw materials such as carbon dioxide and water. This process is called **photosynthesis**. Oxygen is released into the atmosphere during photosynthesis. Carbon dioxide is released from organisms as they extract energy from their food by a process called **respiration**. This carbon dioxide is then available to plants.

Consumers are animals that eat producers or other consumers. They use the oxygen provided by the producers for respiration.

- First-order consumers are **herbivores** that eat the producers.
- Second-order consumers are **carnivores** that eat the herbivores.
- Higher order consumers are carnivores that eat other carnivores. They may include **omnivores** that eat both plants and herbivores or other carnivores.
- **Scavengers** are animals that feed off the remains of animals left by carnivores.
- **Decomposers** are microbes that break down dead plants and animals to extract the nutrients. Bacteria and moulds (fungi) are common decomposers.

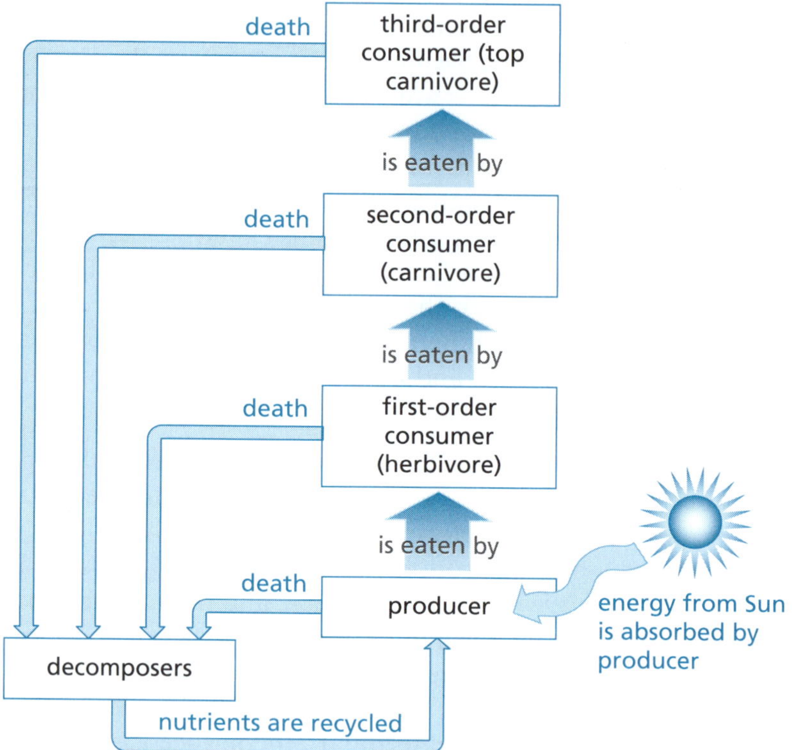

2 Obtaining food is vital for survival. A **food chain** is a form of flow chart that conveniently summarises what different consumers in an ecosystem eat. A food chain begins with a producer such as a green plant or algae. These producers are eaten by a **first-order consumer** or herbivore. The herbivore is then eaten by a second-order consumer or carnivore. In some food chains, higher order consumers eat carnivores lower in the chain.

The diagram on the next page shows a simple food chain for organisms living on a sandy beach and in coastal waters. In a food chain each organism is separated by an arrow. The arrow means '**is eaten by**', and shows the direction of movement of energy. The sandy beach is an arid habitat and most life forms within it are mobile because of the changing tides. At the back of the beach, hardy, pioneering plants bind the sand with their long, branching roots. Leaf-eating insects and crabs live in this upper zone. Lower down the beach in the zone between the tides burrowing crabs and sand worms are found. Burrowing is important to avoid the drying action of the wind and the Sun. As the tide comes in and the beach is partially covered by the seawater, tube worms and fan worms display their tentacles to filter and trap food particles in the water. Small shrimps scavenge decaying debris. Crabs begin to hunt for shrimps and scavenge for other debris. Small fish search out and feed off worms, larvae and various crustaceans such as shrimp.

As the tide ebbs and retreats, the feeding ceases and the larger fish head once more into deeper water. Sea birds can feed off larger fish.

3 In any one ecosystem there are many food chains because of the great variety of feeding relationships. In a freshwater creek community the producers include large water plants as well as microscopic **phytoplankton**. They receive their energy from the Sun. Various first-order consumers (herbivores), such as copepods and snails, rely on these producers for food. In turn, the second-order consumers (carnivores), such as dragonfly nymphs and fish, eat the first-order consumers. Second-order consumers are predators. A **food web** of common organisms of the freshwater creek ecosystem is shown below. You will notice that in this complex web of feeding relationships, some organisms (e.g. fish) can be both first-order and higher order consumers.

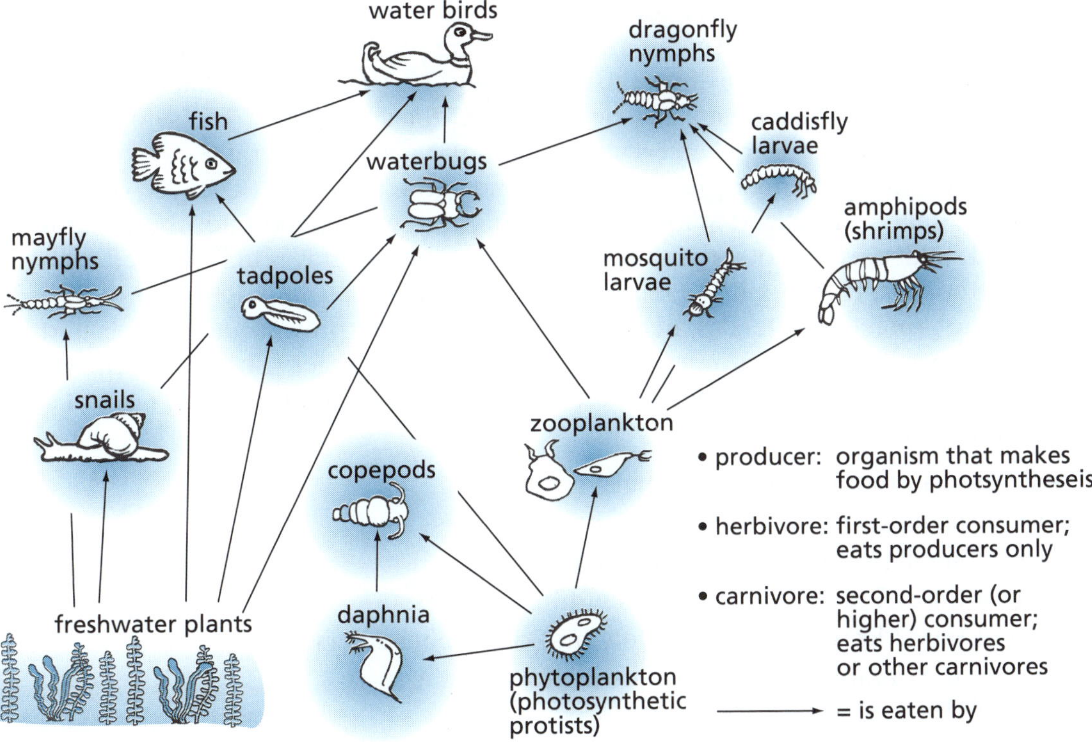

Checklist

Can you:

1 *Classify organisms as producers and consumers?* ☐
2 *Construct simple food chains for a community of living things?* ☐
3 *Analyse food webs and classify organisms as producers and various orders of consumers?* ☐

FOOD CHAINS AND FOOD WEBS

Ecology

REVISION TEST

1 The diagram below shows a simple food chain for a river estuary.

seagrass → sea snails → crabs → egret

a Name the second-order consumer. (1 mark)

b What would happen to the population of organisms in this food chain if large numbers of the producer suddenly died? (2 marks)

c When these organisms die, their remains are decomposed. Name a common decomposer. (1 mark)

2 Read the text about the organisms living in a muddy river estuary and answer the questions that follow.

A rich population of organisms live in the mudflats at the river mouth. The producers in this community include the mangrove trees, sea grasses (*Zostera*) and microscopic phytoplankton. As the leaves of the mangroves die, they fall into the water and the bacteria and microscopic fungi begin the decaying process. Sea grasses can be found growing in the shallow, clear waters of the estuary. These sea grasses make food by photosynthesis. This food is stored inside the plant and when it dies, and bacterial and fungal decomposition begins the rotting process, the tissue particles become a valuable food source for many estuarine organisms. Some of this food finds its way into the sea as the tide runs out and is an important food source for coastal marine organisms. Sea snails eat the sea grasses. The semaphore crab, which burrows into the mud, emerges to feed on the nearby sea grasses and decayed leaves. Mudworms feed on the microbes in the mud and extract whatever decayed organic matter is present. Small prawns and mud whelks live in these waters and filter the water to find the algae and small particles of larger animals that they eat. Clams remain buried in the mud and get their food fragments and oxygen via a siphon which is pushed up to the surface. Young fish use the protected waters as feeding grounds. They eat prawns and protists and when they reach maturity they swim into the deeper ocean waters. Larger fish feed off the smaller fish and the prawns. Birds such as egrets nest in the canopies of mangroves. They eat small fish, crabs and prawns.

a Is *Zostera* a producer or a consumer? (1 mark)

b Are sea snails producers, herbivores or carnivores? (1 mark)

c Are the small fish herbivores or carnivores? (1 mark)

d True or false? Prawns are examples of filter feeders. (1 mark)

e True or false? Prawns are exclusively first-order consumers. (1 mark)

f Identify the organism X in the following food chain. (1 mark)
PROTIST → X → EGRET

3 a Read the following information about a marine rock pool and use it to construct a food web. *Hint 1* (11 marks)

Small herbivorous molluscs feed off the green algae in the rocky pools on a marine rock platform. At high tide, the green phytoplankton flood into the rock pools and are a source of food for barnacles, oysters and tube worms, which are filter feeders. Brittle stars and sea urchins feed off the oysters and barnacles. Very small fish in the rock pool also consume the phytoplankton, and in turn may be eaten by the sea anemones. *Morula* is a sea snail which burrows through the shells of the herbivorous molluscs and eats them.

b Predict what would happen to the numbers of herbivorous molluscs if the populations of green algae suddenly increased due to fertilisers being washed into the water from nearby farms. *Hint 2* (1 mark)

4 A typical leaf-litter food web from an Australian woodland is shown below. The dead leaves are a food source for many organisms. Answer the questions that follow.

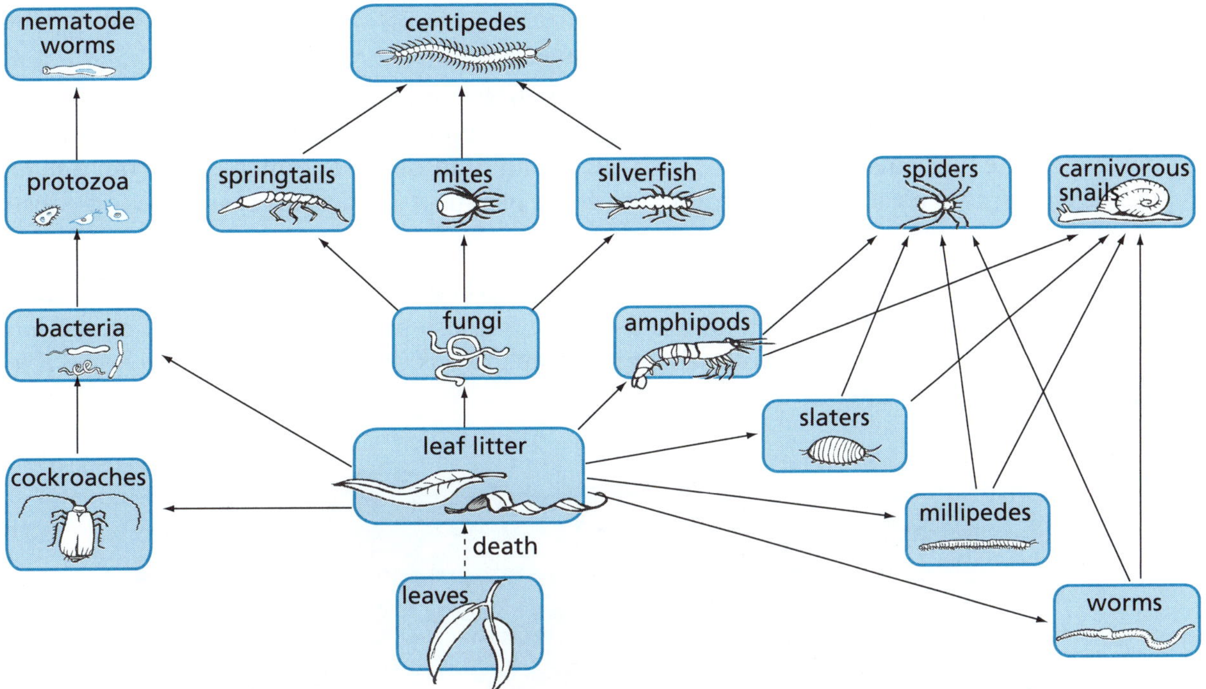

a How many different herbivores eat the leaf litter? (1 mark)

b Identify the organisms that eat amphipods. (2 marks)

c Are carnivorous snails first-, second- or third-order consumers? (1 mark)

d True or false? Centipedes are predators and will prey on almost any invertebrate, including other centipedes. Millipedes are herbivores and scavengers and mainly survive on rotting vegetable or animal matter. (1 mark)

e True or false? The nematode population will decrease if the protozoan population decreases. (1 mark)

f Classify slaters as herbivores or carnivores. (1 mark)

Hint 1: Food webs start with producers at the bottom level and then consumers at higher levels.
Hint 2: Algae increase their growth rates when fertilised.

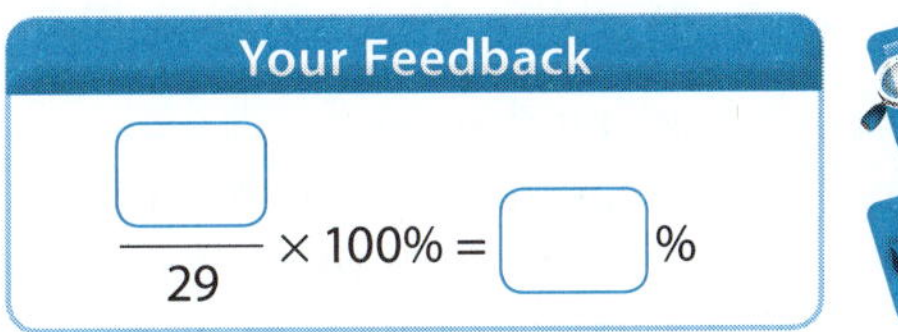

ECOSYSTEMS AND NATURAL CHANGE

Ecology

QUICK REVISION

1 An ecosystem is a community of __________ and non-living things that work together. Ecosystems don't have any particular __________ and can be large or __________. A healthy ecosystem has plenty of species diversity (that is, many __________ types of plants and animals) and is less likely to be critically __________ by human interference, natural calamities or __________ in climate. Organisms are linked in __________ ways in an ecosystem. For instance, if there isn't enough __________ (solar energy) or water, or if the soil doesn't have the right chemical __________, the plants will die. If the __________ die, animals that depend on them will perish. A healthy ecosystem is a balanced system.

In food chain and __________ web diagrams, an arrow indicates the flow of energy. For example, 'worm → bird' shows that the __________ (and matter) moves from the worm to the bird when the bird eats it. As only a __________ amount of the energy passes from one level to the next, there aren't too many arrows between a producer __________ and the top consumer organism.

2 Energy enters an ecosystem when __________ plants trap some of the energy in __________ and use it to produce energy-rich materials. Some of these producer organisms are eaten by __________ organisms. In the following photo the grass is the producer and the horse is a herbivore or first-order consumer.

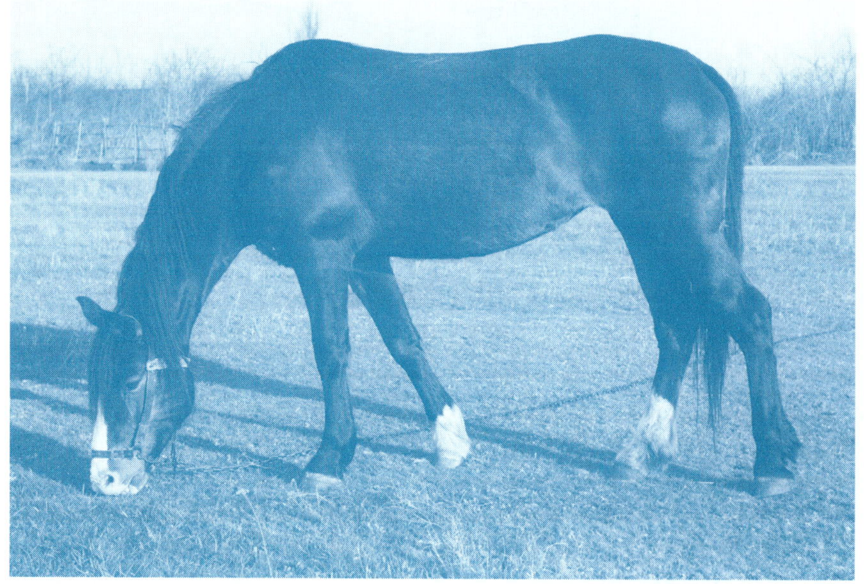

Organisms need __________ to grow, move and maintain their bodily functions. As energy moves along a food chain there is less and less of it available. Towards the top of the food chain the total mass of animals becomes __________. If the organism is large (such as a lion) there are far __________ of them than their prey (such as antelope). On the other hand, if the organism is small (such as a flea) there could be thousands of these __________ by each larger organism (such as a dog). An energy pyramid is a way of showing that there is less and less __________ available to an organism as it moves up the food chain.

3 Organisms causing disease (__________) can affect the numbers of organisms in an ecosystem. Disease can affect the organism directly, or it can attack the __________ (prey) the organism needs to survive. For example, devil facial tumour disease is killing __________ devils (*Sarcophilus harrisii*). This is an unusual cancer as it can spread like a contagious __________ from devil to devil through biting. The __________ system of Tasmanian devils is unable to reject live tumour cells. Unless a cure is discovered soon, Tasmanian devils could become __________ in Tasmania. The photo shows a healthy Tasmanian devil.

If a disease was to kill eucalyptus trees in an area then the __________, which rely on them for food, could be wiped out. Even if not all eucalypts were equally affected, the remaining koalas would have their __________ numbers reduced and, could in turn, be weakened and be more susceptible to __________.

4 A parasite lives in a close association with another life form, its __________, and causes it harm. The __________ relies on its host for its life functions. Viruses and flat worms (Platyhelminthes) are common parasites. The parasite has to be in its host to survive, mature and __________. Parasites rarely kill their __________. A common, well-known parasite of humans and their pets is the hookworm. Hookworms attach themselves to the lining of the small __________ causing diseases and malnutrition, since they eat the __________ and prevent them from being available to the host.

5 The make-up of an ecosystem can __________ from season to season. Plants generally don't produce much __________ (leaves) during winter and __________ trees drop their leaves altogether. Herbivores that rely on succulent green plants will be __________ in numbers. The numbers of __________ which rely on these herbivores for prey will also be reduced. Other animals eat different types of food as the seasons change. Birds can __________ northwards during these times and return during spring. Some animals (such as squirrels, mice and beavers) gather extra food in autumn and store it to __________ later. A few animals (such as bears, rodents and bats) can hibernate to ride out the times when food is __________. Hibernation is a condition of inactivity and low metabolism in __________, where the body temperature __________ and breathing becomes __________. Hibernating animals __________ energy, so they use up their energy reserves and body fat slowly.

Answers **1** living; size; small; different; damaged (upset); changes (variations); different; light; nutrients (minerals); plants; food; energy; small; organism **2** green; sunlight; consumer; energy; less; fewer; supported; energy **3** pathogens; food; Tasmanian; disease; immune; extinct; koalas; population; disease (sickness) **4** host; parasite; multiply (reproduce); hosts; intestine; nutrients (food) **5** vary (alter); foliage; deciduous; reduced; carnivores; migrate (fly); consume (eat); scarce; animals; drops; slower (shallower); conserve (save)

ECOSYSTEMS AND NATURAL CHANGE

Ecology

REVISION SUMMARIES

1 **Ecology** is the scientific study of associations between living and non-living things in the natural world. Ecologists examine these interactions to better understand the abundance and variety of life in Earth's ecosystems. By learning how ecosystems function, our ability to predict how they will respond to changes in the environment is improved. Living organisms in ecosystems are linked in complex relationships so it is not always straightforward to foresee how altering one part of it will affect the rest of an ecosystem.

Ecosystems sustain themselves by taking in **energy** and cycling nutrients. At the first **trophic** level (the feeding position that an organism occupies in a food chain), producers (green plants, algae and some bacteria) use **solar energy** to manufacture organic plant material using photosynthesis. Herbivores (animals that feed exclusively on plants; also called first-order consumers) form the second trophic level. Predators that eat these herbivores comprise the third trophic level. These predators are second-order consumers and are called carnivores if they feed solely on meat. Still higher trophic levels exist. Decomposers, which include bacteria, fungi, moulds, worms and insects, break down wastes and dead organisms, returning nutrients to the soil. These nutrients are then taken up again by producers.

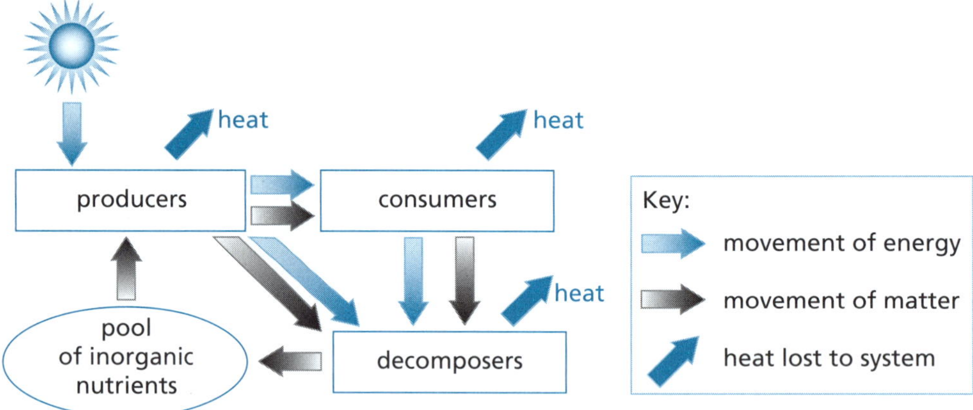

Nutrients (minerals and other inorganic substances) are recycled, but energy only moves one way and is eventually lost to the system, mostly as **heat**. The continual input from sunlight is necessary to maintain the energy of the system. On average only about 10% of the total energy at one trophic level is passed on to the next level. Processes such as respiration, defecation, growth and reproduction reduce the energy movement between trophic levels, as do organisms that die without being eaten by consumers. The nutritional value of the material that is eaten also influences the efficiency of energy transfer.

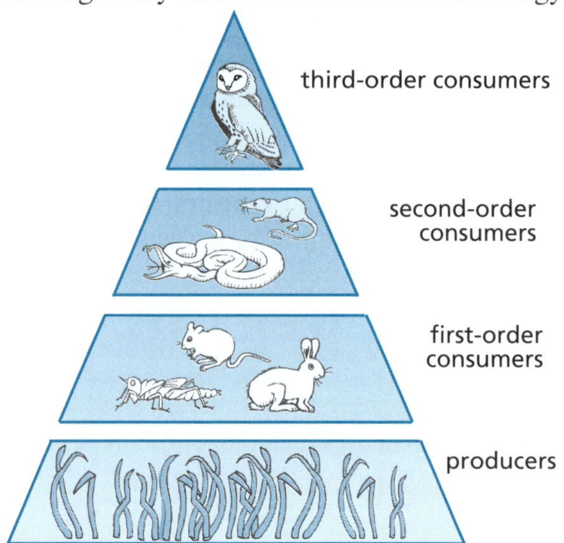

2 An **energy pyramid** is a graphical representation of the **biomass** (the amount of living or organic matter present) at each trophic level in a given ecosystem. The amount of biomass is a measure of the energy stored at each level.

The simplest way to illustrate the change of energy through ecosystems is a food chain where energy passes from one trophic level to the next. Only about 10% of the energy at one level is passed on to the next level.

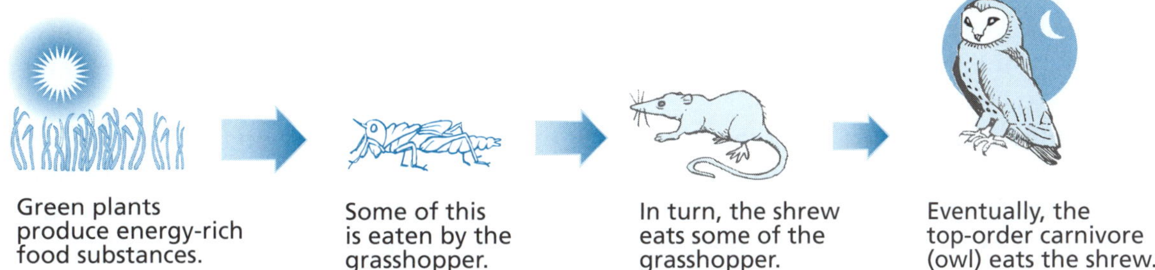

3 Disease in an ecosystem can affect **population numbers**. Diseases can affect individuals directly. For example, a virulent outbreak of the sexually transmitted **viral disease**, *Chlamydia*, is spreading through koala populations. If left unchecked, this disease could result in the koala's extinction. Scientists are working to develop a *Chlamydia* vaccine, and are currently treating visibly ill koalas with medicines. Diseases can affect the food (prey) that animals require. Plant pathogens, for example, can destroy **vegetation communities** and several plant species can be at risk of extinction. The animals relying on these plants for food will either starve or migrate elsewhere, if they can. Higher order carnivores that feed on these animals will be affected too.

Widespread **loss of predators** at the top of a food chain, and harsh environmental conditions, can encourage opportunistic pests and pathogens. For example, if a disease killed off foxes that feed on rabbits, then the rabbit population numbers would increase dramatically. More rabbits mean more grasses and other crops are eaten, leading to food shortages. This can stress the rabbits and, with more rabbits being in closer proximity to each other, diseases can easily spread. This could see a net reduction in the rabbit population that would open up the opportunity for other pest organisms to come in and fill the void.

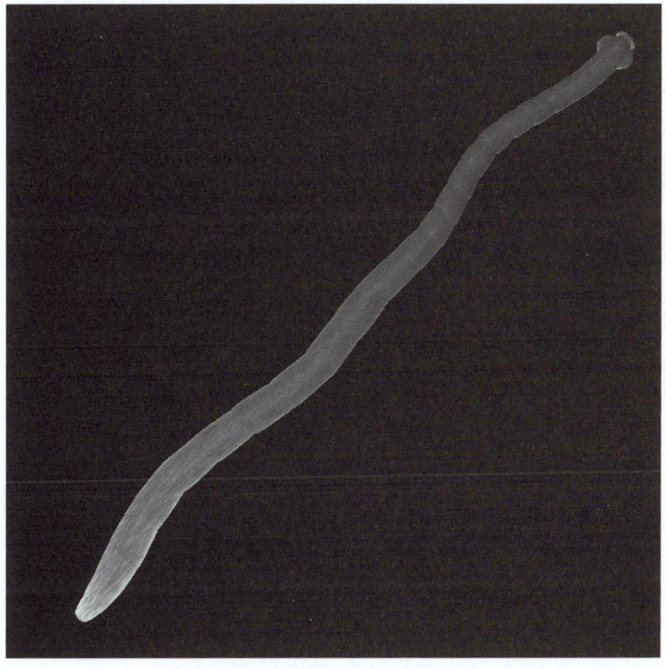

4 A **parasite** is an organism that lives on or in a host and gets its food from or at the expense of its host. It rarely kills its host, but can make the host sick. **Tapeworms** are parasitic flatworms that infect the digestive tract. These are sometimes ingested by consuming undercooked food. Once inside a host, a tapeworm larva can grow into a very big adult. Tapeworms of the genus *Echinococcus* infect and cause the most harm to intermediate hosts such as sheep and cattle. This is often referred to as hydatid disease. Anaemia can develop in some infected people as the tapeworm lives off the food which was meant for the host. The photo shows a tapeworm.

(cont.)

5 Production of energy-rich substances by green plants in land ecosystems generally increases with temperature up to about 30 °C, and declines with higher temperatures. **Productivity** also increases with moisture. On land, productivity is highest in warm, wet zones in tropical forests. In contrast, desert scrub ecosystems have the lowest productivity because their climates are extremely hot and dry. There are also **seasonal effects** on ecosystems, which animals and plants respond to in a number of ways.

- **Deciduous** plants lose all of their leaves for part of the year. They do this to conserve water or to better endure winter weather conditions. Losing leaves in winter may also decrease damage from insects. Evergreens can suffer greater water loss during the winter and they also can experience greater pressure from predators, especially when tiny. Fewer leaves on the trees provide less food for herbivores.
- Some animals can feed on a variety of food sources, so when one is scarce during a particular season, the animals can turn to the others that are available.
- Some birds, insects and other animals are able to **migrate** during wintry months to areas where food sources are plentiful before returning for the following season.
- Some animals survive the lean times by **hibernating** in caves and burrows. Hibernating animals store food as body fat during the end of summer and autumn. During hibernation the animal's body temperature drops very low, almost to that outside, and its heartbeat and breathing slow down. Body fat provides the energy for their bodies all winter.
- Some animals store food in their burrows during times of plenty to provide for when there is not much food around.

Checklist

Can you:

1. *Recognise how energy and nutrients flow in an ecosystem?* ☐
2. *Describe the function of an energy pyramid?* ☐
3. *Account for how disease can affect population numbers?* ☐
4. *Explain what parasites are?* ☐
5. *Describe some responses to seasonal effects on ecosystems?* ☐

ECOSYSTEMS AND NATURAL CHANGE

Ecology

30 MINUTES

REVISION TEST

1 True or false?

a An ecosystem is a natural unit consisting of plants, animals and microorganisms in a region functioning collectively with the non-living physical factors of that environment. (1 mark)

b Sunshine, temperature, rainfall, oxygen and moisture are all abiotic (non-living) factors of an ecosystem. (1 mark)

c Carnivores are an abiotic component of an ecosystem. (1 mark)

d Energy enters an ecosystem from the Sun. During photosynthesis it is transformed into simple sugars, and then it is stored. (1 mark)

e The nutrient carbon cycles in an ecosystem through photosynthesis, cellular respiration and decomposition. (1 mark)

f All ecosystems have an approximately equal size. (1 mark)

g A herbivore is not an example of a consumer. (1 mark)

h The flow of energy through an ecosystem is:
Sun → producers → herbivores → carnivores. (1 mark)

i Decomposers keep the community clean by breaking down dead organisms. (1 mark)

j Energy enters an ecosystem as light energy and is then eventually converted into heat energy and recycled. (1 mark)

k An energy pyramid shows the amount of energy at each trophic level. (1 mark)

l In an energy pyramid energy flows from the top of the pyramid to the bottom. (1 mark)

2 The following diagram shows an example of an ecosystem.

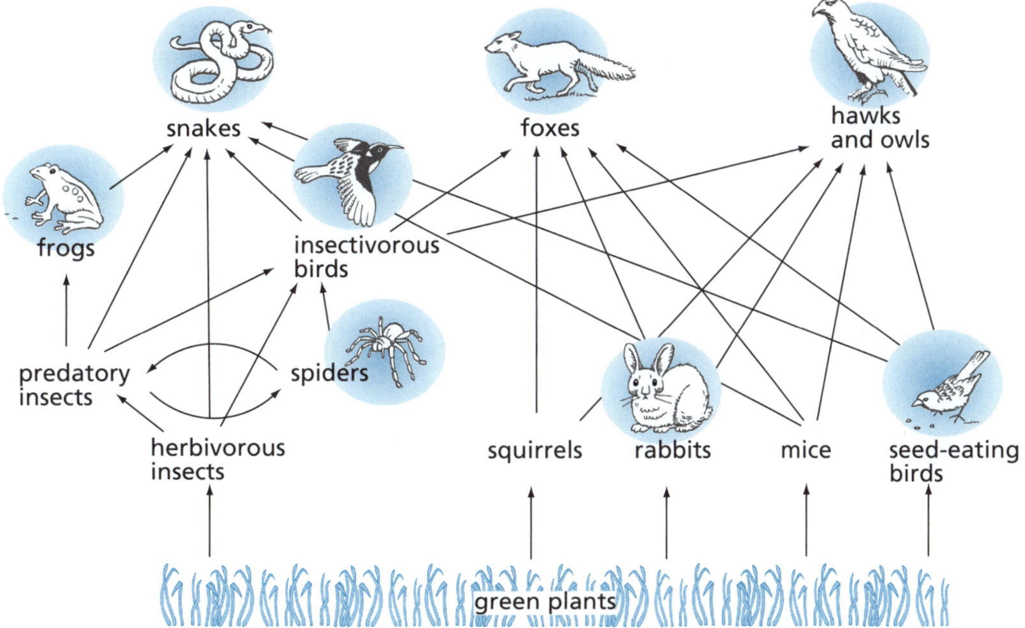

a What do the arrows indicate? (1 mark)

b What is the original source of the energy used by these organisms to live in this ecosystem? (1 mark)

c Which animals compete with hawks for seed-eating birds? (1 mark)

(cont.)

d If a farmer baited the rabbits with poison, which other animals might also be poisoned? *Hint 1* (1 mark)

e If a disease were to wipe out the snakes, what might happen to the number of frogs? (1 mark)

f What is meant by the term *insectivorous bird*? (1 mark)

g Name an organism *not* in the same food chain as snakes. (1 mark)

3 Read the following and answer the questions below.

Many scientists consider that re-introducing wolves into Yellowstone National Park (located primarily in the state of Wyoming in the United States) in 1995, after they had been exterminated from the park for decades through hunting, has caused a trophic cascade. This has been beneficial for the ecosystem. Wolves have sharply reduced the elk population, allowing willows and other plants to grow back in areas adjacent to rivers and streams where the elk had grazed these plants heavily. Healthier willows are attracting birds and other small mammals in great numbers. Songbirds, insects, rodents and other species have come back into these favoured habitat types. And now foxes and coyotes are moving into these areas because there's more prey for them.

a What is meant by the words *a trophic cascade*? *Hint 2* (1 mark)

b Government predator control programs in the early 1900s essentially helped eliminate the grey wolf from Yellowstone. What happened to the number of elk in the national park once the grey wolf was gone? (1 mark)

c How did elk suppress the numbers of birds, insects and rodents? (1 mark)

d How would the outcome be similar if a disease had killed off the wolves? (1 mark)

4 Hydatid disease is a parasitic infection of various animals, caused by a small tapeworm living in dogs, dingoes and foxes, which can be transmitted to humans. Humans can become infected by inadvertently swallowing the tiny and sticky hydatid eggs.

The diagram shows the life cycle of the *Echinococcus granulosus* tapeworm which can infect dogs, particularly in sheep farming areas.

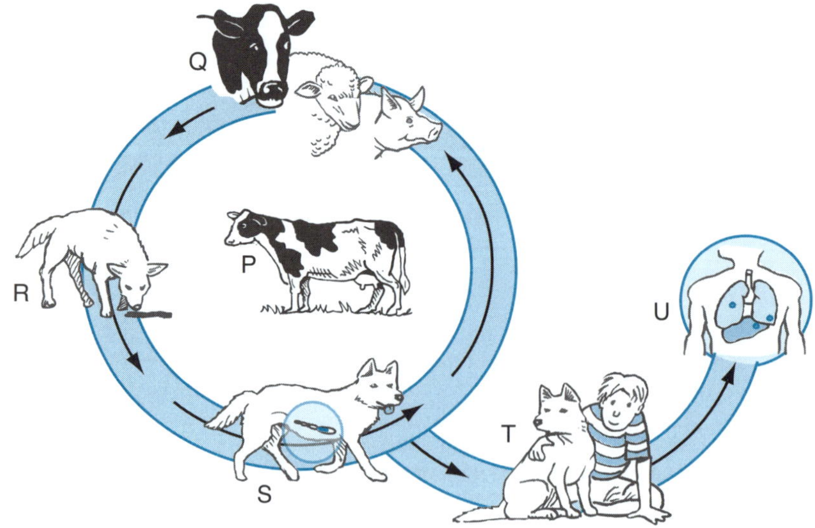

Match the letters in the diagram with each of these statements.

a Dogs become infected when they eat meat from infected stock. (1 mark)

b Cows, pigs and sheep become infected when they eat tapeworm eggs from the faeces of infected dogs that are stuck on grass. (1 mark)

c Hydatid cysts develop inside humans and cause serious health issues. (1 mark)

d Cysts develop in the liver, lungs and heart of infected animals. (1 mark)

e Tapeworms develop inside the dog, which can then shed many tiny eggs onto pastures. (1 mark)

f Humans can pick up tapeworm eggs when playing with infected dogs. (1 mark)

5 For many species, the climate where they live or spend part of the year influences key stages of their annual life cycle, such as migration, blooming, and mating. As the climate has warmed in recent decades, how might these events alter and impact on the survival of the species? In your answer make reference to different regions in a country and migratory species. *Hint 3* (5 marks)

6 Read the following and then answer the questions below.

> Declines in the duration and range of sea ice in the Arctic lead to declines in the great quantity of ice algae, which flourish in nutrient-rich pockets in the ice. These algae are eaten by zooplankton, which are in turn eaten by Arctic cod, a main food source for many marine mammals, including seals. Seals are eaten by polar bears.

a Draw a food chain for the information given. (2 marks)

b Assume that in a particular area there is 6 000 000 kJ of energy stored in the algae and that only 10% of the energy in one trophic level is available to the next level. Estimate the amount of energy available to:

i zooplankton (1 mark)

ii polar bears. (1 mark)

c Explain how declines in ice algae can contribute to declines in polar bear populations. (1 mark)

d Explain, in terms of energy, why food chains rarely extend for more than five or six levels. (2 marks)

Hint 1: The infected, dying and dead rabbits have poison in their system. Which organism(s) are likely to eat these rabbits and so also be poisoned?

Hint 2: Trophic levels are the feeding position in a food chain such as producers, herbivores and carnivores.

Hint 3: Many organisms do specific things at certain times of the year. As they don't have calendars, they rely on the seasons for cues.

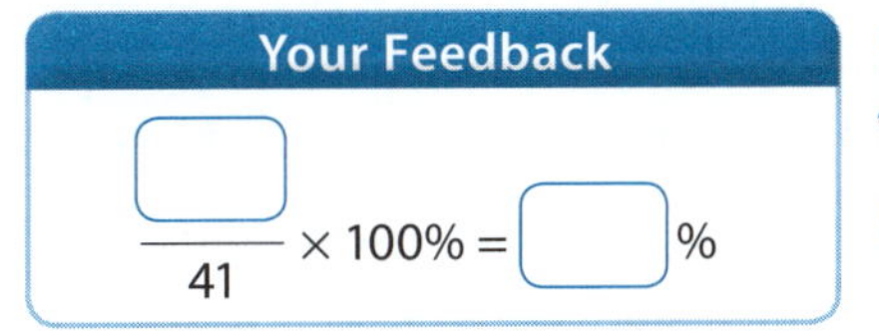

IMPACTS ON ECOSYSTEMS

Ecology

QUICK REVISION

1 The Earth's ____________ are complex ecological environments which have developed over ____________ of years. Altering the intricate components of an ecosystem may affect its ____________. While humans have harmed many environments, by causing ____________ warming, habitat destruction and ____________ pollution, changes have also occurred through natural means.

- Bushfires. The Australian environment has adapted over millions of years to ____________. Some plants have adapted to survive through these conditions by developing thick, protective ____________. Other plants can regenerate through ____________ crowns. Fires also clear undergrowth allowing seeds to ____________, producing new plants.
- Drought. Australia is the driest inhabited ____________ and so droughts can have devastating effects and destroy both ____________ and agricultural ecosystems. With less vegetation, the soil is not held firmly in place and so is vulnerable to ____________ and dust storms. The following photo shows the effects of a severe drought.

- Floods. Plants and animals that can't escape floods are ____________, and those that do survive can die from ____________ through lack of food. Some ecosystems have adapted to regular ____________, especially desert plants which need flood waters to regenerate and ____________. There is also an ____________ in animal activity after floods. As well, floods can replenish the ____________ of its depleted nutrients.
- Cyclones. These usually develop along the tropical ____________ coast of Australia and can travel hundreds of ____________ inland. Strong winds and rain can damage much of the environment. Winds can bring down long-standing trees and damage ____________ habitats. Rains can ____________ the landscape causing further damage. Damaging storm surges can flood low-lying ____________ areas.

2 Humans are responsible for introducing numerous ____________ and plant species to Australia. Many of these survive at the expense of native ____________ and animals. Rabbits, for example,

were originally brought into Australia from Europe as a __________ source. Some escaped in the 19th century, with devastating effects on the Australian ecology. They are suspected of being the most significant known factor in the loss of native __________ in Australia. Rabbits often kill young trees by __________ them. They are also responsible for serious __________ problems, since they eat native plants, exposing the topsoil to erosion. In order to regenerate this topsoil may take many hundreds of __________.

3 Humans have also affected large areas by turning bushland into __________ ('concrete jungles') with very little __________ of plant and animal life. Other areas have been cleared to make room for pastures, crops, mining and other __________. Large areas can be permanently __________ when dams are built. Farming practices require the use of pesticides and __________, some of which may end up in natural waterways many __________ away with damaging consequences.

4 Oil spills can have catastrophic consequences to the __________. When an oil tanker is damaged, __________ of litres of oil can be spilled into the __________ and other waterways. As oil is less dense than __________, and immiscible with it, it floats on the surface and covers many square kilometres. Many __________ organisms (plankton) in these upper water surface layers are destroyed or displaced. These organisms form the basis of coastal marine __________ webs. Destroying plankton in such numbers can collapse the __________ food web. Birds and other animals can be coated with this oil, eventually leading them to a slow __________. Many kilometres of __________ can be covered with this oil, further damaging animals and plants on land. Clean up operations can run for years and cost many millions of __________. The following photo shows workers cleaning up an oil spill on a beach.

Answers 1 ecosystems; billions (millions); survival; global; environmental; bushfires; barks; underground; germinate (sprout); continent; natural; erosion; drowned; starvation; flooding; bloom (flower); increase; soil; north; kilometres; animal; flood; coastal 2 animal; plants; food; species; ringbarking; erosion; years 3 cities; variety (diversity, range); industries; flooded; fertilisers; kilometres 4 environment; millions; sea (ocean); water; small (minute, microscopic); food; marine; death; shoreline; dollars

IMPACTS ON ECOSYSTEMS

Ecology

REVISION SUMMARIES

1 Changes in an ecosystem can result from human activity, changes in climate, introduction of non-native plant and/or animal species and changes in populations. **Climate changes** have different impacts on different ecosystems and species. Changing climate may cause different species and populations to migrate into or out of a region, or become extinct at different rates. Alterations in climate could cause the disappearance of existing ecosystems and the creation of new ecosystems. **Catastrophes** also play a part in impacting ecosystems.

- **Bushfires.** Fires have been shaping the Australian ecosystem for a very long time. Some plants need low-heat fire to set seed, or to open tightly-closed fruits to release new seed. Some can regenerate from underground root stock after being burned to the ground. Other plants can grow new branches from beneath thick, protective bark. Fires produce chemicals in the ash to stimulate new growth, and remove growth-inhibiting toxins from the litter. This encourages animals back into an area. The first inhabitants of Australia learnt the advantages of selective fires, so they burnt broadly and frequently. They benefited from the plant and animal responses to this burning so they could coexist. Many plants have developed characteristics that encourage the spread of fire.

  Coarse, slowly decaying eucalypt litter provides plentiful fuel to support the next fire; the flammable bark that is loosely attached to many trees assists fire to cross natural barriers; and many green leaves contain highly **combustible** oils and resins which act as catalysts, promoting burning before the leaves are fully dry. However, ecosystems such as rainforests, alpine moss beds and peat bogs are susceptible to fire. Some species are also fire sensitive. Other consequences after a fire may include sediment and ash entering waterways degrading water quality, food supplies and breeding habitat for many species.

- **Drought.** Severe, repetitive cycles of drought can potentially cause catastrophic changes in ecosystems, making it unlikely that many species will manage to recover. Australia is prone to drought because the continent is situated under the subtropical high pressure belt which is an area of dry, stable air and cloudless skies. Rainfall over most of the country is generally low and irregular. As streams and rivers dry up during droughts, organisms relying on this water die off. In turn, this causes food shortages for other animals higher up the food chain. There are long-term repercussions for the survival of natural ecosystems as damage is caused by the loss of vegetation and soil erosion. While farmers can protect their animals during drought by bringing in feed and water, it is costly.
- **Floods.** Floods are often caused by tropical cyclones (in the north of Australia) or by severe weather patterns. Some inland floods, such as those of Lake Eyre (in central Australia), can result from flooding rainfall hundreds of kilometres away. This water moves slowly downstream and may take months. On the rare occasions that the lake fills, it is the

largest lake in Australia and a major breeding ground for plants and animals (especially birds). Flooding often causes damage to properties, roads, bridges and railways, and other infrastructure. Farmlands can be severely affected by floods, destroying crops and drowning animals. Floods can also cause erosion of farmlands, coast lines and other environments. However, there are some positive effects of flooding for natural ecosystems: they saturate the soil and fill up ground and underground water supplies, providing a long-term water supply for both native plants and agricultural crops; they cleanse excess salt and chemicals from the soil and the rivers, and help to remove debris and other wastes; and they dump fertile soil on river flats and provide an environment that enables fish, birds and other organisms to breed and multiply. If floods reach the desert, plants can regenerate and start to bloom.

- **Cyclones.** Tropical cyclones occur along the northern Australian coast, but may travel many hundreds of kilometres inland before the energy is spent. Cyclones are dangerous because:
  - destructive winds cause property damage, uproot trees, destroy animal habitats and whip up seas damaging harbours, ships and marine life
  - heavy, persistent rainfall causes flooding which causes further deaths and damage
  - damaging storm surges can cause low-lying coastal areas to be inundated
  - damage to electricity, water and sewerage pipes can lead to disease due to lack of sanitation.

2 Where humans have gone they have often taken plants and animals with them. Cattle, sheep, cats, rabbits and rats were all introduced to Australia, as were many plants (wheat, corn, rice, cotton, ornamental shrubs and flowers). Some of these organisms can survive better in the Australian environment than many **native plants and animals** and, over time, **out-compete** them and become pests. This has led to the extinction of many native plants and animals. For example, the cane toad, *Bufo marinus*, is native to Central and South America. Toads were deliberately introduced from Hawaii to Australia in 1935 to control scarab beetles that were sugar cane pests. While they weren't much good at that, they spread widely and now occur throughout the eastern and northern half of Queensland, into the Northern Territory, and the northern parts of New South Wales. Toads eat almost anything they can swallow and, being tough and adaptable, as well as poisonous throughout their life cycle, have few predators in Australia. **Invasive species** have a major impact on Australia's environment, putting pressure on our unique biodiversity and reducing overall species abundance and variety.

3 Currently Earth's human population is over seven billion and continues to grow at the rate of more than 80 million people per year. This puts tremendous pressure on resources to house, feed and clothe people. The inventions of industry and agriculture, and the extent of these practices around the globe, have allowed humans to commandeer large sections of the useable parts of the Earth. Large tracts of land are cleared by humans for sprawling cities, industries and to grow crops. For example, **clearing forests** for palm oil plantations is a major cause of **deforestation**. Indonesia and Malaysia account for the majority of this deforestation, and more than 50 orangutans die each week because of this devastation. Where they once lived, there are now less than 500 Sumatran tigers and 3000 Sumatran elephants left in the wild.

4 An **oil spill** is a form of pollution and is the deliberate or accidental release of petroleum into the environment, especially marine areas. This oil may come from tankers, offshore platforms, drilling rigs and oil wells. Spills can be very damaging to the environment. Oil spills at sea are

(cont.)

generally much more damaging than those on land, as they can spread for hundreds of square kilometres as an oil film, covering beaches with a thin oil coating. Oil penetrates into birds' feathers and the fur of mammals, reducing their insulating ability and leading to **hypothermia**. The oil also weighs them down, making them liable to sink and drown in the water. Oil can blind animals and affect their breathing, digestive systems and other bodily functions. Animals that rely on scent to find their offspring or parents are confused by the strong oil smell. This causes the young to be rejected and deserted, leaving them to starve and ultimately perish. Without human assistance, the majority of animals caught up in oil spills die. Plankton, larval fish, seaweed, clams, oysters and mussels are all affected by oil spills. This impacts organisms higher up the food chain. The oil spill can harm the entire food web in the area.

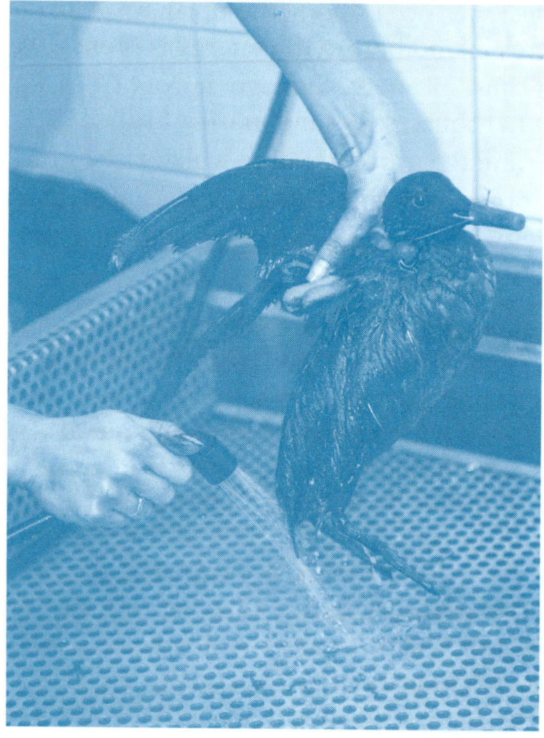

Checklist

Can you:

1. *Describe some of the effects of bushfires, drought, floods and cyclones on an ecosystem?* ☐
2. *Describe the impact of introduced species, such as rabbits or cane toads, on an ecosystem?* ☐
3. *Describe the ways that humans have destroyed natural habitats or harmed ecosystems?* ☐
4. *List some of the damage caused by oil spills?* ☐

IMPACTS ON ECOSYSTEMS

Ecology

30 MINUTES

REVISION TEST

1 Read the following and answer the questions below.

The black-flanked rock wallaby (*Petrogale lateralis*) of Western Australia is endangered. This shy and fearful marsupial would rather starve to death in its burrow than risk being torn apart in the open by feral cats and foxes. This makes it difficult for it to escape its haven and find food or a mate as it becomes an easy target. Inappropriate fires and introduced grazers have also impacted upon populations.

a Suggest two ways the rock wallaby can be brought back from the brink of extinction. (2 marks)

b Several groups have joined forces to build a 5-km long fence around the nature reserve where rock wallabies are found. How might this help to save the population? *Hint 1* (1 mark)

2 List two harmful and two useful effects of bushfire in the Australian environment. (4 marks)

3 Roads are often built through forest areas to reach industries such as mineral and oil mines. The table shows the number of species of birds and mammals in a particular area that have been unfavourably affected by building such a road.

Distance to road (km)	1	2	3	4	5	6	7	8	9
Number of species affected	152	70	39	34	30	18	10	9	8

a Describe what is shown by this table. *Hint 2* (2 marks)

b Suggest two possible reasons why these species are affected. *Hint 3* (2 marks)

c If an ecologist noted around two dozen bird and mammal species with reduced abundance in an area, approximately how far would this area be from the road? (1 mark)

4 Read the following information and answer the questions below.

The European red fox, *Vulpes vulpes*, was introduced into Australia in the mid-1800s for recreational hunting purposes. Today, foxes are found in most areas of the mainland south of the tropics and number more than 7.2 million. Foxes are opportunistic feeders. Overall, about a third of this diet is from farm livestock, a third from pest vertebrates like rabbits and mice and a third from wildlife. The species has been directly implicated in the extinction of the desert rat-kangaroo. Its spread also matches the decline in the distribution of several small and medium-sized ground-dwelling native mammals, ground-nesting birds and reptiles.

a Foxes represent a very successful invasive species. Explain what this means. (2 marks)

b Foxes are opportunistic feeders. What does this mean? (1 mark)

c Explain how foxes cause environmental damage. (1 mark)

d Serious economic damage results from fox predation on farm livestock. Explain this statement. (2 marks)

e Each fox will eat about 400 g a night. How much does this amount to in a year? (1 mark)

f Foxes eat other pest animals, like rabbits and mice, in approximately equal numbers to native animals. However, the number of native animals is decreasing but pest animals are not. Suggest a reason for this. *Hint 4* (1 mark)

g Suggest three ways humans can control fox populations. (3 marks)

(cont.)

IMPACTS ON ECOSYSTEMS *(continued)*

Ecology

REVISION TEST

5 Read the following and then list three conditions that plants must endure to survive in this desert landscape. (3 marks).

> Lake Eyre covers one-sixth of the Australian continent and holds some of the rarest and least exploited ecosystems in the world. During the rainy season the rivers from the north-east flow into it. Typically a 1.5-m flood occurs every three years, a 4-m flood every 10 years and it fills only a few times each century. The water in the lake soon evaporates, drying by the end of the following summer. Few plants can live around this desert saltpan.

6 a What is an oil spill? (1 mark)

b Suggest a reason why oil spills in the sea are generally far more damaging than those on land. (2 marks)

c List three effects oil spills have on birds and mammals. (3 marks)

d Certain insoluble polymers are able to turn spilled oil into a floating, porous, semi-solid or rubber-like mass. How are these of benefit in cleaning up oil spills? (2 marks)

e One serious oil spill occurred from the tanker *Exxon Valdez* in 1989. It is estimated that around 40.9 million litres of oil escaped into the sea. Assuming the average oil spread at the surface was 2.9 L/ha (litres per hectare), calculate the area covered by this oil spill (correct to the nearest thousand square kilometres; 1 ha = 0.01 km^2). (2 marks)

7 Read the informations below and answer the questions that follow.

> The dodo (*Raphus cucullatus*) is an extinct flightless bird, related to pigeons, which lived on the island of Mauritius. It evolved in isolation from significant predators and was entirely fearless of humans. It fed on nuts, seeds, bulbs and roots as well as fallen fruits. The bird became easy prey for hungry sailors in the early 1600s. These seafarers later brought with them many domesticated animals and other invasive species. For example, dogs, pigs, cats, rats and crab-eating macaques raided dodo nests eating the eggs, and competed for other restricted food resources. Humans also destroyed the dodo's home forests. Within a little over half a century the dodo was extinct.

a The dodo birds were perfectly evolved for their environment. Explain this statement. (2 marks) b State three factors that led to the downfall of the dodo. (3 marks)

c How did the dodo call attention to the problem of human involvement in the disappearance of an entire species? (1 mark)

8 Read the following information and answer the questions below.

Some ecosystems have adapted to regular flooding and need this water to function. The river red gums (*Eucalyptus camaldulensis*) that grow along the Murray River are well suited to the floods that used to occur regularly. These trees need periods of partial flooding where the trunk may be inundated for months. The floods disperse the seeds to high ground and drop fertile soil around the gums' roots. The seeds germinate, take root and grow before the next flood submerges the new tree. After the Murray River was dammed, and with the increased use of weirs and irrigation, the river flow became reduced so that floods no longer take place. In some areas, the majority of river red gums are stressed, dead or dying from a lack of water. A government project aims to simulate a major natural flood every three to four years, to help the threatened river red gums.

a What caused the river red gums to become stressed? (2 marks)

b How will simulating a natural flood help their survival? (1 mark)

Hint 1: What might the purpose of the fence be?
Hint 2: Relate the distance from the road to the number of species affected.
Hint 3: You will need to suggest possible causes.
Hint 4: Think of how fast pest animals and native animals can breed.

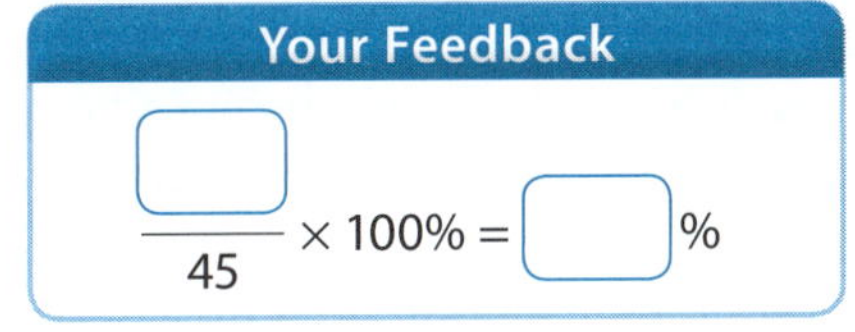

HISTORICAL DEVELOPMENT OF ATOMIC THEORY

Atomic theory

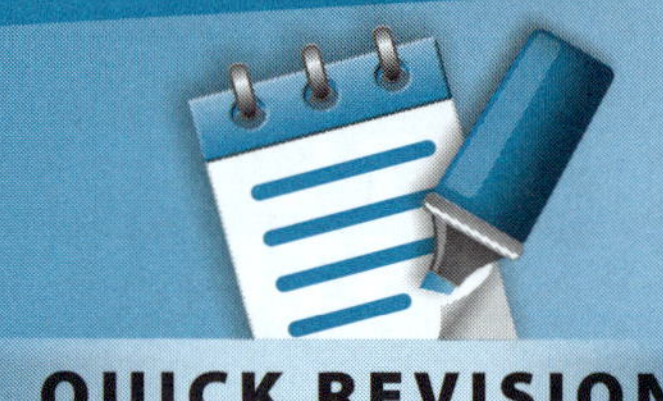

QUICK REVISION

1 The Greek philosopher ____________ (~460 BC) reasoned that if you break a piece of matter into two, and then continue to break it over and over, you would eventually reach the ____________ component that could not be broken down any further. This smallest component was called *atomos* meaning 'not able to be ____________'. Still, the prevailing views of Aristotle that matter could be divided ____________ remained for 2000 years. The Renaissance in Europe, however, led to the emergence of critical thinking strategies and experimentation which ultimately showed that Democritus was correct.

2 The English scientist John Dalton re-established the idea of atoms in his atomic theory of 1803. Dalton believed that the smallest particles of matter were ____________ and that they could not be divided into anything ____________ or destroyed. He also proposed that atoms could not be ____________. Dalton believed that each element had its own unique atom with a specific weight. Light elements like hydrogen would have ____________ atoms. Heavy elements like gold would have ____________ atoms. He conducted experiments to try to determine the relative atomic weights of elements, and he simplified the study of elements by giving each element its own unique ____________. Dalton also proposed that compounds were ____________ of a specific number of atoms of each element. A painting of John Dalton is shown below.

3 In the latter part of the 19th century many experiments were conducted on electrical discharges in low pressure gases. From these experiments the English chemist JJ Thomson discovered a small ____________ particle called the electron. Thomson was able to show that this particle was ____________ than an atom of hydrogen. He reasoned that ____________ were subcomponents of atoms. He developed an atomic theory in which electrons were embedded in a sphere of

__________ charge. This model required that the total number of negative charges from the electrons was balanced by the total positive __________ of the sphere. Heavy atoms had many more electrons than __________ atoms and consequently the heavy atoms had a greater positive charge than lighter atoms. In all cases the atoms were __________.

4 The development of the atomic theory was greatly influenced by the discovery of __________. Elements such as uranium emitted particles or rays. One of these particles was called an alpha particle which had a __________ charge. Ernest Rutherford conducted a series of experiments in which alpha particles were fired at thin __________ of metals such as gold. The results, however, did __________ support Thomson's atomic model. Rutherford found that __________ alpha particles passed through the gold foil without deflection. However, some alpha particles were __________ from their path. His analysis of this data showed that the gold atom had a __________ zone at its core. Rutherford called this dense core the __________. In 1911 Rutherford developed a new atomic model in which an atom had a central __________ nucleus surrounded by lightweight __________ negative electrons. Rutherford proposed that the nucleus contained small positive particles called __________ that he had discovered previously. In a neutral atom the number of positive protons __________ the number of negative electrons. This model of the atom was called a __________ system model as it resembled the way that planets orbit the Sun.

5 Rutherford's atomic model had its critics. Physicists said that orbiting electrons would not be stable, and so in 1913 a new model of the atom was developed by Niels Bohr. Bohr's atomic model retained the central __________ nucleus, but had the electrons arranged in stable energy levels or __________ at specific distances from the nucleus. There was also a __________ number of electrons that could be accommodated in each shell. This maximum electron population was related to the shell __________ (n) and given by the formula $2n^2$. The first shell ($n = 1$) is called the __________ shell and has $2(1)^2 = 2$ electrons. The second shell (L shell) has $2(2)^2$ = __________ electrons. Bohr's atomic model was very successful. Following the discovery by James Chadwick of a small, neutral particle called the __________, the model of the atom was further refined. The neutrons along with the protons occupied the nucleus. The neutrons were of similar __________ to the protons. Thus, most of the mass of the atom was in its nucleus.

Rutherford's atomic model

Answers 1 Democritus; smallest; divided; infinitely 2 atoms; simpler; created; light; heavy; symbol; composed 3 negative; lighter; electrons; positive; charge; lighter; neutral 4 radioactivity; positive; foils; not; most; deflected; dense; nucleus; positive; orbiting; protons; equalled; solar 5 positive; shells; maximum; number; K; 8; neutron; mass

HISTORICAL DEVELOPMENT OF ATOMIC THEORY

Atomic theory

REVISION SUMMARIES

1 The idea that matter is made up of particles can be traced back to the ancient Greeks. **Democritus** (~460 BC) made assumptions based on his observations of the natural world. Democritus hypothesised that matter was composed of incredibly tiny, indestructible particles which joined together to make materials. He called these particles *atomos* which means 'not able to be sliced apart or divided'. This idea of **atoms** was radical for his time and was opposed by the existing ideas of Aristotle, who had said that matter was 'infinitely divisible'. Democritus' idea of the atom remained untested until the Renaissance (about 1500 AD) when the new scientific thinkers and experimenters began to question Aristotle's ideas. The **atomic theory** began to emerge again.

2 By 1803, **John Dalton**, an English educator and mathematician, realised that he could use the old idea of atoms to explain many of the recent experimental observations on matter. Some of the important ideas of Dalton's atomic theory are:

- all matter is made of atoms
- atoms are **indivisible** and they cannot be destroyed or created
- all atoms of the same element are exactly alike but different from the atoms of all other elements.

Dalton also created **diagrammatic symbols** for elements and conducted many experiments to measure the relative weights of the atoms of different elements. Dalton believed that compounds were composed of atoms of each element in a fixed ratio. Within a few years the atomic theory became one of the cornerstones of modern chemistry.

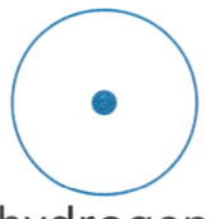
hydrogen

carbon

oxygen

phosphorus

3 In the late 19th century, the English scientist **JJ Thomson** discovered that there were smaller particles than atoms. He called these particles **electrons** and proposed that they were components of atoms. He proved that electrons had a negative charge. He produced a model of the atom in which the negative electrons were embedded in a sphere of equal positive charge. For example, an atom with six electrons would also have six positive charges so the whole atom was neutral. This atomic model from 1897 is often called the **plum pudding model** as it resembled a pudding containing dried fruits.

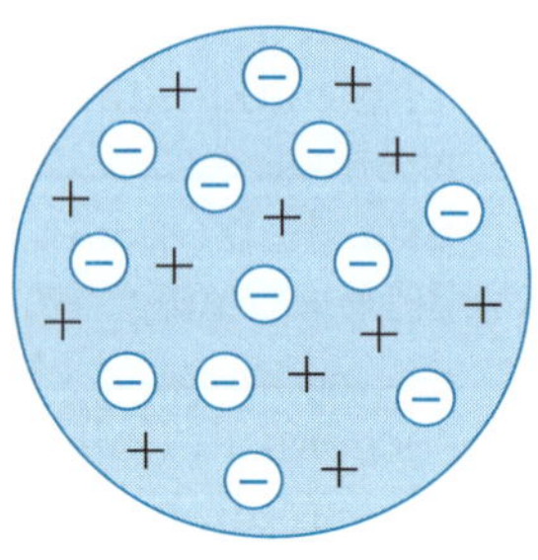

4 The discovery of **radioactivity** in the late 19th century allowed scientists to use radiation to probe the atom. One type of radiation called **alpha particles** was used by the New Zealand scientist **Ernest Rutherford** to investigate the internal structure of gold atoms. Alpha particles have a positive charge. When they are fired at thin gold foils, most alpha particles pass through without any deflection. Some alpha particles, however, were scattered at very large angles and a few almost totally rebounded. Rutherford applied a mathematical analysis to the data collected and showed that the atom had an extremely small yet dense central nucleus. The remainder of the atom, where the electrons were located, was largely empty. Later experiments proved that the **nucleus** also contained positive particles that were called **protons**. A proton had the opposite charge to an electron. Rutherford proposed that neutral atoms contained equal numbers of positive protons and negative electrons. The proton number is unique to each element. For

example, all carbon atoms have six protons and all oxygen atoms have eight protons. It is the number of protons in the nucleus of the atom that defines what element it is. In 1911 Rutherford developed an atomic model, often called the **solar system model**, in which electrons orbited around a central nucleus of protons, just like planets orbit the Sun.

5 In 1913 the Danish scientist **Niels Bohr** published a new model of the atom. In this model the electrons occupied stable energy levels (or **shells**) around the central nucleus. Electrons were only stable if they occupied these shells. Shells were numbered from the innermost shell ($n = 1$) outwards from the nucleus. Bohr showed that the innermost electron shell (K shell) could hold only two electrons. The next shell ($n = 2$) outwards (L shell) could contain a maximum number of eight electrons. The M shell ($n = 3$) held a maximum of 18 electrons and the N shell had a maximum of 32 electrons. So, an atom of oxygen, which had a total of eight electrons, had two electrons in the K shell and six electrons in the L shell. The maximum electron population (MEP) in any shell is given by the formula:

$$MEP = 2n^2$$

The Bohr model was further modified in 1932 when the English scientist **James Chadwick** discovered that the nucleus also contained small neutral particles that were named **neutrons**. These neutrons had a similar mass to the protons. The number of neutrons varied from one atom to another. For example, most oxygen atoms contain eight neutrons as well as eight protons. However, there are some other types of oxygen atoms that have more or fewer neutrons and these forms are called **isotopes**.

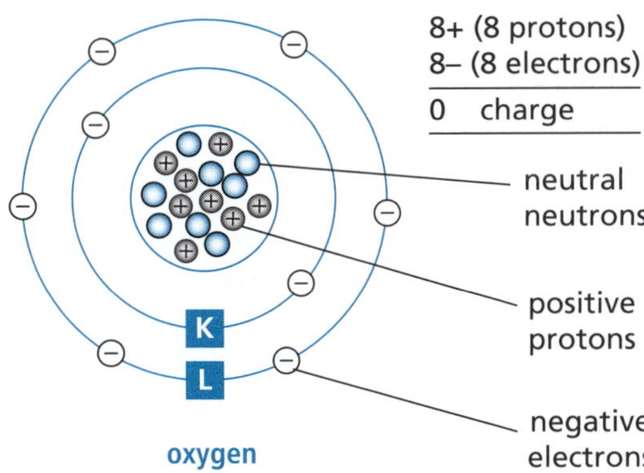

Checklist
Can you:

1 *State the contribution Democritus made to our understanding of matter?* ☐
2 *State the major points of John Dalton's atomic theory?* ☐
3 *Describe JJ Thomson's model of the atom?* ☐
4 *Describe Ernest Rutherford's model of the atom?* ☐
5 *Describe Niels Bohr's model of the atom?* ☐

HISTORICAL DEVELOPMENT OF ATOMIC THEORY

Atomic theory

REVISION TEST

1 The diagram below shows a model of the sodium atom.

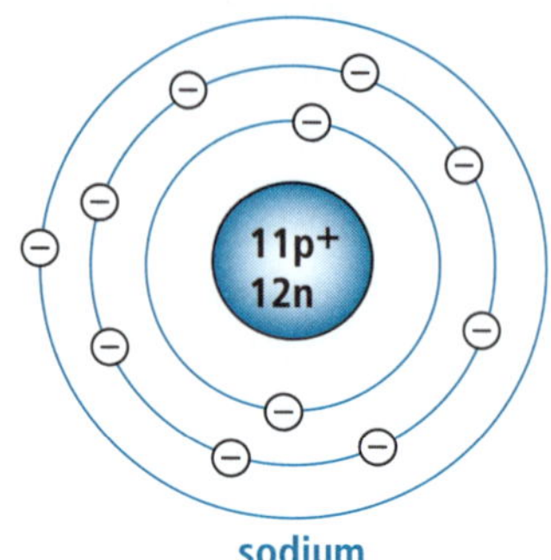

sodium

a Explain why there are 11 electrons in this atom. (1 mark)

b True or false? The L shell of the sodium atom contains a maximum number of electrons. (1 mark)

c Discuss whether or not this model shows the nucleus in a realistic scale relative to the size of the atom. *Hint 1* (1 mark)

d What is the name of the nuclear particle discovered by James Chadwick? (1 mark)

2 Name these scientists.

a Who proposed that compounds were composed of fixed numbers of atoms of each element? (1 mark)

b Who proposed the solar system model of the atom? (1 mark)

c Who proposed the shell model of the atom? (1 mark)

d Who conducted experiments with metal foils and alpha particles? (1 mark)

e Who proposed that an atom consisted of a positive sphere with embedded electrons? (1 mark)

3 Use Bohr's atomic theory to draw a model of an atom of beryllium which has the following subatomic components: four protons; four electrons; five neutrons. (3 marks)

4 Use the formula for the maximum electron population of each electron shell to calculate the maximum number of electrons in the following shells.

a M shell (1 mark)

b N shell (1 mark)

c O shell (1 mark)

5 Which of the following statements is not true? (1 mark)

A Protons have a positive charge.

B Electrons have a negative charge.

C Neutrons and protons have the same charge.

D Neutrons have no charge.

6 The diagram on the next page shows an experiment in which a gold foil is irradiated with a beam of alpha particles from a radioactive source. The experiment also uses a special screen coated with zinc sulfide. When an alpha particle strikes the screen a flash of light is observed.

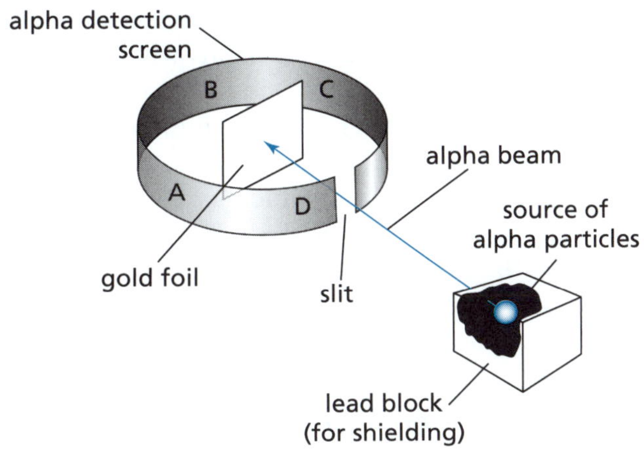

- **a** Where on the screen (A, B, C or D) are the most alpha particles detected? *Hint 2* (1 mark)
- **b** Where on the screen (A, B, C or D) are the least alpha particles detected? (1 mark)
- **c** What conclusions about the atom were made by Rutherford from such an experiment? (2 marks)

7 The diagram below shows the dimensions of a typical atom and its nucleus. The dimensions are in the units of picometres (pm) where one million million picometres equals one metre.

- **a** Use the information to calculate what percentage the nuclear diameter is of the atomic diameter. (2 marks)
- **b** A two-dimensional scale model of the atom was constructed on a grassy field. A circle of diameter 200 m was constructed on the grass. This circle is the size of the atom. Calculate the diameter (in millimetres) of the nucleus on this scale. *Hint 3* (2 marks)

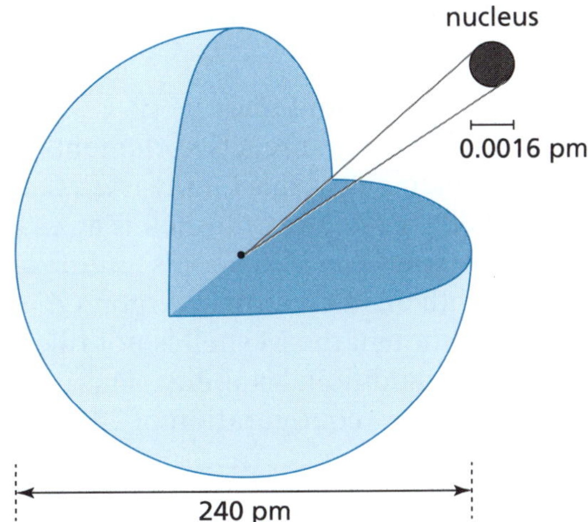

Hint 1: The Rutherford experiment showed that the nucleus was extremely small relative to the size of the atom.
Hint 2: The atom is mostly empty space.
Hint 3: Use your result in part a to work out the nuclear diameter in metres and then convert your answer to millimetres.

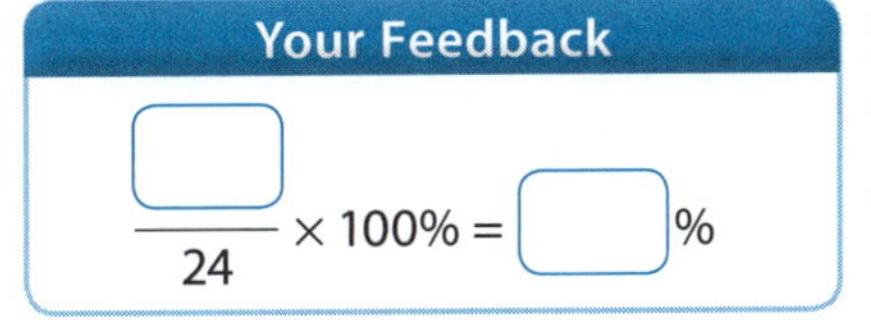

PAGE 228
PAGE 252

QUICK REVISION

1 Electrons are located in __________ around the central nucleus. The shell closest to the nucleus (K shell) can hold up to __________ electrons whereas the L shell can contain up to __________ electrons. The number of electrons in the atom of an element depends on the atomic __________ (Z) of that element. The atomic number is the number of __________ in the nucleus. The first element of the periodic table is __________. It has an atomic number of one. This means it has one __________ in the nucleus. In order to maintain the neutral charge of the atom there is one __________ in the K shell. Helium has an atomic number of two and so has __________ electrons in the K shell. For heavier elements with more protons and electrons, the L shell starts to fill up. For oxygen (Z = 8) there are two electrons in the K shell and __________ electrons in the L shell. The shell model below shows the electrons in the K and L shells of a neon atom (Z = 10).

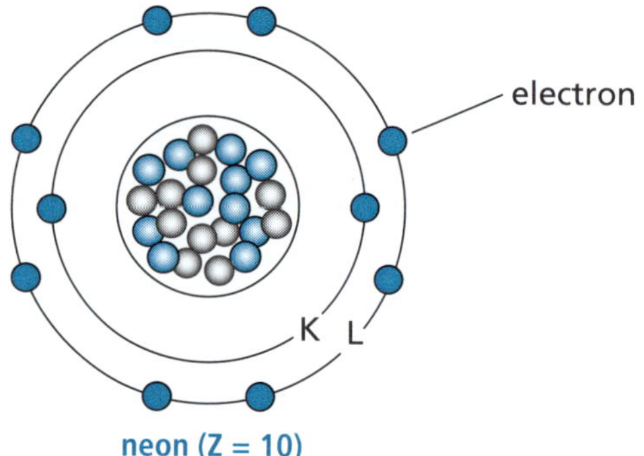

neon (Z = 10)

Magnesium has an atomic number of 12. This means that the 12 electrons are arranged with two electrons in the K shell, eight electrons in the L shell and __________ electrons in the M shell. This arrangement is called the electron __________ and in the case of magnesium it is written as 2,8,2. In this example the last two electrons occupy the M shell as the L shell can only contain a __________ of eight electrons.

2 Some elements, such as sodium (Z = 11) and fluorine (Z = 9), are very reactive elements. Other elements are very __________. The most unreactive elements are a family of gases known as the __________ gases. The first member of this family is __________ (Z = 2). The two electrons of helium __________ the K shell. A full electron shell is a __________ configuration. The second member of this family is neon (Z = 10). Neon's __________ configuration is 2,8. In this case the L shell is __________ with eight electrons. Argon (Z = 18) is the third noble gas and its electron configuration is 2,8,8. Although the M shell is not filled, the presence of __________ electrons (an octet) gives the atom additional stability. This __________ stability is shown by krypton (Z = 36) which has an electron configuration of 2,8,18,8. Apart from helium, all the noble gases contain eight electrons in their __________ shell.

Reactive elements can gain or __________ electrons in chemical reactions. Magnesium, for example, can lose the __________ electrons in the outer shell to produce a magnesium ion (Mg^{2+}). This ion has a stable __________ configuration of 2,8. Chlorine atoms can achieve stability by __________ an electron to form a __________ charged chloride ion (Cl^{-}). The electron configuration of this ion is 2,8,8. Positive ions are called __________ and negatively charged ions are called __________.

3 Different __________ have different numbers of protons. For example, oxygen atoms always have eight protons whereas neon atoms have 10 protons. Elements can vary in the number of

__________ found in the nucleus. Oxygen atoms, for example, normally have eight neutrons in the nucleus, but some oxygen atoms contain nine or 10 neutrons. Most neon atoms have 10 neutrons but some have 11 or 12 neutrons. These different forms of an element are called __________.

The number of protons in the nucleus of an atom is called its __________ number (Z). Every element has its own unique atomic number. The __________ number (A) of an atom is the total number of protons and neutrons in the nucleus. Thus, the number of neutrons in a nucleus is the __________ between A and Z. Each isotope has its own nuclear symbol. The nuclear symbol contains the __________ of the element as well as the Z and A values as subscripts and superscripts respectively. ${}^{16}_{8}O$ is the nuclear symbol for the oxygen-16 isotope, while ${}^{17}_{8}O$ is the nuclear symbol for oxygen-17 which has one more __________ than oxygen-16.

4 It is important not to confuse the term mass __________ with the atomic weight of an element. The mass number is the sum of the number of __________ and neutrons in the nucleus. The atomic weight of an element, however, is the __________ atomic mass of the element compared with the __________ isotope. The atomic weight is measured in atomic mass units (u) and includes the mass of all the __________ as well as protons and neutrons. As most elements consist of isotopic forms, the atomic weight is normally not a __________ number. For example, the atomic weight of chlorine is 35.45 u as __________ chlorine consists of chlorine-35 and chlorine-37 isotopes.

The atomic __________ of hydrogen is 1.008 u as it consists of hydrogen-1, hydrogen-2 and an extremely small amount of radioactive hydrogen-3. Hydrogen gas has a very __________ density but it is highly explosive in the presence of air and a spark or flame. In 1937 the German passenger airship *Hindenburg* caught fire and was destroyed while attempting to dock with its mooring mast at the Lakehurst Naval Air Station in New Jersey, United States. While helium gas (atomic weight 4.003 u) is almost as light as and __________ than hydrogen, it was too expensive and not as plentiful at the time. Hence the operators of the *Hindenburg* took the risk of using hydrogen.

Answers **1** shells; two; eight; number; protons; hydrogen; proton; electron; two; six; two; configuration; maximum **2** unreactive; noble; helium; fill; stable; electron; full; eight; octet; outer (outermost); lose; two; electron; gaining; negatively; cations; anions **3** elements; neutrons; isotopes; atomic; mass; difference; symbol; neutron **4** number; protons; relative; carbon-12; electrons; whole; natural; weight; low; safer

1 The electron shell model of the atom is a simple model that shows how electrons are arranged around the nucleus of an atom. This electron arrangement is called the **electron configuration**. Hydrogen (H) is the simplest element. It has one proton in its nucleus and one electron in the first electron shell (K shell). Helium (He) has two protons and two neutrons in its nucleus, and two electrons in the K shell. The K shell is now full and so lithium (Li) with three electrons has two of these in the K shell and one in the second shell (L shell) which is further from the nucleus. The electron configuration for lithium is written as 2,1 which means two electrons in the K shell and one electron in the L shell. The table below shows the electron configuration of the first 18 elements based on their proton number. The proton number of an element is also called the **atomic number** and it is symbolised by the letter Z. In a neutral atom the electron number is also equal to Z. Beyond argon (Z = 18) the electron configurations become more complex due to the existence of subshells. Potassium (Z = 19) has an electron configuration of 2,8,8,1 rather than 2,8,9. These issues will be discussed in the Year 10 book.

Element	Z	electron configuration
hydrogen	1	1
helium	2	2
lithium	3	2,1
beryllium	4	2,2
boron	5	2,3
carbon	6	2,4
nitrogen	7	2,5
oxygen	8	2,6
fluorine	9	2,7

Element	Z	electron configuration
neon	10	2,8
sodium	11	2,8,1
magnesium	12	2,8,2
aluminium	13	2,8,3
silicon	14	2,8.4
phosphorus	15	2,8,5
sulfur	16	2,8,6
chlorine	17	2,8,7
argon	18	2,8,8

Sodium (Z = 11) has an electron configuration of 2,8,1. The L shell can only hold a maximum of eight electrons and so the last electron must occupy the M shell.

2 Elements vary in their **chemical reactivity**. The most stable and unreactive elements are a family of gases known as the **noble gases**. These gases can be found in the right hand column of a periodic table. The following table shows the electron configurations of the noble gases. Apart from helium (which has a filled K shell) the remaining noble gases all have eight electrons in their outermost shell. The presence of an **octet** (group of eight) of electrons creates stability. Krypton, for example, has 18 electrons in the M shell. This shell is now full. The remaining electrons form a stable octet in the N shell. In xenon, 18 electrons are also found in the N shell and eight electrons in the O shell, rather than all 26 electrons occupying the N shell. Electrons are more stable when they form groups of 8 or 18.

Element	Z	electron configuration
helium	2	2
neon	10	2,8
argon	18	2,8,8
krypton	36	2,8,18,8
xenon	54	2,8,18,18,8

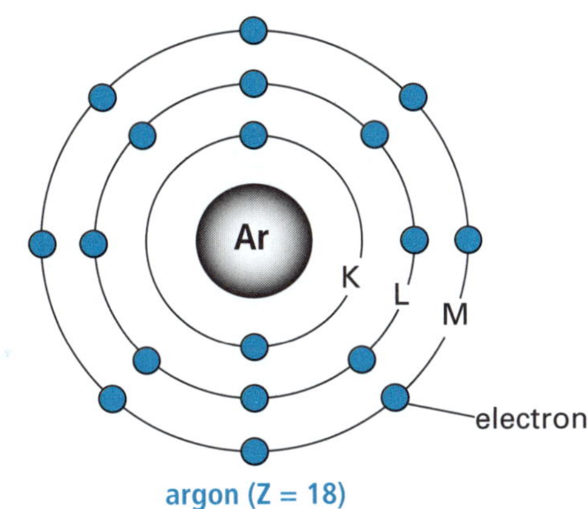

argon (Z = 18)

Reactive elements can achieve a stable octet of electrons in their outer shell by losing or gaining electrons. Consider sodium (Z = 11) which has an electron configuration of 2,8,1. Sodium readily loses the single outer electron during chemical reactions. The atom that is left after this loss of an electron is positively charged. This charged atom is called an **ion**. The symbol for the sodium ion is Na^+. This is a very stable ion as its electron configuration is 2,8. Sodium ions have a stable octet in the outer shell. Crystals of table salt always contain sodium ions.

Fluorine (Z = 9) is a very reactive element. During chemical reactions fluorine readily gains an electron to form a negative fluoride ion (F^-). The fluoride ion also has an octet in its outer shell (2,8) which makes it a stable ion. All positive ions are called **cations**. All negative ions are called **anions**. It is the transfer of electrons to or from an atom that produces ions; protons are not transferred.

3 The atoms of some elements can have variable numbers of neutrons. For example, 20% of boron atoms have five neutrons and 80% have six neutrons. These different forms of the same element are called **isotopes**. The more neutrons an element has in its nucleus the heavier it is.

(cont.)

The **mass number** (A) of an element is:

$$A = \text{number of protons} + \text{number of neutrons}$$

Boron (B) has an atomic number of 5 (Z = 5). The mass number for the isotope with five neutrons in its nucleus is 10 (five protons + five neutrons). This isotope is referred to as B-10, because it is a boron atom (chemical symbol B) with a total of 10 protons and neutrons in its nucleus. The mass number of the boron isotope with six neutrons in its nucleus is 11 (five protons + six neutrons). This isotope is referred to as B-11.

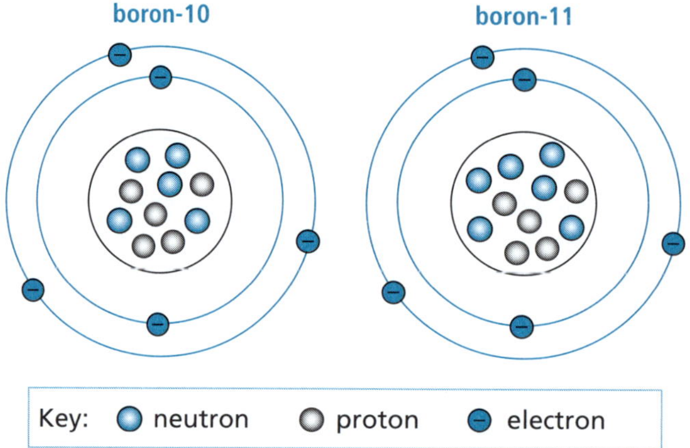

The nuclear symbols for the isotopes of any element (E) can be represented as ${}^{A}_{Z}E$. The two isotopes of boron are represented symbolically as ${}^{10}_{5}B$ and ${}^{11}_{5}B$.

For any isotope of an element, the number of neutrons can be found by subtracting Z from A, that is, neutron number = A – Z.

4 Atoms of different elements have different **atomic weights**. Most of the mass of an atom is due to the mass of the protons and neutrons, as electrons are very much lighter than protons or neutrons. Isotopes of an element also have different masses as they vary in the number of neutrons present. Chemists have established a **relative mass scale** for atoms based on the carbon-12 isotope. This isotope of carbon is assigned an atomic weight of 12 u. The symbol u is used for atomic mass unit. The atomic weight of other elements is relative to this carbon-12 standard. The atomic weight of hydrogen is 1.008 u (see the periodic table on the inside front cover of this book). This atomic weight is not a whole number because natural hydrogen contains three isotopes of different mass. The most abundant of these isotopes is hydrogen-1 (H-1) but smaller amounts of H-2 (deuterium) and H-3 (tritium) exist. Similarly the atomic weight of natural carbon is 12.01 u. The most common isotopic form of carbon is C-12 but C-13 and C-14 also exist in smaller amounts.

Checklist

Can you:

1. *Write electron configurations for the first 18 elements?* ☐
2. *Explain the unreactive nature of the noble gases?* ☐
3. *Explain what an isotope is?* ☐
4. *Explain why the atomic weights of elements are not whole numbers?* ☐

ATOMIC STRUCTURE

Atomic theory

REVISION TEST

1 Write the electron configuration of each of the following elements.

a fluorine (Z = 9) (1 mark)

b aluminium (Z = 13) (1 mark)

c sulfur (Z = 16) (1 mark)

2 Use the following nuclear symbols to determine the number of protons, neutrons and electrons in each element.

a $^{51}_{23}V$ (3 marks)

b $^{238}_{92}U$ (3 marks)

3 The diagram below shows the nuclei of two different elements.

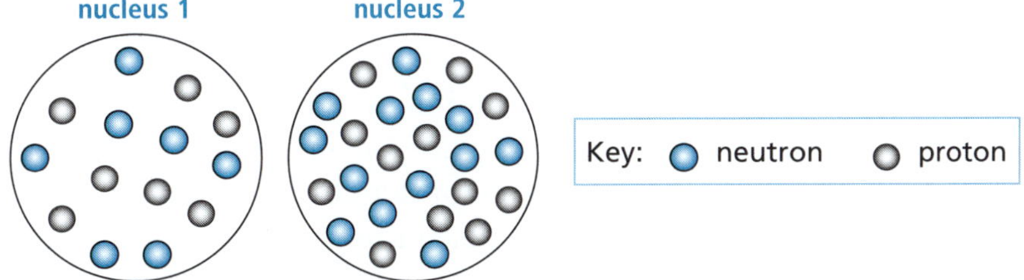

a Use the periodic table on the inside cover of this book to identify these elements. *Hint 1* (2 marks)

b Write the nuclear symbols for each of these two elements. (2 marks)

4 The element indium (In) has an atomic weight of 114.8 u. Natural indium is composed of two isotopes: indium-113 and indium-115.

a What is the difference between indium-113 and indium-115? (1 mark)

b Explain which isotope is in the greater proportion in a sample of natural indium. *Hint 2* (2 marks)

5 Xenon is a colourless gas. It is sometimes used in car headlights. The light produced is bluish-white. Xenon has an atomic number of 54.

a Write the electron configuration of this element. (1 mark)

b Explain the chemical stability and low reactivity of this element. (1 mark)

(cont.)

6 The following diagram shows information about the isotopes of lithium and their relative proportions in natural lithium.

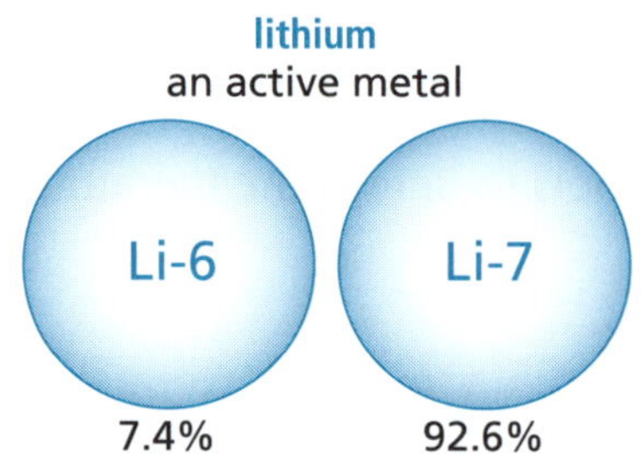

a Find lithium in the periodic table on the inside cover of this book and determine the number of protons in its nucleus. (1 mark)

b Write the electron configuration of lithium. (1 mark)

c Write the nuclear symbols for each of the lithium isotopes. (2 marks)

d Assuming the Li-6 atom has an atomic weight of 6 u and the Li-7 atom has an atomic weight of 7 u, calculate the average atomic weight of a sample of natural lithium. *Hint 3* (2 marks)

e During chemical reactions, lithium loses an electron to achieve a stable electron arrangement. Write the symbol of the lithium ion formed and write its electron configuration. (2 marks)

7 True or false?

a Chloride ions (Cl^-) have a stable octet of electrons in their outer shell. (1 mark)

b In chemical reactions magnesium ($Z = 12$) loses one electron to form a stable magnesium ion (Mg^+). (1 mark)

c The fluoride ion (F^-) has the same electron configuration as the neon ($Z = 10$) atom. (1 mark)

d The standard used for the atomic weight scale is the hydrogen-1 isotope. (1 mark)

e The mass number of an atom is the sum of the number of protons and electrons. (1 mark)

Hint 1: Count the number of protons.
Hint 2: Which of the isotopes has a mass number closer to the atomic weight?
Hint 3: Use the percentage data in the diagram to work out the sum of the weighted averages of each isotope's mass.

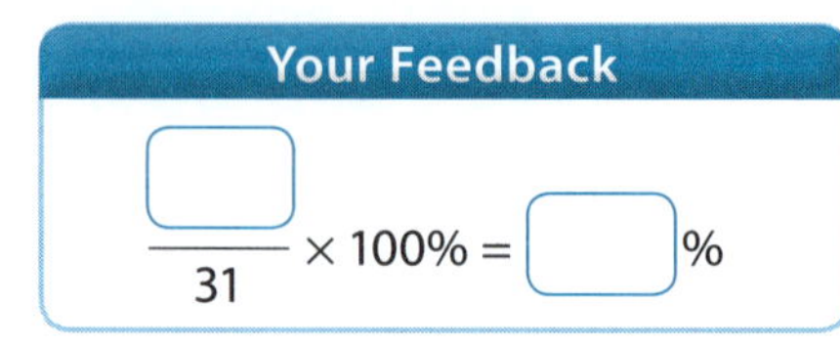

RADIOACTIVITY

Atomic theory

QUICK REVISION

1 Some atoms are unstable. Particles and __________ are emitted from their nuclei to try and make them more __________. A chain of decays may take place until a stable nucleus is __________. This is called radioactivity. In the process the __________ become smaller. There is no way to exactly predict when a particular unstable nucleus will __________ in this way, but it is easy to predict the decay pattern of a __________ number of radioactive nuclei. The rate at which nuclei decay is proportional to how many __________ there are. The activity of a sample of __________ material (that is, a very large number of unstable nuclei) can be measured in disintegrations per second.

With Frederick Soddy, Ernest Rutherford noticed that __________ such as uranium and thorium changed to __________ elements during radioactive decay. This was an incredible idea at the time, since the alchemists' notions of changing cheap metals into gold or other precious metals was shown to be fake. In 1899 Rutherford described two distinct types of radiation: alpha rays and __________ rays. In 1903 he found a new type of radiation that had a much higher penetrating power. He called these gamma __________. Rutherford also observed that a sample of radioactive material took the same amount of time for half the sample to __________, no matter how much of the sample he had.

2 There are three common types of radioactive decay: __________, beta and gamma. The difference between them is the particle or energy emitted by the __________ during the decay process.

- Alpha particles (α particles) consist of two __________ and two neutrons bound tightly together. They can travel through air for only a few centimetres and can be stopped by the thickness of a sheet of paper.
- Beta particles (β particles) are high-speed (and high energy) __________ formed when neutrons break down. They can travel a few __________ through air and can be stopped by a sheet of aluminium or a lump of wood.
- Gamma rays (γ rays) are high energy __________ and are not particles. They are not charged and can travel kilometres through air. It takes thick concrete blocks or around 3 cm of __________ to stop them.

3 Radiation cannot be detected by human __________. Since radioactivity affects the atoms that it passes by, it can be easily monitored using a variety of methods. Several devices have been invented to __________ radiation. The most famous is the Geiger __________. It detects the emission of alpha particles, beta particles or gamma rays by the ionisation produced in a low-pressure __________ inside a Geiger-Müller tube. Invented in 1908, these are popular instruments due to their wide and highly visible use as hand-held radiation detection instruments. The photo shows a portable Geiger counter.

(cont.)

RADIOACTIVITY *(continued)*

Atomic theory

QUICK REVISION

In 1896, Henri Becquerel discovered that uranium compounds darken a __________ plate. This effect can be used to __________ how much radiation has struck the film. For example, workers in the nuclear industry wear badges containing __________ that is developed periodically (every month), in much the same way as old-style photographic film. (This is like developing a black-and-__________ X-ray film.) This gives an indication of the radiation each worker has been __________ to. More modern body __________ contain a sheet of radiation-sensitive aluminium oxide or lithium fluoride crystals sealed in a light and moisture-proof __________. When radiation falls on these crystals electrons absorb energy. Later these electrons can be made to discharge their __________. The energy of these electrons, which is released as visible light, is a measure of the __________ dose.

4 The half-life is the time it takes for a substance undergoing decay to __________ by half. A half-life can range from billions of __________ to less than a thousandth of a __________, depending on the substance. The half-life of a substance always remains the same, but the quantity of the substance remaining after each half-life gets __________ and smaller.

The element actinium (Ac) has no stable __________, and around 31 known radioactive isotopes. One of these is ^{225}Ac with a __________-life of 10 days.

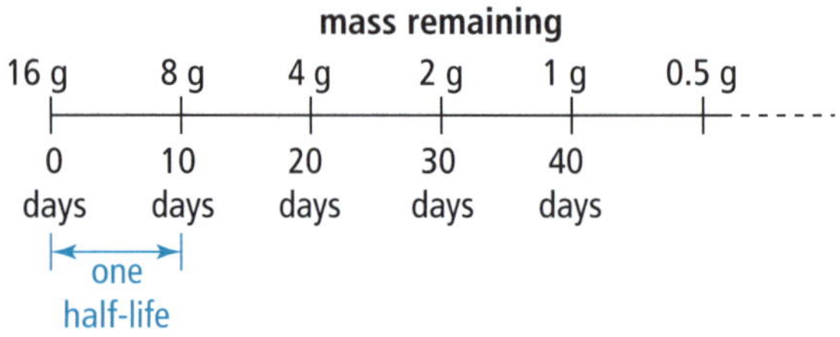

mass remaining

16 g	8 g	4 g	2 g	1 g	0.5 g
0 days	10 days	20 days	30 days	40 days	

one half-life (0 days to 10 days)

After 10 days, half of the __________ amount of actinium remains. After a further 10 days, half of this remaining amount has decayed away, so a quarter of the original __________ remains. If there was originally 16 g of actinium, after 10 __________ there would be 8 g remaining. After 50 days __________ g of actinium is left in the sample.

Answers **1** radiation; stable; reached (formed, produced); nuclei; decay; large (huge); nuclei; radioactive; elements; different (other); beta; rays; decay **2** alpha; nucleus; protons; electrons; metres; radiation; lead **3** senses; detect; counter; gas; photographic; measure (detect); film; white; exposed; badges; packet (container, capsule); energy; radiation **4** decrease; years; second; smaller; isotopes; half; original; amount; days; 0.5 (½)

RADIOACTIVITY

Atomic theory

REVISION SUMMARIES

1 The neutrons and protons that make up the nucleus of an atom interact with each other, and there are strong forces keeping the nucleus together. **Radioactive decay** occurs when the nucleus of an unstable atom spontaneously loses energy by emitting ionising particles. An atom with one type of nucleus might decay into an atom with a nucleus of different energy, or into a different nucleus containing a different number of protons and neutrons. There are many isotopes which are unstable and emit some kind of radiation.

Several important events led to the discovery of radioactivity.

- In 1896 French scientist **Henri Becquerel** discovered that a piece of uranium mineral produced an image on a photographic plate in the absence of light. He explained this event as being due to spontaneous emissions by the uranium.
- **Pierre and Marie Curie**, who coined the word 'radioactivity', found that uranium ore was considerably more radioactive than the pure element. Believing that the ore contained radioactive components additional to the uranium, they discovered two new radioactive elements which they named polonium and radium.
- Becquerel and the Curies together received the Nobel Prize in Physics in 1903 for their discoveries in radioactivity.
- In 1899 New Zealand-born chemist and physicist **Ernest Rutherford** discovered and named **alpha and beta radiation**. He was awarded the Nobel Prize in Chemistry in 1908.
- In 1900 French chemist **Paul Villard** discovered a new type of radiation. This had a much greater penetrating power than either alpha or beta radiation. Rutherford studied this third type of radiation and called it **gamma ray**.

2 The most common types of radiation are called alpha, beta and gamma radiation.

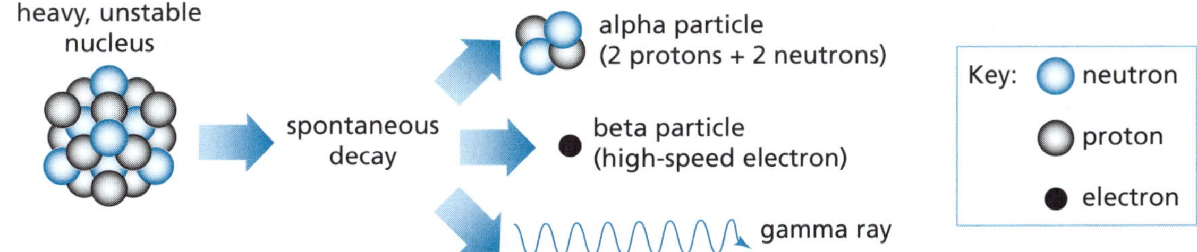

Radiation	Charge	Particles or radiation	Penetration	Other features
alpha (α)	+2	two protons and two neutrons bound together	stopped by just a few centimetres of air, or a piece of paper	relatively slow and heavy; mass is more than 7000 times the mass of a beta particle; due to large charge, alpha particles strongly ionise other atoms; deflected by electric and magnetic fields
beta (β)	–1	high speed electrons	penetrate several metres of air, several millimetres of plastic or less of very light metals	fast and light; ionise atoms that they pass, but not as strongly as alpha particles; deflected by electric and magnetic fields
gamma (γ)	0	high energy electromagnetic radiation	high penetrating power; concrete or a thick sheet of metal, such as lead, needed to reduce penetration significantly	waves, not particles; no mass and no charge; do not directly ionise other atoms; not deflected by electric and magnetic fields

(cont.)

3 Our senses cannot detect radiation, so various devices have been constructed that can.

- A **Geiger counter** consists of a Geiger-Müller tube filled with an inert gas (usually helium, neon or argon) that briefly conducts an electrical charge when a nuclear particle or photon of radiation ionises the gas. The tube amplifies each ionisation event and outputs a current pulse that is further amplified and displayed on the meter. The Geiger counter only registers broad ionisation so it cannot distinguish between the various types of ionising radiation.

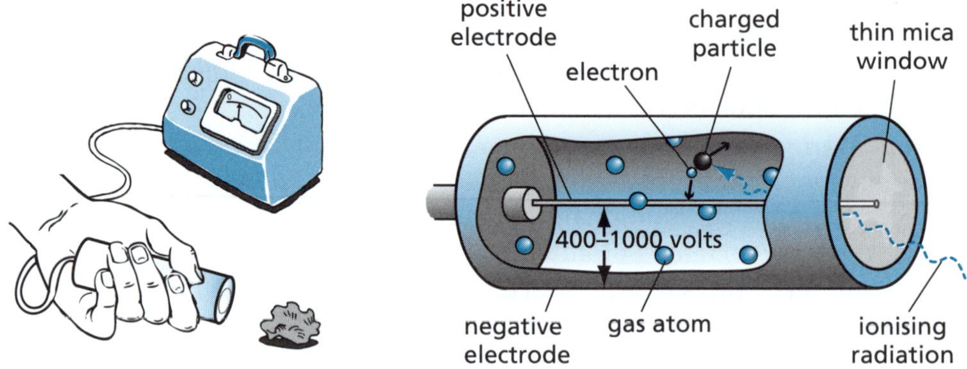

- Radioactivity will darken **photographic film**, and this effect can be used to measure how much radiation has fallen on the film. Photographic film contains a silver bromide emulsion which darkens when exposed to light or nuclear radiation.
- **Body badges** or ring badges are worn by workers using radioactive materials so they have a fair idea as to how much radiation they have been exposed to in the course of their work. In the past these contained a small amount of unexposed photographic film that was later developed. Modern badges contain a sheet of radiation-sensitive aluminium oxide or lithium fluoride crystals sealed within a dark and dry container. When these atoms are exposed to radiation, electrons are able to trap this radiation energy. Later, the energy can be released as visible light. This is measured to establish radiation dosage.

Some older technologies include the following.

- **Spark counter**: a radioactive source ionises the air between the gauze and the wire and sparks jump.

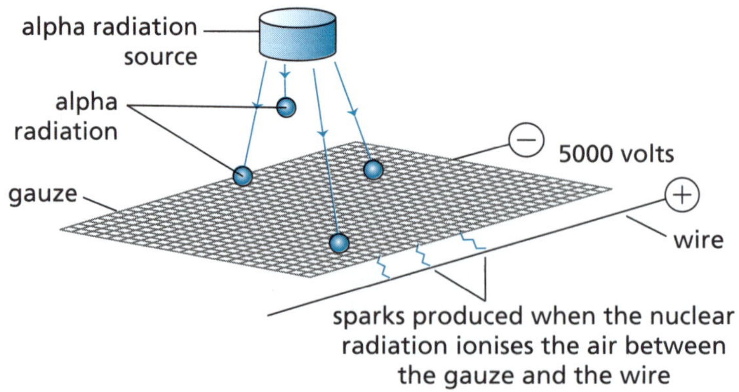

- **Gold-leaf electroscope**: radioactivity ionises the air and conducts electricity. When an electroscope is charged, the charges repel and the gold leaf sticks out. As the charge leaks away the electroscope discharges and the gold leaf falls.

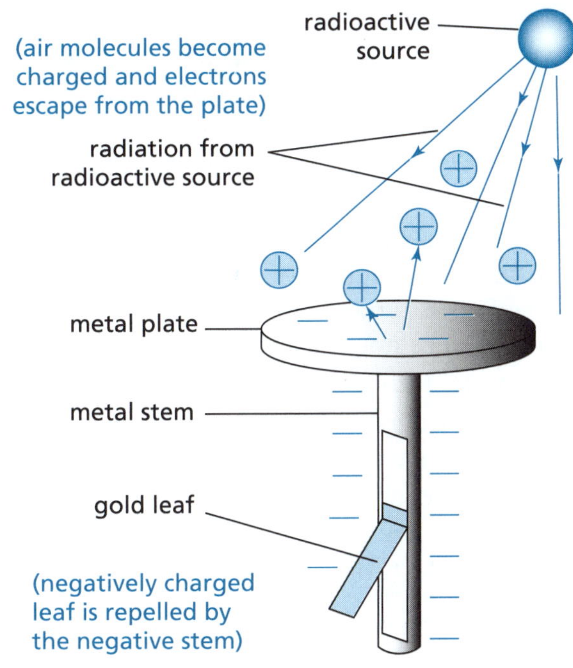

Scientists are always trying to develop more efficient, sensitive and effective ways to detect and measure radiation.

4 It is not possible to guess when a particular atom might decay, but it is possible to determine how long it takes for half the nuclei to decay in a sample of a radioactive isotope. This is known as the **half-life**. One way to determine the half-life is to measure the time it takes for the radioactive count rate from a sample containing the isotope to fall to half its initial level. Different radioactive isotopes have different half-lives. For example, the half-life of carbon-14 is 5730 years, but the half-life of beryllium-11 is just 13.81 seconds. Different isotopes have their own specific half-life. Half-life does not depend on the initial mass of matter nor on temperature or pressure; it is **constant** for each particular nucleus.

Checklist

Can you:

1. *Describe what radioactivity is and outline some of the key events in the discovery of radioactive elements?* ☐
2. *Describe the nature and properties of alpha, beta and gamma rays?* ☐
3. *List some devices for detecting radioactivity?* ☐
4. *Define 'half-life' and use this in plotting/interpreting half-life graphs?* ☐

RADIOACTIVITY

Atomic theory

30 MINUTES

REVISION TEST

1 Coins can be used to model a sample of radioactive nuclei. This is because when they lie on a flat surface they can be in one of just two states: they can lie with the Queen's image facing up or with her image facing down. In this model the image facing down represents a decayed nuclei.

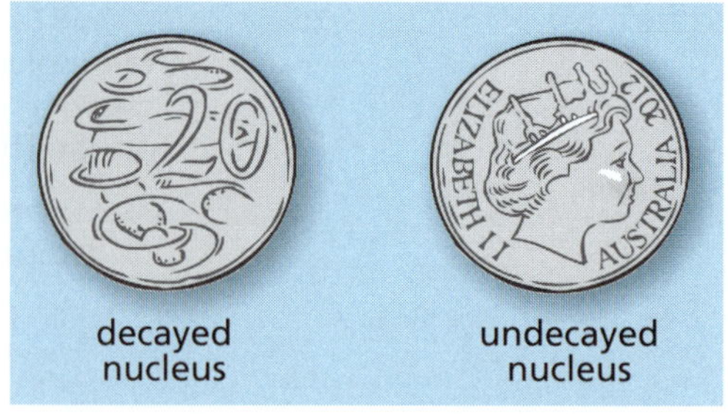

Louisa took a handful of 20c coins and tossed them onto a table. She counted the total number of coins and how many had 'decayed'. She removed all the 'decayed' coins from the sample. She repeated the steps until there were no coins left.

a Explain how coins can model radioactive nuclei. *Hint 1* (2 marks)

b Before tossing the coins can Louisa predict which particular coin will fall face up or face down? Explain how this relates to nuclear decay. (2 marks)

c Her friend, Sarah, stated that each of Louisa's throws represents going through a half-life. Is Sarah correct? Explain. (2 marks)

d Does a radioactive substance last forever? (1 mark)

2 The diagram shows the penetrating power of the three types of radiation.

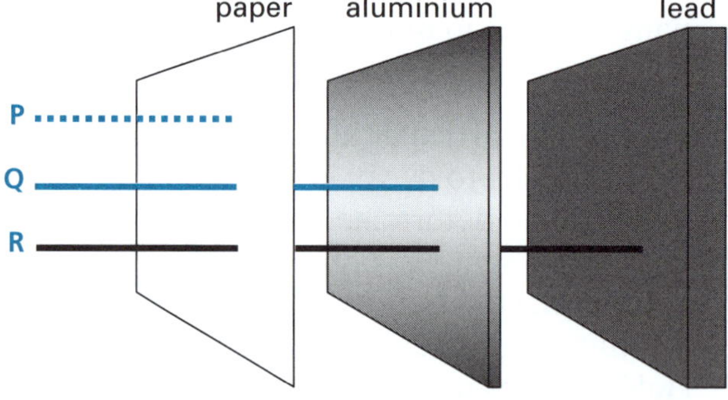

Identify the different types of radiation labelled P, Q and R. (3 marks)

3 The following diagram shows the path followed by alpha, beta and gamma radiation as they pass through a magnetic field.

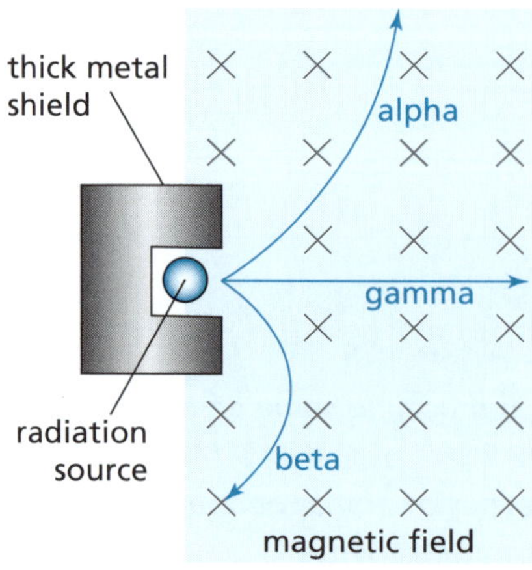

a Why are gamma rays not deflected? (1 mark)

b Why are alpha particles deflected in one direction and beta particles in the opposite direction? (2 marks)

c Why are beta particles deflected more than alpha particles? *Hint 2* (2 marks)

4 Which of the following statements about half-life is true? (1 mark)

A It varies with the mass of material.

B It alters with changes in the temperature.

C It changes for different isotopes.

D It is affected by changes in pressure.

5 P, Q, R and S are elements which form compounds PR, P_2S and QS. It is known that PR and P_2S are radioactive while QS is not a radioactive compound. Determine whether each of the following substances is radioactive, not radioactive or if radioactivity can't be determined.

a P_3 (1 mark)

b P_2R (1 mark)

c QR_3 (1 mark)

d R_2S (1 mark)

6 The timeline shows the radioactive decay for an isotope.

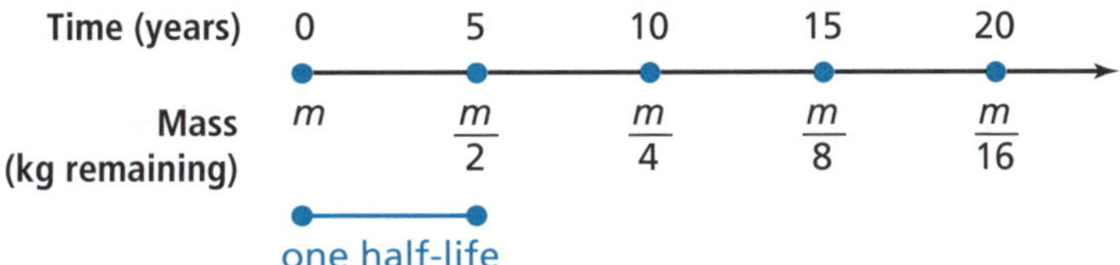

If the decrease in mass of the radioactive matter between 15 and 20 years is 7 kg, what was its initial mass, *m*? (2 marks)

7 The graph shows the radioactive decay for a particular isotope.

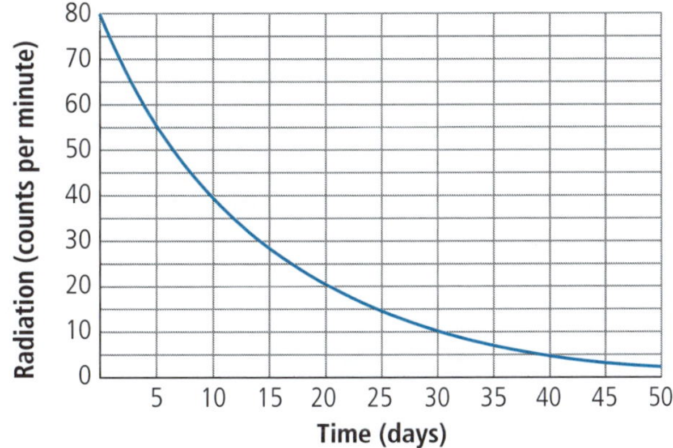

a What is the radiation count after 40 days? (1 mark)

b How long does it take for the radiation counts to halve? What is this time called? (2 marks)

c How many half-lives does it take for the radiation count to drop to 20 counts per minute? (1 mark)

(cont.)

8 The graph shows the decay of a certain radioactive substance.

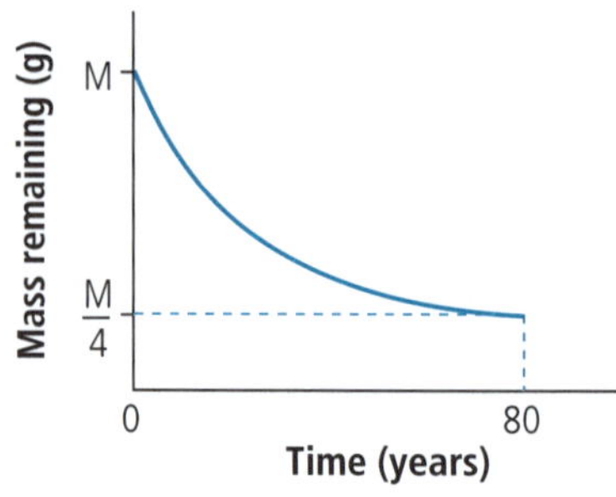

True or false?

a The half life of this substance is 40 years. (1 mark)

b The rate of decay of this substance after 40 years is less than the initial rate of decay. *Hint 3* (1 mark)

c The amount of matter that decays during the first half-life equals the amount of matter that decays during the second half-life. (1 mark)

9 Technetium-99m is a widely used radioactive tracer isotope in nuclear medicine.

a Use the table of values to draw up a graph of the radioactive decay for technetium (Tc-99m). Let the horizontal axis have values from 0 h to 15 h. (4 marks)

Time (h)	Percentage Tc-99m remaining (%)
0	100
1	89.1
2	79.4
3	70.8
4	63.1
5	56.2
6	50.1

Time (h)	Percentage Tc-99m remaining (%)
7	44.7
8	39.8
9	35.5
10	31.6
11	28.2
12	25.1

b What is the half-life of this isotope of technetium? (1 mark)

c Extrapolate your graph to determine what percentage of Tc-99m is present after 15 hours. (1 mark)

d This isotope is used in humans for radioactive imaging. Why do you think it is important that the half-life be short? *Hint 4* (1 mark)

Hint 1: A scientific model is a depiction of an object or system.
Hint 2: Consider their masses.
Hint 3: One way to show rate of decay (how fast a substance decays) is to look at the slope of the curve.
Hint 4: Do you want radioactive substances hanging around inside your body longer than is absolutely necessary?

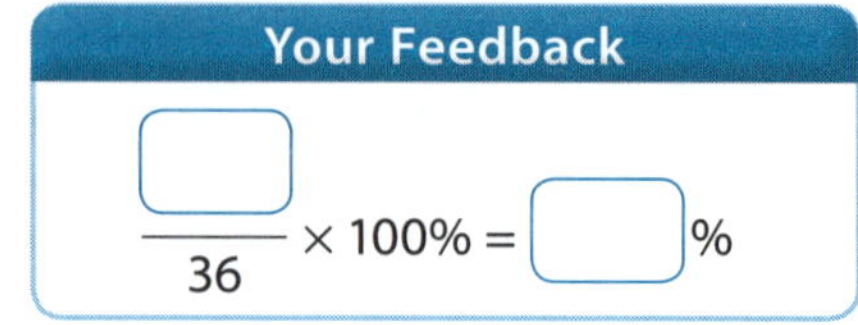

APPLICATIONS OF RADIOACTIVITY

Atomic theory

QUICK REVISION

1 Since their discovery over a __________ ago, radioactive __________ have been put to a number of uses. Geologists use radiometric dating to __________ how long ago rocks were produced, and to deduce the ages of __________ enclosed within those rocks. Radiometric dating is a method of dating geological specimens by determining the proportions of particular radioactive __________ present in a sample. In order to determine the age of a sample of material, one needs to know:

- how much of this radioactive material was __________ initially in the object
- how much (or what proportion) of the material now __________, or how much of the original material has decayed and the __________-life of the material.

As soon as any living organism dies, it __________ taking in carbon. The ratio of carbon-12 to carbon-14 at death is the __________ for all living thing. However, the radioactive carbon-14 __________ and is not replaced. The carbon-14 decays with a half-life of 5730 years, while the quantity of non-radioactive carbon-12 remains __________ in the sample. By calculating the __________ of carbon-12 to carbon-14 in the sample, and comparing it to that in a living organism, it is possible to determine the __________ of a once living thing fairly accurately.

2 There are a number of applications of radioactivity in medicine.

- Medical __________ are sterilised using radioactive isotopes. These strongly penetrating beams can __________ all forms of life, including fungi, bacteria, __________ and spores that might exist on the surface of an instrument.
- Radioactive isotopes can be used to kill cancer __________ inside the body. This is called radiotherapy.
- X-rays are used to look at __________ and to detect some diseases. Dentists use X-rays to diagnose problems with teeth, such as __________.
- A radioactive element can be used as a tracer whose __________ around the body can be followed. For example, a tracer can be used to check for blocked or damaged kidneys. As it moves through the __________ the tracer can be monitored with a radiation counter or on a __________ film.

3 Radiation has a range of uses in many facets of modern life.

- Gamma rays can be used to kill __________ in food, prolonging its shelf life. Although it kills microbes, this radiation does not make the __________ radioactive.
- X-ray imaging can be used to investigate paintings, making it possible to see artwork that has been __________ over.
- X-ray detectors are used to detect X-rays from distant __________ in astronomy. They are also used by customs officers to examine shipping __________ for prohibited and restricted goods.
- Tracers can be used to identify __________ in underground water and gas pipes.
- Radioactive carbon can be used to determine the age of ancient documents, wooden artefacts and other objects containing __________.
- Smoke alarms contain a weak source of americium-241 that emits alpha __________. These can detect smoke entering the alarm. The range of these alpha particles is very __________ through air and so does not pose a health __________ to people who have them in their homes.

Answers **1** century; isotopes (elements); estimate (determine, calculate); fossils (materials); isotope; present; remains; half; stops (ceases); same; decays; constant; ratio (proportion); age **2** instruments; kill (destroy); viruses; cells; bones; cavities; pathway; kidney; photographic **3** microorganisms (microbes); food; painted; objects (stars, nebulae); containers; leaks; carbon; particles; short; risk

APPLICATIONS OF RADIOACTIVITY

Atomic theory

REVISION SUMMARIES

1 There are two main techniques for dating **fossils**.

- **Absolute dating** is used to determine the actual age of the fossil. Most absolute dating methods make use of radioactive elements that occur naturally in different types of minerals and organic matter.
- **Relative dating** compares the age of one object with another, and determines whether it is older or younger. Relative dating methods do not determine the actual age of the object.

There are a number of dating techniques using radioactive elements.

- Radioactive **uranium-238** is present in many different rocks and minerals. Some of these atoms can split and the pieces travel apart very quickly, damaging the rock or mineral and leaving fission tracks. The number of tracks increases with time depending on the rock's uranium content. The age of a sample is found by measuring the uranium content and the concentration of the fission tracks.
- **Potassium–argon dating** can be used to date fossils found between different layers of volcanic ash and rock. Radioactive potassium-40 in these layers decays to argon-40. The age of the layers can be determined by measuring the proportions of argon gas and potassium within them. A fossil can then be given an approximate age from this.
- **Argon–argon** dating converts the stable potassium-39 in a rock sample into argon-39. The proportions of argon-39 and argon-40 within the sample are then measured, allowing the age of the sample to be determined.
- Non-radioactive carbon-12 accounts for 98.9% of all carbon, while radioactive **carbon-14** comprises less than 1%. Carbon-14 decays to nitrogen-14 with a half-life of 5730 years. When alive, plants and animals integrate all isotopes of carbon into their tissues, and continually replenish it. After death carbon-14 begins to decay. By measuring how much carbon-14 is left, the age of the fossil can be determined. **Carbon dating** can be used to date material that is up to about 50 000 years old. Of course, carbon-14 dating is only useful for material containing carbon, such as linen, cotton, bones, wood, fossils, sea shells, seeds, coal, tombs and diamonds.

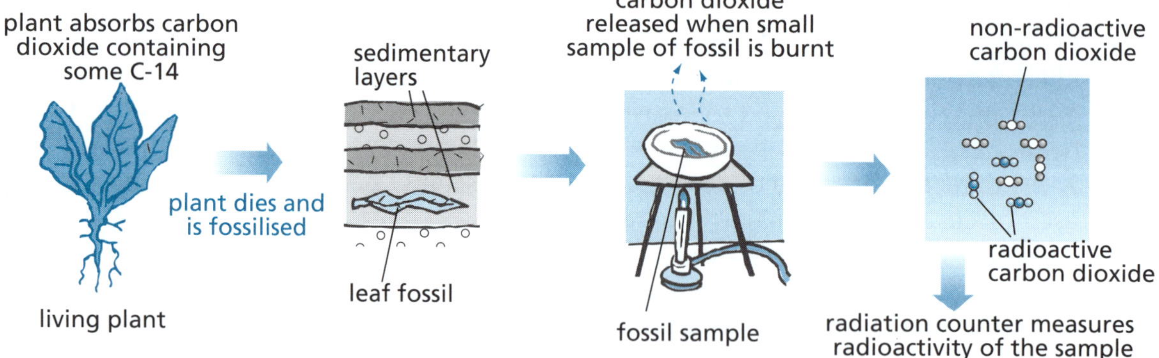

Carbon dating has become an important tool for scientists such as archaeologists and anthropologists.

2 Because of its radioactivity, a **radioisotope** can be detected simply by the radiation that it emits. In medicine, this can be used to take a 'picture' of various body parts. After a patient takes the radioactive substance, orally or by **injection**, a special camera can be used to trace the path of the radioactive isotope through the body and determine where it ends up. This is called scanning. For example, a bone scan can show up fractures (breaks) which may not be visible on normal X-rays.

APPLICATIONS OF RADIOACTIVITY

Atomic theory

REVISION SUMMARIES

Nuclear medicine uses radiation to collect **diagnostic** information about the functioning of specific organs in a person. These diagnostic procedures are now routine. Radiotherapy can be used to treat some medical conditions, such as cancer, by using radiation to weaken or destroy targeted cells. The radioisotope that produces the radiation can be localised in the required organ, just as it is for diagnosis. The radioactive element may be used as it follows its path though the body, or it might be attached to a suitable biological compound. A number of radioactive materials are used in medicine.

- Radioactive rays can be used to treat skin diseases.
- Radiation from cobalt-60 and iodine-131 is used to diagnose and treat thyroid disorders.
- Phosphorus-32 is used to study the metabolism of plants, nucleic acids and phospholipids; to treat certain types of leukaemia; and in identifying the location of some tumours during surgery.
- Sodium-24 in the form of sodium chloride (NaCl) is used to study both electrolytes and blood.
- Lutetium-177 and yttrium-90 are used to treat cancer.
- Technetium-99m is used to image the skeleton, heart muscle and many other organs. This is the most widely used isotope in nuclear medicine.

Recently, Australian and Chinese researchers jointly developed a portable battery-operated plasma flashlight that emits reactive plasma particles. The flashlight is capable of killing some of the strongest bacteria, and can be used to treat patients in natural disasters and war zones. They can help heal wounds and selectively kill cancer cells. The flashlight works by making reactive species, such as oxygen ions. These highly reactive particles can oxidise the pathogens' cell membranes and damage its DNA.

3 Radiation is also used in **industry**.

- Sodium-24 is an industrial tracer used to find leaking underground pipes.
- Radiation is used in the manufacture of paper, plastic and metal sheets to control the thickness.
- Radioactive isotopes of carbon and hydrogen are used to examine the path of nutrients into plants.
- Radioisotopes are used to approximate the wear in bearings.
- Just as X-rays show a break in a bone, gamma rays show flaws in metal castings or welded joints.
- Some old watches and clocks have dials that were painted with a radiation-emitting compound allowing them to be seen in the dark. These are no longer manufactured.
- Gamma irradiation is used extensively to **sterilise** wool, food and medical products. Cobalt-60 is the chief isotope used, since it emits energetic gamma rays.
- **Food irradiation** destroys food-borne bacteria and parasites, and extends the **shelf life** of some foods. Food irradiation won't improve the food if it is already spoiled. Irradiation does not make food radioactive nor does it create abnormal or harmful chemicals in the food. Irradiation does not affect the nutritional value of food.

Checklist

Can you:

1 *Outline how radioactive isotopes are used to date fossils?* ☐

2 *Describe some applications of radioactivity in medicine?* ☐

3 *Describe some applications of radioactivity in industry?* ☐

APPLICATIONS OF RADIOACTIVITY

Atomic theory

REVISION TEST

1 The film badge dosimeter is worn by people who may become exposed to radioactive materials. It is used for monitoring total exposure to ionising radiation.

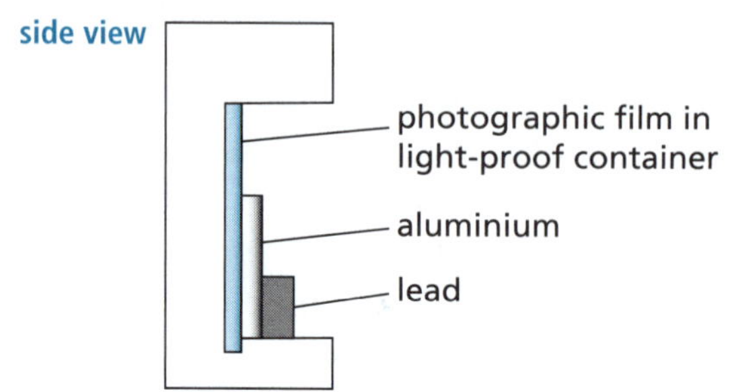

a Why do people working with radiation wear such a device? (2 marks)

b Why does the badge contain a thickness of aluminium and lead? *Hint 1* (2 marks)

c Would this dosimeter detect exposure to alpha particles? Explain. (2 marks)

2 In order to date many older fossils, scientists examine layers of igneous rock or volcanic ash above and below the fossil. They can date these igneous rocks using elements that are slow to decay, such as uranium and potassium.

a Why use elements that are slow to decay? (2 marks)

b The diagram shows a cross-section through a cliff where the ages of volcanic ash have been determined. How old are the fossils shown? (2 marks)

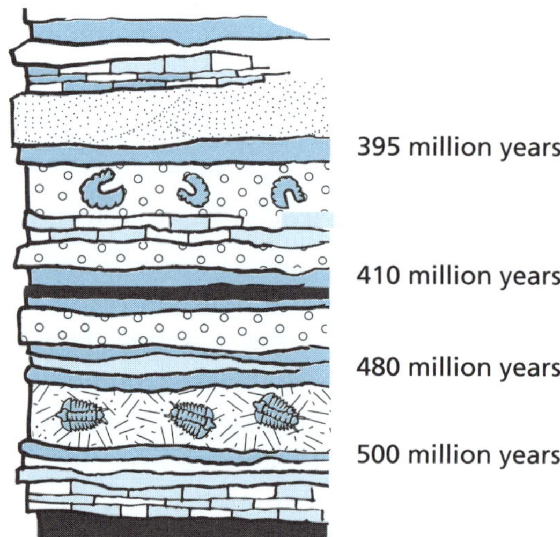

3 Large-scale irradiation facilities use very high doses of gamma radiation for sterilisation and are used for disposable medical supplies such as syringes, cannulas, gloves, clothing and instruments, many of which would be damaged by heat sterilisation.

a What is sterilisation? (1 mark)

b Heat is often used to sterilise substances. Why is heat not used to sterilise all materials? *Hint 2* (1 mark)

c Why are gamma rays used for sterilisation purposes? (1 mark)

d Food irradiation is the last step before packaged food is sent to the retail market. Why is it the last step? (2 marks)

4 The short half-life of carbon-14 means it cannot be used to date extremely old fossils. Explain this statement. *Hint 3* (1 mark)

5 The diagram shows the use of a source of beta particles during the manufacture of paper.

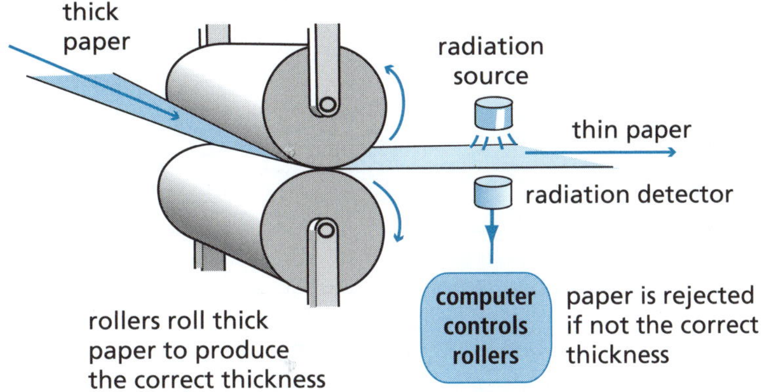

a Why couldn't an alpha particle source be used? (1 mark)

b Should the source of these particles have a long or short half-life? Explain. (3 marks)

c As beta particles pass through the paper, the reading in the detector should remain unchanged. Why? (1 mark)

d Suppose the radiation received by the detector decreases. What does this mean? What happens next? (3 marks)

6 Leaks in underground industrial pipes can be detected by adding a radioactive isotope (called a tracer) to the liquid or gas in the pipe, as illustrated in the diagram below.

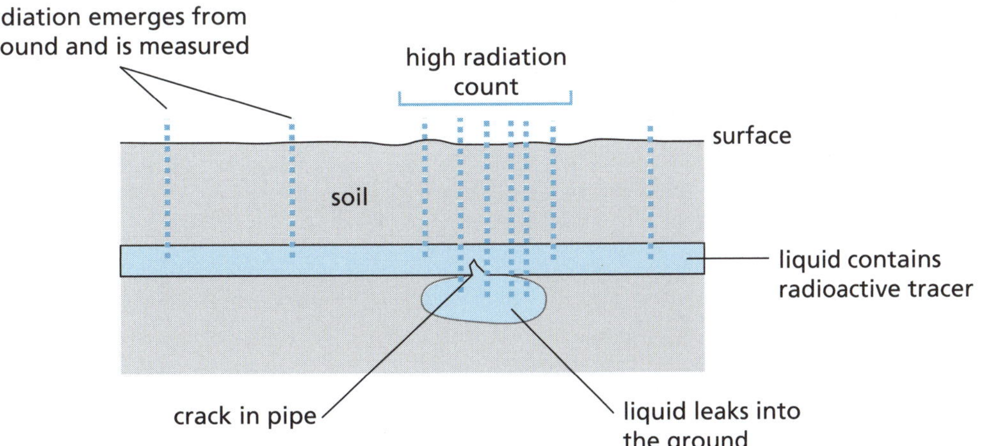

a Describe how underground leaks would be detected. (3 marks)

b Would the radioactive source used emit alpha, beta or gamma radiation? Explain. (3 marks)

c Why should the substance emitting the gamma radiation have a short half-life? (2 marks)

(cont.)

7 Phytoextraction is a method that uses plants to extract radioactive metal contaminants from the soil by accumulating them in plant tissue. The graph below shows the results of an experiment removing two such contaminants, caesium-137 and strontium-90, from the soil.

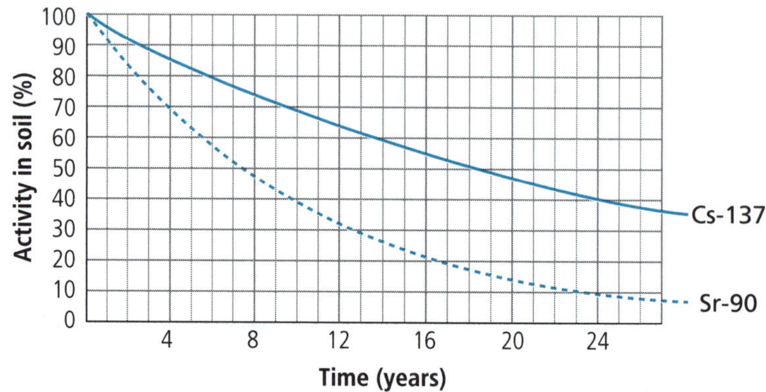

a Approximately how long will it take to remove 50% of the Sr-90 from the soil? (1 mark)

b How much *longer* will it take to remove 50% of Cs-137, compared to Sr-90? (1 mark)

c After how many years will the activity of the strontium drop to half of that shown by the caesium? (1 mark)

d The half-life of Sr-90 is 28.8 years and the half-life of Cs-137 is 30.2 years. Comment on the effectiveness of plants in removing these radioactive contaminants from the soil. *Hint 4* (4 marks)

Hint 1: Consider the penetrating power of different radiations.
Hint 2: Consider the effect of heat on some substances.
Hint 3: Consider how much is left after a long period has passed.
Hint 4: You need to compare how much of each of these radioactive materials are removed with the half-life of each substance.

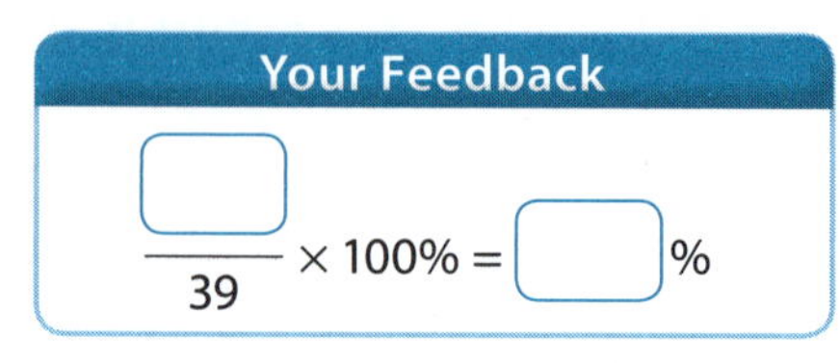

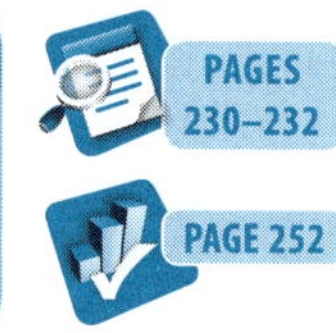

OBSERVING AND MODELLING CHEMICAL REACTIONS

Chemical reactions

QUICK REVISION

1 The evaporation of water is an example of a __________ change, as the water vapour that forms is just a different physical __________ of water. However, the electrolysis of water is a __________ change because new substances (hydrogen and __________) are formed. There are various observations that can be made that indicate whether or not a chemical __________ has occurred. When dilute hydrochloric acid is added to limestone, the mixture is observed to __________ as bubbles stream from the limestone. The limestone gradually __________ in the acid. These observations indicate a __________ reaction has occurred. When solutions of sodium hydroxide and hydrochloric acid are mixed, there is no visible __________ but the mixture gets quite hot. The generation of __________ energy is an indicator of chemical change. When solutions of barium hydroxide and sulfuric acid are mixed a white solid (or __________) is formed. The mixture also gets warmer. These observations __________ a chemical change. Fires and explosions are also indications that a chemical change has occurred. Some chemical reactions are quite slow, particularly if the __________ is low. Corrosion of iron is usually called __________. Rusting is a chemical change and it is accelerated if air and water are in contact with the iron. The presence of salt in the water also increases the rate of rusting.

2 Compounds are formed when two or more elements combine together. Such reactions are called __________ reactions. When magnesium __________ is heated in the presence of oxygen molecules, a chemical change occurs in which a white powder called magnesium oxide is formed. Considerable amounts of __________ and light energy are also released in this reaction. Another example of a synthesis reaction is the reaction of nitrogen and oxygen molecules in the __________ during a lightning storm. The energy provided by the lightning flash causes the __________ and oxygen molecules to react to form nitric oxide molecules.

Decomposition reactions involve a compound breaking down into its __________ elements or into simpler compounds. When silver oxide crystals are heated they break down into __________ metal and oxygen gas. The oxygen gas can be detected using the rekindling of a __________ splint of wood test. Lead chloride can be decomposed by electrolysis. The lead chloride is first melted and electrodes inserted into the molten material. When electricity is passed through the melted material the lead chloride decomposes to form lead metal and __________ gas.

3 Chemical synthesis and decomposition reactions can be visualised with the aid of atomic __________ kits. These kits typically contain coloured balls of __________ diameters. Each coloured ball represents an atom of a different __________. These balls can be joined together using plastic or metal rods that fit into small __________ in the surface of the balls.

In the reaction between sulfur and __________ to form sulfur dioxide, the sulfur is represented by a single atom. Oxygen is a diatomic __________ and so two oxygen atoms are joined together with a small rod which represents the __________ bond. The product molecule (sulfur dioxide) consists of a sulfur atom and two oxygen atoms. The molecule is bent with the __________ atom in the middle. Once these three models are constructed, we can see that one sulfur atom reacts with __________ oxygen molecule to form one sulfur dioxide molecule. The models show both mass and __________ conservation has occurred during the reaction. The law of mass __________ is a fundamental law of chemistry.

Answers 1 physical; state; chemical; oxygen; change (reaction); effervesce; dissolves; chemical; change (reaction); heat; precipitate; indicate; temperature; rusting 2 synthesis; metal; heat; air; nitrogen; component; silver; glowing; chlorine 3 model; different; element; holes; oxygen; molecule; chemical (covalent); sulfur; one; atom; conservation

1 Chemical reactions are occurring all the time in our natural and built environment. The **rusting** of iron structures, the digestion of food, the **explosion** of fireworks and the production of glucose via photosynthesis in green plants all involve **chemical changes**. In a chemical change new substances are formed. When you perform chemical reactions in a laboratory experiment, there are a variety of **observations** that you can make that indicate a chemical change has occurred.

- **Gas bubbles** form in the reaction mixture—this is called effervescence.
- The reaction mixture gets hotter or colder.
- The reaction mixture **changes colour**.
- An insoluble substance (or **precipitate**) forms.
- Flames are produced or an explosion occurs.

Some chemical reactions are very slow so they are hard to observe. For example, the rusting of an iron nail may initially be hard to see. However, if a time-lapse video is made of the nail over a period of several days then the formation of the rust layer on the surface of the nail can be seen.

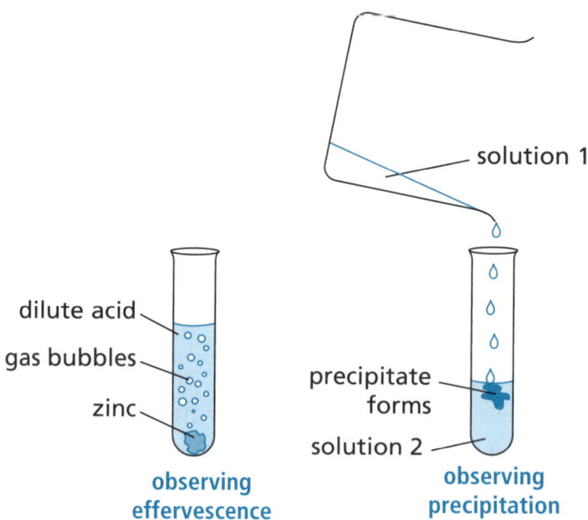

2 A common type of chemical reaction is the reaction between two elements to form a compound. This is called a **synthesis reaction**. The elements are called **reactants** and the compound formed is called the **product**. An example of this type of reaction is the formation of water (the product) from oxygen and hydrogen (the reactants). If oxygen gas and hydrogen gas are mixed together in a test tube at room temperature, no reaction is observed. However, if a flame from a lit candle is introduced into the mixture of gases there is a loud pop sound as the gases react explosively. On the walls of the test tube some small droplets of water can be seen.

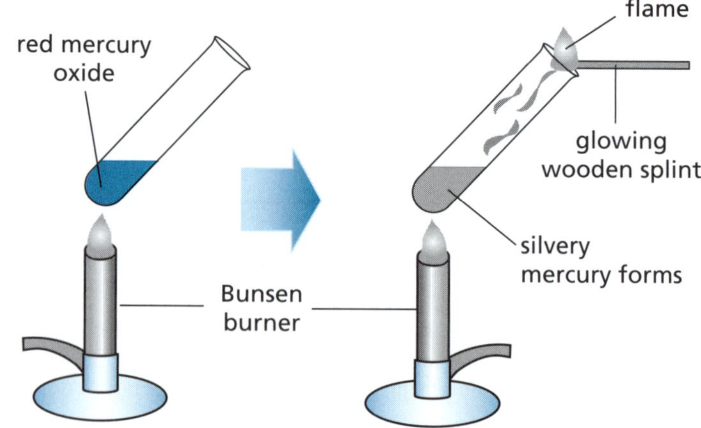

Another common type of reaction is the **decomposition** of a compound into its component elements. The reaction usually occurs when the compound is heated strongly. For example, when heated the red compound mercury oxide decomposes and turns into liquid mercury and gaseous oxygen. The oxygen cannot be seen but if a glowing splint of wood is placed in the mouth of the test tube, the wood is rekindled by the oxygen that is produced.

3 Chemists use **molecular model kits** to create models of compounds. These models are helpful in understanding a chemical reaction. Model kits consist of plastic balls of different colours and sizes which represent the atoms of different elements. Plastic connecting pieces fit into holes in these balls and these connectors represent **chemical bonds**. Some older kits use wooden balls and metal springs as bonds.

The reaction between hydrogen and oxygen to form water can be modelled. A hydrogen molecule (H_2) consists of two atoms of hydrogen joined together by a chemical bond. Similarly oxygen molecules (O_2) consist of two oxygen atoms joined together. Water molecules (H_2O) consist of two hydrogen atoms joined to a central oxygen atom. This molecule is bent. It is obvious that the chemical reaction between oxygen and hydrogen molecules cannot occur with just one molecule of each gas. The **law of mass conservation** states that mass is neither created nor destroyed in chemical reactions, that is, that mass is conserved in a chemical change. In this case the number of atoms of each element in the reactants must be the same as their number in the products. The diagram below shows how many molecules of each element are required to satisfy this fundamental law of chemistry. Two hydrogen molecules are required for each oxygen molecule. As a result, two water molecules form.

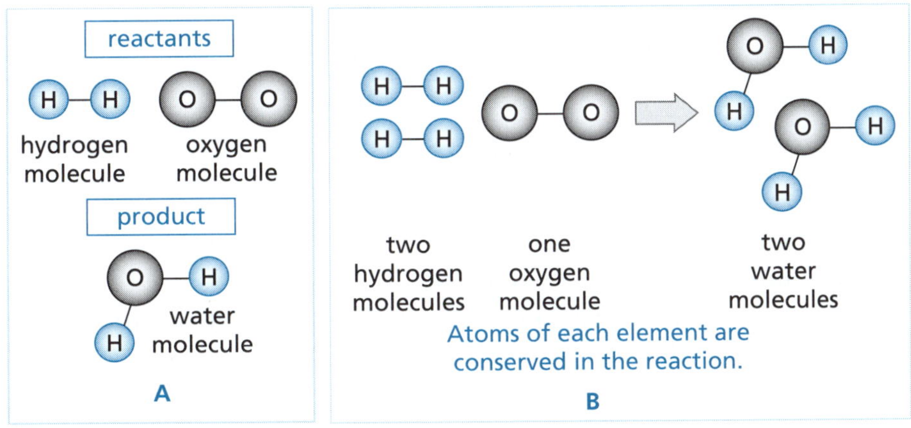

Checklist

Can you:

1 *Identify the indicators of chemical change?* ☐
2 *Explain the difference between reactants and products in a chemical change?* ☐
3 *Explain how molecular models can be useful in understanding chemical change?* ☐

OBSERVING AND MODELLING CHEMICAL REACTIONS

Chemical reactions

REVISION TEST

1 Elements and compounds are known as pure substances. A student used coloured plastic balls to create models of different types of pure substances. These models are shown in the diagram.

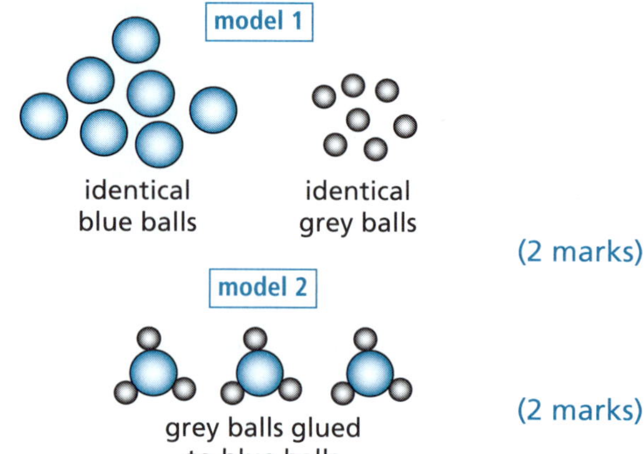

- **a** Identify the types of pure substances that models 1 and 2 represent. (2 marks)
- **b** Using the same coloured balls, demonstrate the synthesis reaction that leads to the formation of the substance shown in model 2. *Hint 1* (2 marks)

2 The following diagram shows the apparatus used to heat magnesium ribbon in air.

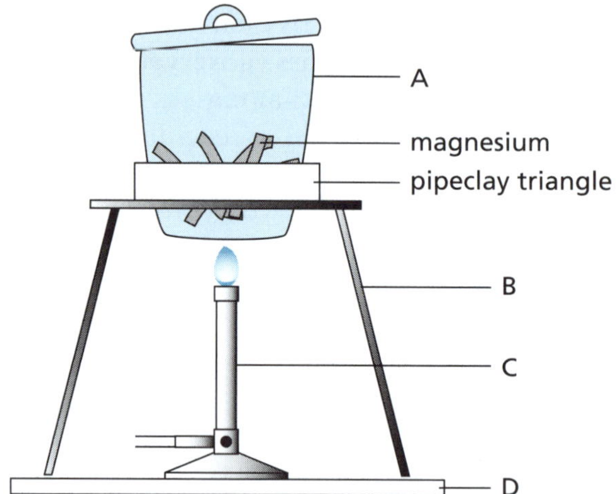

- **a** Classify this reaction as a synthesis or decomposition reaction. (1 mark)
- **b** Identify the equipment labelled A, B, C and D. (4 marks)
- **c** When heated the silvery magnesium ribbon turns into a white powder. Explain why this is an example of a chemical change. (1 mark)
- **d** Draw models of the reactants and products. *Hint 2* (3 marks)

3 State whether the following reactions are physical or chemical changes.

- **a** melting ice (1 mark)
- **b** cooking toast (1 mark)
- **c** boiling an egg for breakfast (1 mark)
- **d** baking a muffin (1 mark)
- **e** burning alcohol in air (1 mark)
- **f** burning magnesium in air (1 mark)
- **g** adding copper sulfate solution to sodium hydroxide solution to form a precipitate (1 mark)
- **h** dissolving copper sulfate in water (1 mark)
- **i** mixing alcohol and water (1 mark)
- **j** shaking olive oil with water in a stoppered test tube (1 mark)

4 The following diagram shows models of molecules in a chemical reaction involving a gas called acetylene.

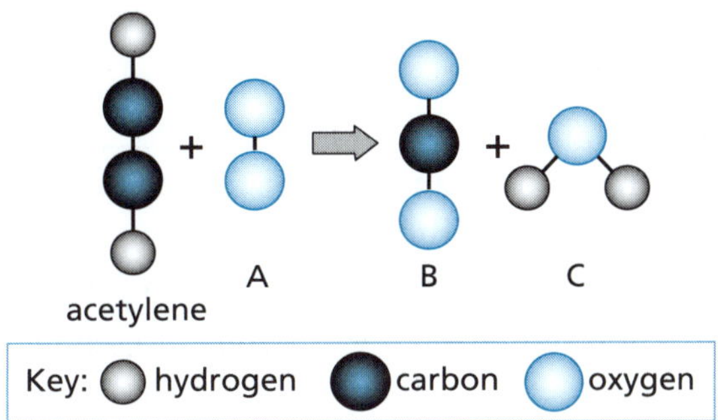

a Which of the following is the correct formula of acetylene molecules: CO, CH, C_2H_2, C_2O_2? (1 mark)

b Name the molecules A, B and C. *Hint 3* (3 marks)

c i This model equation does not demonstrate the law of mass and atom conservation. Explain why. (1 mark)

ii Redraw the diagram to show the correct number of molecules of reactants and products so that the law of mass conservation and atom conservation is demonstrated. (4 marks)

5 This diagram shows an experiment in which a colourless solution is poured into a pale blue solution. Identify the observations that suggest a chemical change has occurred. (2 marks)

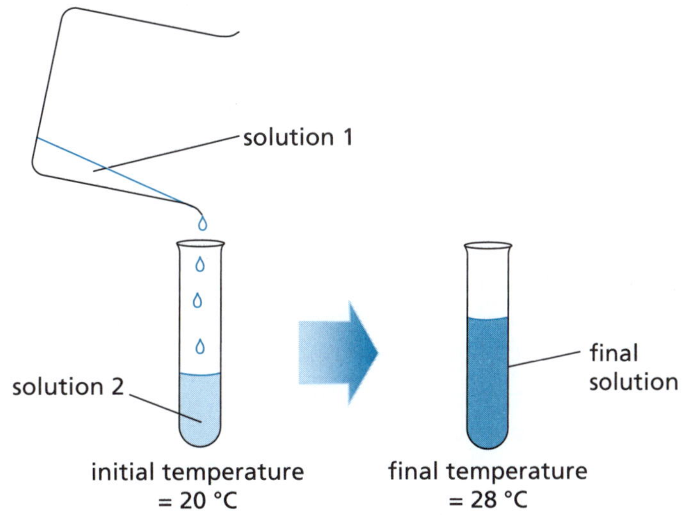

Hint 1: How many green and grey balls will it take to make the compound in model 2?

Hint 2: Oxygen is diatomic. Magnesium oxide is a simple 1:1 binary compound.

Hint 3: Molecule C is dihydrogen oxide. What is its more common name?

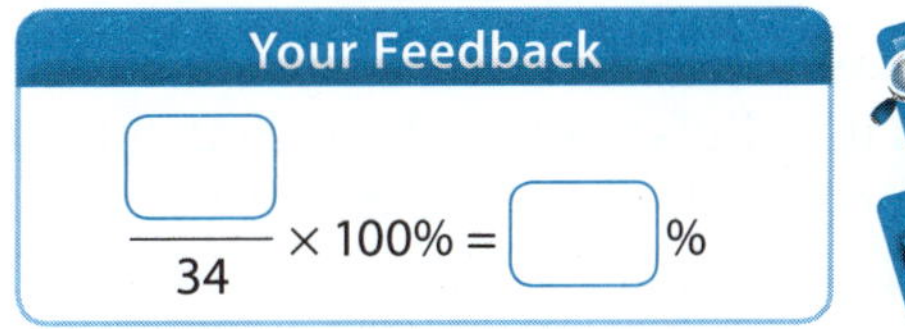

PAGE 232

PAGE 252

CHEMICAL EQUATIONS

Chemical reactions

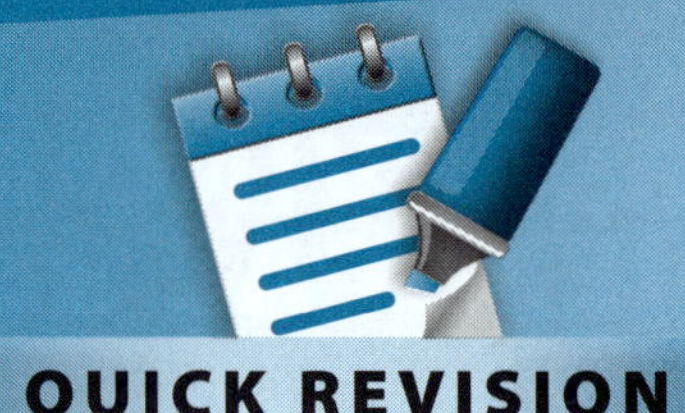

QUICK REVISION

1 In a chemical reaction the __________ combine to form new products. This can be expressed in a __________ equation or a symbolic equation. The name or formula of each reactant is written __________ followed by an arrow. The arrow represents the words '__________ to form'. Following the arrow are the names or formulae of the __________.

For example, powdered aluminium reacts rapidly with __________ to produce a white powder called aluminium oxide. Considerable amounts of __________ in the form of heat and light are also produced. The word equation for this reaction is:

$$\text{aluminium} + \text{oxygen} \rightarrow \text{aluminium oxide}$$

In order to write a __________ equation, the chemical formulae of reactants and products must be known. The symbol for __________ is Al and the formula for oxygen gas is __________ rather than O because oxygen in the air is a diatomic molecule. The formula of aluminium __________ is Al_2O_3. The next step is to write these formulae in an equation and then check for conservation of __________.

$$Al + O_2 \rightarrow Al_2O_3$$

It is obvious that the atoms of aluminium and oxygen are __________ equal on both sides of the equation. The equation has to be balanced by inserting __________ in front of the formulae. The final balanced equation is:

$$4Al + 3O_2 \rightarrow 2Al_2O_3$$

Each side of the equation has __________ aluminium atoms and __________ oxygen atoms.

2 Metals have quite variable __________ reactivities. Gold is a very __________ metal. Calcium and magnesium are reactive metals. Potassium is __________ reactive. When metals are added to solutions of dilute acids the __________ of the reaction varies considerably. Gold will not react at all, while magnesium will __________ vigorously. The three common dilute acids used in a school laboratory are hydrochloric acid (HCl), sulfuric acid (H_2SO_4) and __________ acid (HNO_3). When dilute HCl and dilute H_2SO_4 react with magnesium, __________ gas is produced. An effervescence is observed as the reaction occurs at the __________ of the magnesium. The magnesium __________ in the acid and is converted into an __________ compound which is called a salt. The salt is usually __________ in the water of the acid solution. In the case of hydrochloric acid, the salt is called magnesium __________ ($MgCl_2$). In the case of __________ acid the salt is called magnesium sulfate ($MgSO_4$). Both salts can be recovered in crystalline form by slow __________ of the water.

The balanced equations for these reactions are:

$$Mg + 2HCl \rightarrow MgCl_2 + H_2$$

$$Mg + H_2SO_4 \rightarrow MgSO_4 + H_2$$

Dilute nitric acid behaves differently to the other two acids. Instead of hydrogen gas, most metals (except magnesium) react to form __________ oxide (NO) gas. Water and a __________ are also formed. These salts are called nitrates. For example, copper metal slowly dissolved in dilute nitric acid forms __________ nitrate ($Cu(NO_3)_2$).

3 The leaves of green __________ contain special molecules called chlorophyll. Chlorophyll is found in cellular organelles called chloroplasts. The chlorophyll molecule __________ certain frequencies of sunlight and uses the energy from this light to power a biochemical process called __________. During this process water molecules and __________ dioxide molecules react to form __________ gas and glucose molecules. Glucose is an example of a group of molecules

CHEMICAL EQUATIONS

Chemical reactions

QUICK REVISION

called sugars. The chemical ____________ of glucose is $C_6H_{12}O_6$. Chemical energy is stored in this molecule in its ____________ bonds. Green plants are classified as producers in a food web because of their ability to use ____________ energy to produce nutrient molecules. Some of the oxygen formed is used by the plant. The remainder is released into the ____________.

When consumers eat producers, the energy-rich nutrient molecules are transferred to the ____________ of the consumer. In these cells a reaction called ____________ respiration occurs. In this process, glucose reacts with oxygen to produce carbon dioxide and ____________. Considerable amounts of energy are also released. This ____________ is used by an organism to power its bodily processes and to stay alive.

Shown below is a molecular model of glucose. Note that five of the six carbon atoms form a hexagonal ring.

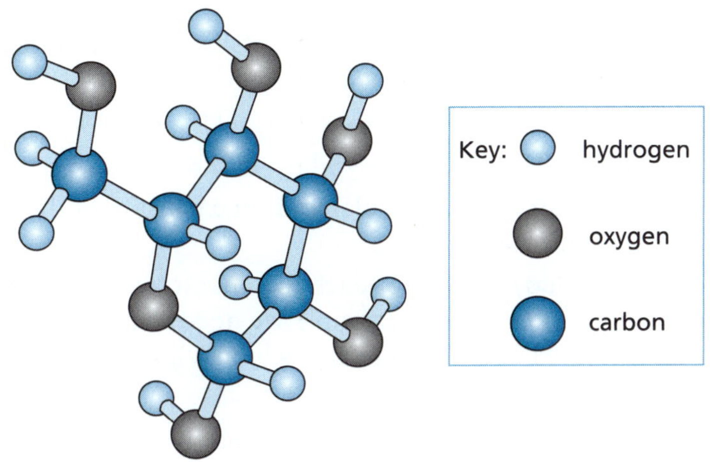

Answers 1 reactants; word; first; react; products; oxygen; energy; symbolic; aluminium; O_2; oxide; atoms; not; coefficients (numbers); four; six 2 chemical; unreactive; very; rate (speed); react; nitric; hydrogen; surface; dissolves; ionic; dissolved; chloride; sulfuric; evaporation; nitric; salt; copper 3 plants; absorbs; photosynthesis; carbon; oxygen; formula; chemical; solar; air; cells; cellular; water; energy

CHEMICAL EQUATIONS

Chemical reactions

1 The conversion of reactants into products can be expressed as a **word equation** and as a **symbolic equation**.

A word equation has the format:

reactants → products

where the arrow stands for 'react to form'.

Example: magnesium + hydrochloric acid → magnesium chloride + hydrogen gas

A **balanced** symbolic equation shows the symbols and formulae of the reactants and products. It also shows conservation of atoms and therefore **conservation of mass**.

The first step in writing a balanced symbolic equation is to write the correct **chemical formula** of each substance. Using the example above: magnesium = Mg; hydrochloric acid = HCl; magnesium chloride = $MgCl_2$; and hydrogen = H_2.

The next step is to write the formulae of the reactants on the left of the arrow and the products on the right. Check the number of atoms of each element on each side of the arrow.

$$Mg + HCl \rightarrow MgCl_2 + H_2$$

$$Mg = 1; H = 1; Cl = 1 \rightarrow Mg = 1; H = 2; Cl = 2$$

By checking the number of atoms of each element on each side of the equation you can see that that H and Cl atoms are not balanced.

Write numbers (where needed) in front of each formula to ensure that the **atoms balance** on both sides of the equation. In the above example we need to place a 2 in front of HCl. This will then balance the number of H and Cl atoms so they match the number in the products.

$$Mg + 2HCl \rightarrow MgCl_2 + H_2$$

The figure to the right shows how the hydrogen gas that is produced could be collected in a larger test tube inverted over the reaction test tube. The lighter hydrogen forces air out of the large tube. In order to test whether hydrogen has been collected (hydrogen is colourless and doesn't smell), bring a lighted taper close to the mouth of the inverted test tube. A loud pop is heard. Afterwards, some moisture might be noticed around the inside rim of the test tube. This is because hydrogen has reacted with oxygen to form water. The equation for this reaction is:

$$2H_2 + O_2 \rightarrow 2H_2O$$

In fact the name hydrogen comes from the Greek *hydros* (water) and *gène* (producing) because hydrogen forms water when it is ignited in oxygen.

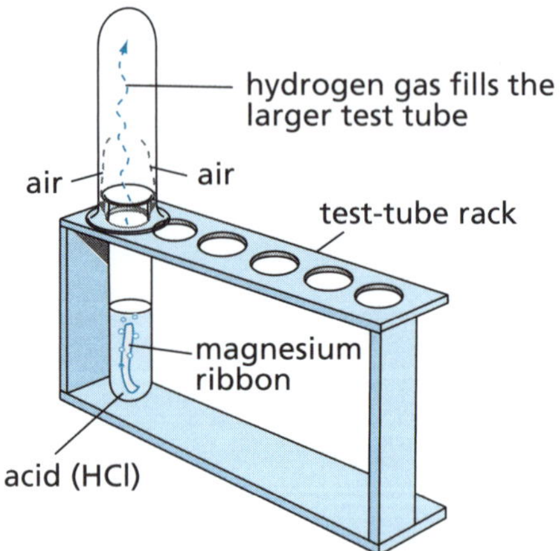

2 In order to practise writing balanced equations, let us examine some more reactions of acids with metals.

Some metals (e.g. potassium and sodium) are very **reactive** while others (e.g. lead and copper) are not very reactive. For dilute solutions of sulfuric acid (H_2SO_4) and hydrochloric acid (HCl) with reactive metals, **hydrogen gas** is produced as the metal reacts and dissolves in the acid. A solution of an ionic compound is formed. This **ionic compound** (or salt) can be recovered as a solid by evaporation of the water as shown in the diagram on the next page.

Dilute nitric acid (HNO_3) behaves differently with different metals. In most cases nitric oxide (NO) gas or even nitrous oxide (N_2O) are produced. With magnesium, however, hydrogen gas is produced. Water and a solution of a salt are also formed in these reactions. The names of the salts in all these examples are dependent on the acid used.

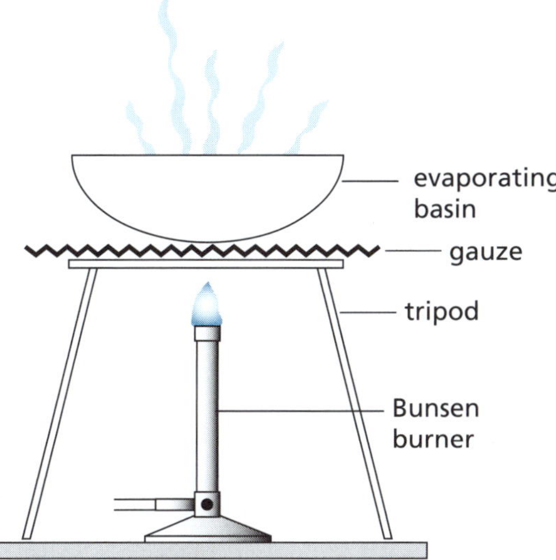

- Hydrochloric acid forms salts called **chlorides**. For example, barium chloride is formed when barium reacts with dilute hydrochloric acid.

  barium + hydrochloric acid → barium chloride + hydrogen

  $$Ba + 2HCl \rightarrow BaCl_2 + H_2$$

- Sulfuric acid forms salts called **sulfates**. For example, potassium sulphate is formed when potassium reacts with dilute sulphuric acid.

  potassium + sulfuric acid → potassium sulfate + hydrogen

  $$2K + H_2SO_4 \rightarrow K_2SO_4 + H_2$$

- Nitric acid forms salts called **nitrates**.

3 Two important chemical reactions in natural systems are **photosynthesis** and **cellular respiration**. The leaves of green plants contain a green molecule called **chlorophyll**. This molecule absorbs solar energy which powers the photosynthetic process. The word equation for photosynthesis is:

carbon dioxide gas + water → glucose + oxygen gas

Glucose ($C_6H_{12}O_6$) is a simple sugar molecule. The symbolic equation for photosynthesis is:

$$6CO_2 + 6H_2O \rightarrow C_6H_{12}O_6 + 6O_2$$

Glucose molecules can be used as a source of chemical energy for the plant or any consumer that eats the plant.

Cellular respiration involves the reaction of nutrient molecules, such as glucose, with oxygen. Energy is released in the process. Cells use this energy to power the biochemical processes that occur in the tissues and organs of the body. The word equation for cellular respiration is:

glucose + oxygen → carbon dioxide + water

The balanced symbolic equation for cellular respiration is:

$$C_6H_{12}O_6 + 6O_2 \rightarrow 6CO_2 + 6H_2O$$

Checklist

Can you:

1. *Write word equations and balanced symbolic equations?* ☐
2. *Write balanced symbolic equations for reactions of metals with dilute acids?* ☐
3. *Write word and balanced symbolic equations for photosynthesis and cellular respiration?* ☐

CHEMICAL EQUATIONS

Chemical reactions

REVISION TEST

1 The following diagram shows a model of a chemical reaction.

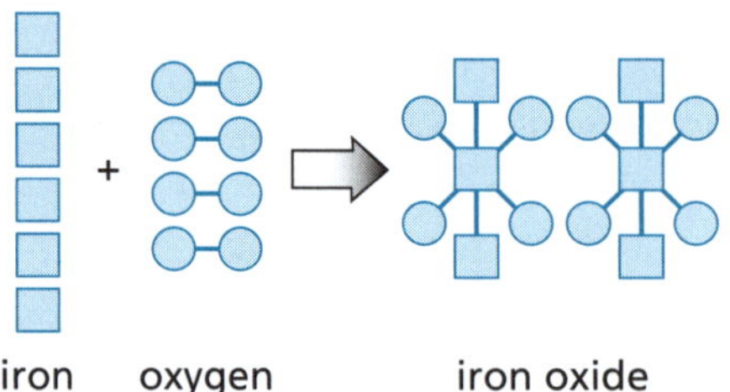

a Write a word equation for this reaction. (1 mark)

b Write a balanced symbolic equation for this reaction. (3 marks)

2 Read the following passage about sugar and then answer the questions.

> Photosynthesis is the process plants use to change light energy into chemical energy. It is the most important chemical process in the world. Plants store this energy in the chemical bonds that hold a glucose molecule together. Scientists have not been able to reproduce this process in the laboratory. In 1804, the Swiss physicist and botanist Nicholas Theodore de Saussure reduced this process to a chemical equation that looks deceptively simple:
>
> $$6CO_2 + 6H_2O + \text{light energy} \rightarrow C_6H_{12}O_6 + 6O_2$$
>
> This process has two major steps. The first involves light being absorbed by the green pigment chlorophyll and providing enough energy to split water into its components, one of which is oxygen. The second part of the reaction involves the hydrogen atoms being picked up by special escort molecules and, together with the remaining energy, passing down a chemical pathway until the hydrogen reacts with carbon dioxide to form sugars such as glucose. The special escort molecules then cycle back to pick up more hydrogen atoms.
>
> The world's green plants produce 160 000 million tonnes of sugar every year. This chemical reaction produces far more product than any other. Sugars are the main storage products of chemical energy, but are also the main energy source for all plant and animal cells. Even though it is a stable chemical product, the moment energy is required by plant or animal cells, it is immediately available from sugar.

Which of the following statements are true and which are false?

a Photosynthesis uses light energy. (1 mark)

b Nicholas Theodore de Saussure was a Dutch botanist. (1 mark)

c Chlorophyll in green plants absorbs light. (1 mark)

d Sugar is a storage product of plants. (1 mark)

e Green plants produce 160 000 million tonnes of sugar every day. (1 mark)

f The products of photosynthesis are carbon dioxide and water. (1 mark)

g Green plants are a source of oxygen which is formed as a by-product of photosynthesis. (1 mark)

h Chlorophyll is used up in the photosynthetic process. (1 mark)

i Glucose sugar has 6 carbon atoms, 12 hydrogen atoms and 12 oxygen atoms in its structure. (1 mark)

j The water used by the plant in the first step of photosynthesis is the same water produced in the products. (1 mark)

3 The following diagram shows a model of a chemical reaction.

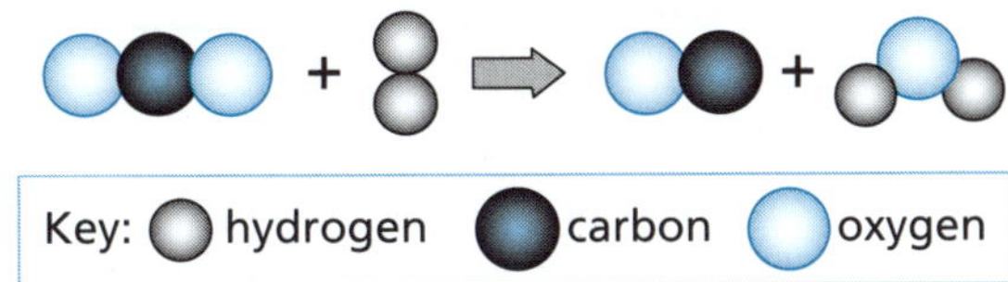

Key: hydrogen carbon oxygen

a Name the reactants and products. *Hint 1* (4 marks)
b Use symbols to write the chemical formulae for each reactant and product. (4 marks)
c Write a balanced symbolic equation for this reaction. (1 mark)

4 The following diagram shows a model of an acetic acid molecule. Acetic acid is found in vinegar.

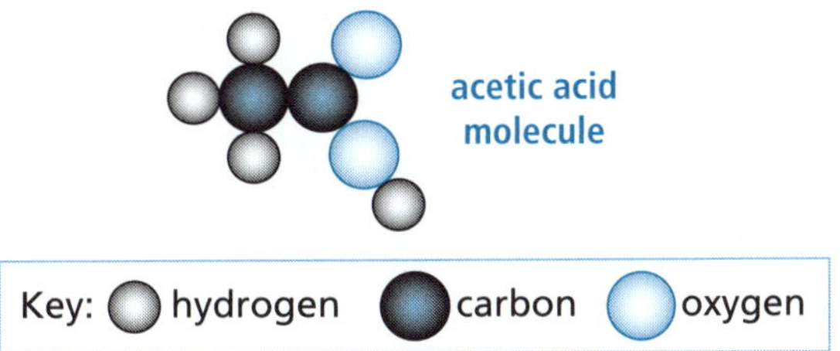

Key: hydrogen carbon oxygen

a Write the chemical formula of acetic acid. *Hint 2* (1 mark)
b An alternative way of writing the chemical formula is CH_3COOH. Suggest a reason why this is a more common way of writing the formula of this compound. (1 mark)
c Sodium metal reacts with acetic acid to produce a colourless gas and a salt called sodium acetate ($NaCH_3COO$).
i The gas that is produced explodes when a flame is present. Identify this gas and write its chemical formula. *Hint 3* (2 marks)
ii Write a balanced equation for this reaction. (4 marks)

5 Name the salts formed in each of the following reactions.
a Calcium metal reacts with dilute nitric acid. (1 mark)
b Cobalt metal reacts with dilute sulfuric acid. (1 mark)
c Aluminium metal reacts with dilute hydrochloric acid. (1 mark)

6 Balance the following equations by inserting coefficients where required.
a $Na_2SO_4 + C \rightarrow Na_2S + CO$ (2 marks)
b $NaCl + H_2SO_4 \rightarrow Na_2SO_4 + HCl$ (2 marks)

Hint 1: Use the prefixes 'di' and 'mon' as needed in the names of two of these substances.
Hint 2: Add up the number of atoms of each element.
Hint 3: This test is sometimes called the pop test. The gas is made up of more than one atom of that element.

Your Feedback

$\frac{\square}{38} \times 100\% = \square\%$

PAGES 232–233
PAGE 252

NEUTRALISATION

Chemical reactions

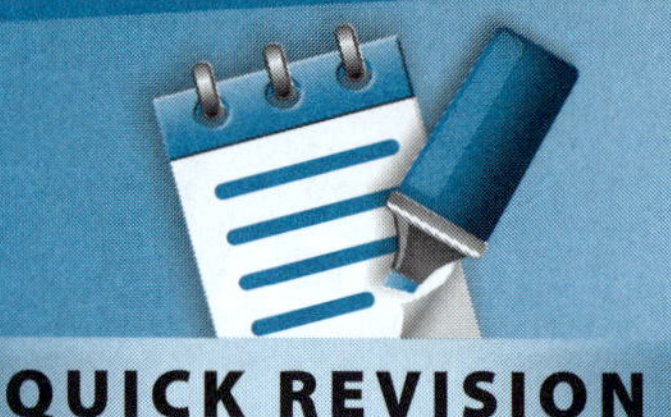

QUICK REVISION

1 Acids have been known since ancient times. Certain common substances, such as ________ fruits (lemons, limes and oranges), sour milk and ________ (obtained when wine is allowed to ferment) tasted sour. Today we recognise that ________ are substances that, in a solution, can break free a hydrogen ion (H^+). This ________ ion attaches itself to a water molecule to form a hydronium ion (H_3O^+). The hydronium ion is just a ________ molecule with an extra hydrogen ion attached to it. It is often used to determine how acidic a chemical compound is. When an ________ is dissolved in water, the greater the number of hydronium ions produced, the higher is the acidity. For example, hydrochloric acid is a strong acid and produces many ________ ions.

$$HCl + H_2O \rightarrow Cl^- + H_3O^+$$

Bases are substances that can ________ with acids to destroy these properties. Soluble bases are called ________. They dissolve in water to release ________ ions (OH^-).

A number of common acidic and basic substances can be found around the home.

Common acidic substances	Common basic substances
car battery acid	caustic soda (drain ________)
vinegar (________ acid)	ammonia solution (window cleaner)
lemon juice	laundry detergents
soft drinks	baking soda
soda water	limestone
bleach	lime

2 Neutralisation is a reaction between an ________ and a base. When bases neutralise acids, a solution of a salt is formed. A ________ is an ionic compound that results from the neutralisation reaction between an acid and a base. The general word ________ for a neutralisation reaction is:

$$\text{acid} + \text{base} \rightarrow \text{salt} + \text{water}$$

Many metal oxides also react with acids. The general word equation for this is:

$$\text{acid} + \text{\_\_\_\_\_\_\_\_ oxide} \rightarrow \text{salt} + \text{water}$$

Acids can also act on carbonates. Marble is a metamorphic rock consisting of calcium ________, and is commonly used for sculpture and as a building material. The general word equation for the reaction between an acid and a carbonate is:

$$\text{acid} + \text{carbonate} \rightarrow \text{salt} + \text{water} + \text{carbon dioxide}$$

For this reason it is important that marble bench tops be ________ and polished, otherwise acidic juices in food substances can break down the marble. Strong detergents, corrosive liquid or scouring powders can also ________ the surface and some cleaning chemicals will destroy the sealer.

3 A chemical indicator is a substance that gives a visible ________ change in the presence or absence of a chemical species, such as an acid or an alkali in a solution. A common example is litmus, which produces a blue colour in an ________ solution. If acid is slowly added, the solution remains blue until all the alkali has been ________, and then the colour suddenly ________ to red. This is useful in a laboratory as it gives an indication of neutralisation.

NEUTRALISATION

Chemical reactions

QUICK REVISION

Chemists use a number, on a pH scale, to represent how ________ or basic a solution is.

- Neutral ________ have a pH = 7.
- Acidic solutions have pH < 7. The ________ the number, the more acidic the solution.
- ________ solutions have pH > 7. The higher the number, the more basic the solution.

pH indicators have many applications in biology and analytical ________.

The following photo shows indicator paper being used to determine the pH of a solution. The results indicate the solution is basic with a pH of about 10.

Answers 1 citrus; vinegar; acids; hydrogen; water; acid; hydronium; react; alkalis; hydroxide; cleaner; acetic 2 acid; salt; equation; metal; carbonate; sealed; damage 3 colour; alkaline (basic); neutralised; changes (alters); acidic; solutions; lower; Basic; chemistry

1 The ancient Greeks noticed that certain substances were **sour-tasting**, such as vinegar. The word vinegar comes from the Old French words *vin* (wine) and *aigre* (sour). **Acid** substances did not only taste sour (don't taste laboratory acids; some are poisonous and corrosive), but also could change the colour of certain chemicals and **corrode** metals. By contrast, **bases** could counteract the effects of acids. It was not until the French chemist Antoine Lavoisier (1743–1796) classified acids and bases that a theoretical organisation of these compounds started. His idea was that all acids contained a common fundamental substance that was responsible for their acidity.

Common acid	Features
hydrochloric acid (HCl)	This strong acid has many industrial uses, including making some plastics and removing rust or iron oxide scale from iron or steel before further processing. Dilute hydrochloric acid in the human stomach helps to digest food.
sulfuric acid (H_2SO_4)	This strong acid is commonly used in industry. It is used in the refining of metals and to make a wide range of substances, including fertilisers. Used with steel it can slow down corrosion.
nitric acid (HNO_3)	Nitric acid is a strong acid, used mainly to produce fertilisers. It is also used to make nylon, manufacture explosives and make compounds used to propel liquid-fuelled rockets.
acetic acid (CH_3COOH)	Acetic acid is a weak acid. It is used to make cellulose acetate mainly for photographic film and polyvinyl acetate for wood glue, and synthetic fibres and fabrics. Common vinegar is 5% acetic acid and 95% water.
citric acid ($C_6H_8O_7$)	This weak acid is found in citrus fruits. It is a natural preservative/conservative and is also used to add an acidic, or sour, taste to foods and soft drinks.
carbonic acid (H_2CO_3)	This weak acid is a solution of carbon dioxide in water (carbonated water) and is found in soft drinks. Carbonic acid is an intermediate molecule in the transport of carbon dioxide out of the body to the lungs. The world's oceans have absorbed around half of the carbon dioxide released through the burning of fossil fuels, forming a dilute solution of carbonic acid.
phosphoric acid (H_3PO_4)	This weak acid is found in soft drinks. In more concentrated solutions it is used for rust removal and to prepare steel surfaces for painting.

Common base	Features
sodium bicarbonate ($NaHCO_3$)	This is a weak base, commonly known as baking soda or bicarb of soda. It is used in cooking. Taken orally as an antacid it treats acid indigestion and heartburn. It is also used in some dry chemical fire extinguishers.
sodium hydroxide (NaOH)	This strong base, also called caustic soda, is used in the manufacture of pulp and paper, textiles, drinking water, soaps and detergents and as a drain cleaner. Being caustic, it can decompose proteins and lipids in skin or other living tissues, causing chemical burns to the skin.
ammonium hydroxide (NH_4OH)	This base is one of a number of anti-microbial processing aids used with meat and poultry to guarantee the safety of these foods before they are distributed to consumers. A dilute solution of ammonium hydroxide is used in household ammonia, a cleaning agent found in many window cleaners.
magnesium hydroxide ($Mg(OH)_2$)	Magnesium hydroxide is also known as milk of magnesia. It is a laxative and antacid. It is usually used with another base, aluminium hydroxide ($Al(OH)_3$), to neutralise stomach acids. Magnesium hydroxide is also used as an anti-perspirant underarm deodorant.
calcium hydroxide ($Ca(OH)_2$)	Calcium hydroxide is used as a flocculent in water and sewage treatment, forming fluffy solids which allow smaller particles to be removed from water, and so make it clear. In the sugar industry it is used to separate sugar from sugarcane or sugar beets.

Soluble metal hydroxides are called alkalis, and react with acids in solution.

NEUTRALISATION

Chemical reactions

2 **Neutralisation** reactions occur when acids and bases react together. The general word equation for this is:

acid + base → salt + water

An example is:

hydrochloric acid + sodium hydroxide → sodium chloride + water

The following diagram shows a simple atomic model of this reaction.

H Cl + Na O H ⇨ Na Cl + H O H

HCl NaOH NaCl H_2O

Neutralisation can also occur when acids react with oxides. The general word equation for this is:

acid + metal oxide → salt + water

An example is:

sulfuric acid + calcium oxide → calcium sulfate + water

$$H_2SO_4 + CaO \rightarrow CaSO_4 + H_2O$$

Most metal oxides are **insoluble** solids.

Acids also react with and neutralise carbonates. The general word equation for this is:

acid + carbonate → salt + water + carbon dioxide

An example is:

nitric acid + copper carbonate → copper nitrate + water + carbon dioxide

$$2HNO_3 + CuCO_3 \rightarrow Cu(NO_3)_2 + H_2O + CO_2$$

Limewater can be used to test for carbon dioxide. The colourless limewater (calcium hydroxide solution) turns a **milky white** when carbon dioxide is bubbled through it.

3 **Indicators** are chemical substances (dyes) that **change colour** in the presence of an acid or base. They can determine whether a solution is acidic, neutral or basic, and can be used to observe neutralisation.

The following table contains some useful indicators.

Indicator	Colour in acidic solutions	Colour in neutral solutions (water)	Colour in basic solutions
litmus	red	purple	blue
methyl orange	red	orange	yellow
bromothymol blue	yellow	green	blue
phenolphthalein	colourless	colourless	crimson

How acidic or basic a solution is can be represented as a number on the **pH scale**.

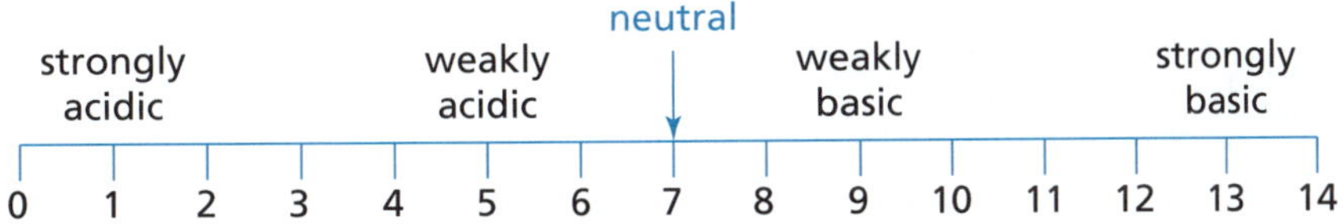

Universal indicator is made from a mixture of several indicators. It exhibits a variety of colours over a wide pH range, so it can be used to establish a rough pH of a solution.

(cont.)

pH range	0–4	4–5	5–6	7–8	8–9	9–10	10–14
Universal indicator colour	red	orange	yellow	green	blue-green	blue-violet	violet

Hydronium ions form when an acid is present in water. The more hydronium ions, the more acidic is the solution. The concentration of hydronium ions (H_3O^+) in a solution is a measure of the pH of the solution. As the acid dissolves in a solution it releases hydrogen ions (H^+). Each hydrogen ion joins with a water molecule to form a hydronium ion.

$$H^+(aq) + H_2O(l) \rightarrow H_3O^+(aq)$$

where *aq* means 'aqueous' (or in a water solution) and *l* means 'liquid'.

The pH of a solution can also be measured using a pH meter. The following photo shows a pH meter and the pH electrode which is placed in the solution.

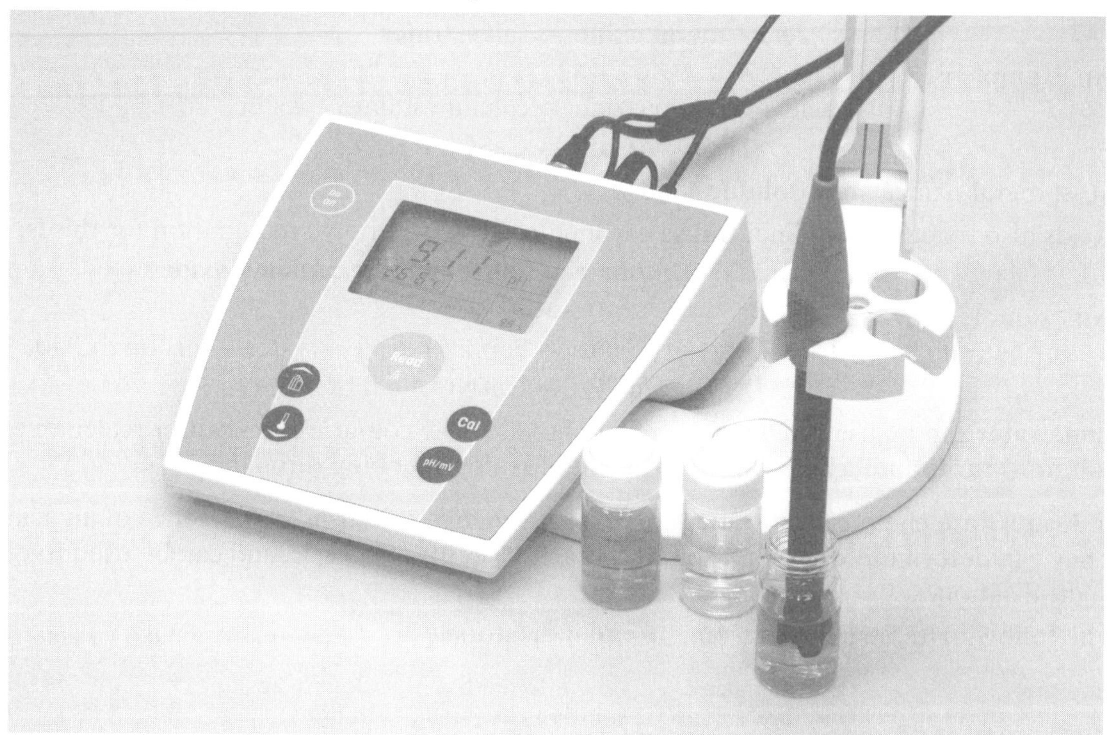

Checklist
Can you:

1 *Name and give some uses for common acids and bases?* ☐
2 *Describe the reaction between acids and hydroxides, acids and oxides, and acids and carbonates?* ☐
3 *Describe how indicators allow us to observe neutralisation?* ☐

NEUTRALISATION

Chemical reactions

1 Name the type of reaction occurring in the following diagram. (1 mark)

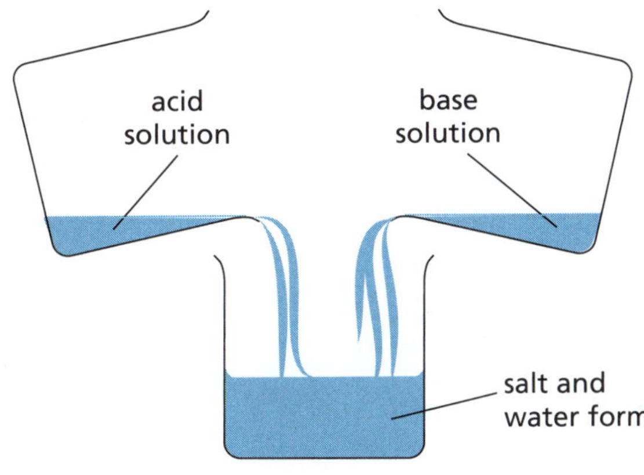

2 Read the following text and answer the question.

> Jim: There are many different kinds of chemicals that can be called a salt.
>
> Geoff: No, there is only one kind. We use it in cooking and put it on our fish and chips.

Who is correct? Explain. (3 marks)

3 **a** Name the salt produced when each of the following acids and bases are mixed. *Hint 1*

- **i** hydrochloric acid and barium hydroxide (1 mark)
- **ii** nitric acid and aluminium hydroxide (1 mark)
- **iii** sulfuric acid and calcium hydroxide (1 mark)
- **iv** sulfuric acid and sodium hydroxide (1 mark)

b The formulae for these substances is given below in unbalanced equations. Balance each of these reactions.

- **i** $HCl + Ba(OH)_2 \rightarrow BaCl_2 + H_2O$ (1 mark)
- **ii** $HNO_3 + Al(OH)_3 \rightarrow Al(NO_3)_3 + H_2O$ (1 mark)
- **iii** $H_2SO_4 + Ca(OH)_2 \rightarrow CaSO_4 + H_2O$ (1 mark)
- **iv** $H_2SO_4 + NaOH \rightarrow Na_2SO_4 + H_2O$ (1 mark)

4 Acids can react with a number of metals as follows:

acid + metal → salt + hydrogen

Write word equations for these symbolic equations.

- **a** $2HCl + Zn \rightarrow ZnCl_2 + H_2$ (2 marks)
- **b** $H_2SO_4 + Fe \rightarrow FeSO_4 + H_2$ (2 marks)

5 Clear, colourless limewater easily absorbs carbon dioxide from air, giving a milky solution. The reaction is:

$$Ca(OH)_2 + CO_2 \rightarrow CaCO_3 + H_2O$$

- **a** What is the chemical name for limewater? (1 mark)
- **b** Is limewater an acidic, basic or neutral solution? (1 mark)
- **c** What causes the milky colour? (1 mark)
- **d** What type of acid–base reaction produces carbon dioxide? (1 mark)

(cont.)

6 An insoluble metal oxide (copper oxide) is reacted with a dilute acid (sulfuric acid) to form a soluble salt. The diagram shows the experiment.

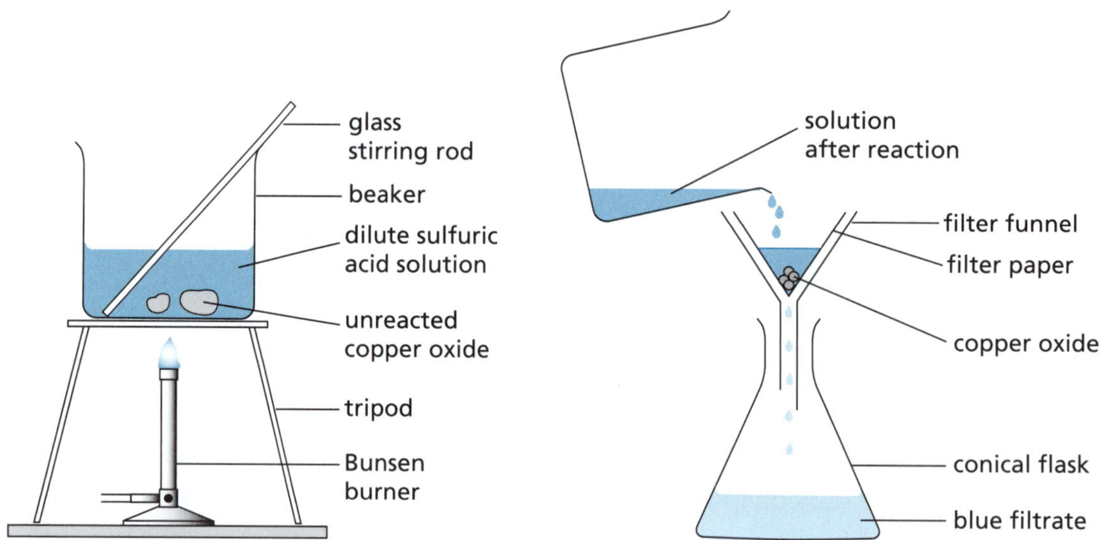

- a Describe the two parts of this experiment, as shown in the diagram. (2 marks)
- b Suggest a reason for heating the solution. (1 mark)
- c How can you tell when the reaction is complete? *Hint 2* (1 mark)
- d Why is there copper oxide in the filter paper? (1 mark)
- e Name the compound in the filtrate causing the blue solution. (1 mark)
- f Write a word and symbolic equation for the reaction. The chemicals involved are: sulfuric acid (H_2SO_4), copper oxide (CuO), copper sulfate ($CuSO_4$) and water (H_2O). (2 marks)

7 Corals are tiny marine animals. Individual coral polyps secrete calcium carbonate to form a hard skeleton. Colonies of polyps form coral reefs. Corals live near the surface of the ocean and are sensitive to changes in temperature, sea level and the amount of carbon dioxide that is dissolved in the ocean.

- a What acid is produced when carbon dioxide dissolves in water? (1 mark)
- b Why are corals more sensitive than many other marine organisms to the amount of dissolved carbon dioxide? *Hint 3* (1 mark)
- c How is the amount of carbon dioxide in the atmosphere related to the amount of carbon dioxide dissolved in the ocean? (1 mark)

8 True or false?

- a Lemon juice is classified as being a base. (1 mark)
- b Tomato juice has a pH of about 4. By adding water the juice becomes less acidic. (1 mark)
- c Acids and bases don't react with each other. (1 mark)
- d Acids have a sour taste, are corrosive and change litmus from blue to red. (1 mark)
- e Vinegar, fruit juice and soft drink are examples of weak acids. (1 mark)
- f A neutral solution has a pH of 7. (1 mark)
- g A base that dissolves in water is known as an alkali. (1 mark)
- h If a substance becomes more acidic its pH increases. (1 mark)

9 The following diagram shows the location of some substances on a pH scale.

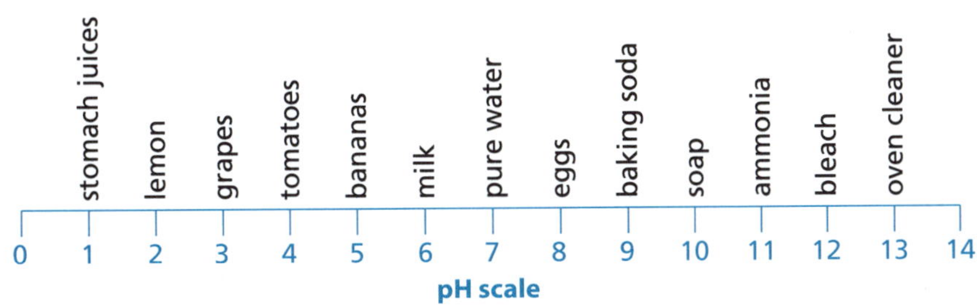

- **a** Name a substance that is more alkaline than pure water but less alkaline than soap. (1 mark)
- **b** If bleach is diluted with plenty of water would its pH increase, decrease or stay the same? (1 mark)
- **c** Which one of these substances is not acidic: bananas, grapes, milk, ammonia? (1 mark)

10 The following chart shows the colour of some indicators in different pH solutions.

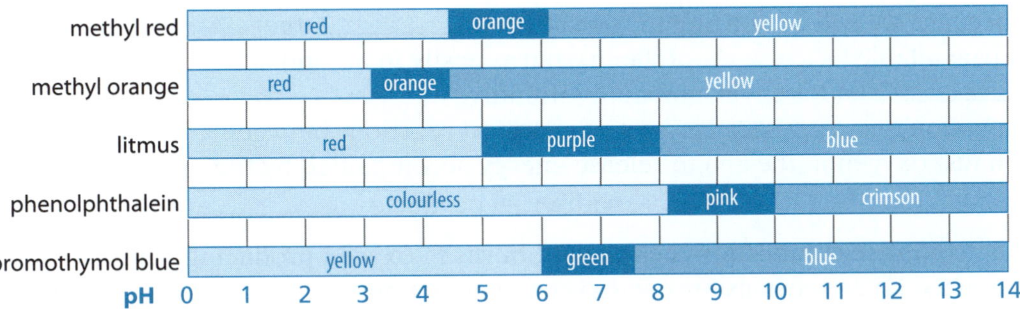

- **a** State a difference between methyl red and methyl orange shown on this chart. (1 mark)
- **b** A solution has a pH of 9.5. What colour would the solution turn if a few drops of bromothymol blue are added? (1 mark)
- **c** A solution turns crimson when phenolphthalein is added. State a possible pH range for the solution. (1 mark)
- **d** A solution is yellow when some methyl orange is added, and red in litmus. What colour would it be with phenolphthalein? (1 mark)
- **e** A strongly acidic solution is slowly diluted until its pH reaches 4. Which indicator would be best to test for this point? (1 mark)

Hint 1: Look at the name of the acid and the metal in the base.
Hint 2: What are some of the features that describe whether a reaction is occurring?
Hint 3: Think of the reaction of acids on carbonates.

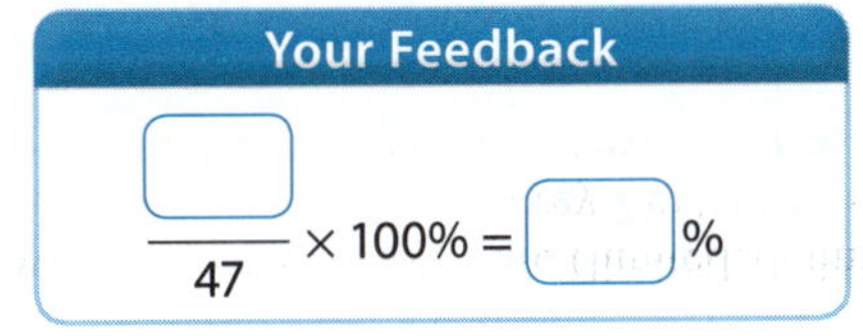

PAGES 233–234
PAGE 252

COMBUSTION

Chemical reactions

QUICK REVISION

1 Wood was one of the first fuels used by ________ nearly 2 million years ago, and it is still the main energy source for most of the world. For a long time in human history ________ obtained from plants or animal fat were the only ones available. ________, from burning wood, has been used for around 8000 years, especially for smelting metals. Coal was first used as a fuel 3000 years ago in China. As forests started to become depleted some three centuries ago ________, derived from coal, increased in importance. Coal became more commonly used as a power source with the development of the steam engine, and it was later used to drive ships and trains. In the 20th and 21st centuries, the chief use of ________ is to generate electricity, providing almost ________ of the world's electrical power supply.
Renewable fuels are fuels produced from ________ resources. Renewable fuels, such as ethanol, are clean burning fuels produced from waste agricultural products. These agricultural crops can be grown ________ and over. The most common renewable fuels are ________ and biodiesel. Petrol and diesel are non-renewable fossil fuels that have a ________ life span as practical transport fuels. Biofuels (liquid fuels derived from other materials such as waste ________ and animal matter) will become essential transport fuels over the next few decades. This is because as ________ becomes more advanced biofuels will be cheaper to produce, and because we will be starting to run short of ________ fuels. A good fuel is one which is readily available, ________ (cheap), burns easily in air and at a moderate rate, produces a lot of ________ and is environmentally friendly.
Combustion or ________ is a set of chemical reactions between a fuel and an ________ (often just oxygen in the air) to release energy accompanied by the production of heat.

fuel + oxidiser → products + ________

2 During complete combustion the reactant burns in oxygen producing a limited number of products. Hydrocarbons burn in oxygen, yielding only ________ dioxide and water. Incomplete combustion will only occur when there is not enough ________ to allow the fuel to react completely to produce carbon dioxide and water. In these instances, carbon ________ and ________ (soot) particles generally form. This can contribute to air pollution. Air pollution occurs in many different ways. Factory and vehicle ________, unburnt fuels, and deliberately lit ________ are all sources of air pollution. Nature also contributes through ________ fires, volcanic eruptions and gases from decaying plants and ________. Many fuels that remain unburnt in incomplete combustion contaminate the smoke with noxious ________ matter and gases. Partially oxidised compounds can produce harmful and toxic ________ monoxide. The quality of combustion can be improved by designing more efficient burners and internal ________ engines. Further improvements are achievable by catalytic after-burning devices (such as ________ converters). Such devices are required by ________ for cars, and may be necessary in large combustion appliances, such as thermal ________ stations, to reach legal emission standards.

3 An exothermic reaction is one where heat is produced or ________ is given out. Burning a piece of paper or a log of wood generates ________ and light. A chemical reaction where energy is ________ is called endothermic. Melting ice is an example, as is the ________ of water. Some compounds can decompose when heat is added.

Answers **1** humans (people); fuels; Charcoal; coke; coal; half; renewable; over; ethanol; finite (limited, definite); plant (vegetation); technology; fossil (non-renewable); inexpensive; energy; burning; oxidiser; energy **2** exhausts; fires; bush; animals; carbon; oxygen; monoxide; carbon; particulate (solid); carbon; combustion; catalytic; law; power **3** energy; heat; absorbed; evaporation

COMBUSTION

Chemical reactions

REVISION SUMMARIES

1 A **fuel** is any material that stores energy which can later be extracted in a controlled manner to perform useful work. Most fuels used by humans undergo **combustion**, where a **flammable** substance releases energy after it **ignites** and reacts with oxygen in the air. Fuels can be gaseous, liquid or solid, and the mixture can be ignited with a heat source, spark or flame.

Combustion is a chemical reaction where a fuel combines with an oxidiser (such as oxygen in the air) to release energy:

fuel + oxidiser → products + energy

Examples of combustion include:

- burning a candle
- burning natural gas in a stove or a barbeque
- burning LPG (liquefied petroleum gas), diesel or petrol in a car
- exploding hydrogen in air.

An **ignition source** is needed for a reaction to occur, as combustion reactions don't occur spontaneously. For example, methylated spirits is used as a **solvent** and as fuel for spirit burners and camping stoves. It essentially consists of ethanol. Although methylated spirits is flammable it does not instantly burst into flames on contact with air. A spark or burning match is needed. Once the combustion begins, the heat energy produced is enough to keep the reaction going.

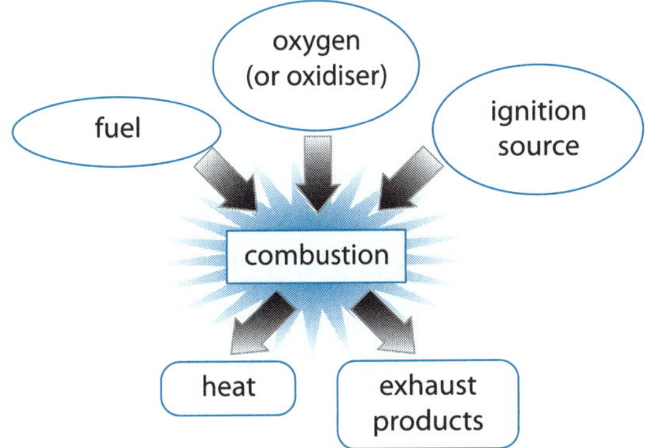

Fuels can be classified according to whether they are solid, liquids or gases at room temperatures. Vapours and gaseous fuels form combustible mixtures with air more readily, so ignite more readily than solid or liquid fuels.

In order to **extinguish** a fire, one or more of the following needs to be removed:

- unburnt fuel
- the ignition (heat) source
- the oxygen supply.

2 **Pollutants** are constantly being released into the environment. A pollutant is a substance or energy that has undesired effects, or that negatively affects the usefulness of a resource. Pollution occurs in many different ways. **Bushfires**, **volcanoes**, wind-blown dust and marsh gas (or methane, CH_4) from swamps and animals wastes are all sources of natural pollution. Humans also contribute to pollution. **Vehicle exhausts**, chimney exhausts from factories and dumping unwanted wastes into water courses are all sources of pollution from humans. Pollution also includes energy sources. There is noise pollution and heat pollution that can adversely affect the surroundings.

A number of common materials cause air pollution.

- **Sulfur dioxide** results from burning coal or oil containing sulfur compounds, or from the smelting of sulfide ores. This poisonous gas can dissolve in rain water producing acid rain. Statues and stone buildings are at risk of damage, as are forests and lakes. Plant and animal life in rivers and streams are affected if the pH drops too much.

  sulfur + oxygen → sulfur dioxide

(cont.)

- **Nitrogen oxides** (such as NO and NO_2) form inside high-temperature furnaces and engines. NO_2 can produce poisonous ozone gas, which is a component of photochemical smog. NO_2 can also dissolve in rain water, forming acid rain.
- Tiny solid particles (**aerosols**) can remain suspended in the air for lengthy periods. Examples include ash and soot, which can affect people with lung conditions.
- Carbon monoxide is a toxic pollutant released during the combustion of hydrocarbons (fuels consisting only of hydrogen and carbon). When hydrocarbons are burnt in an abundant supply of air (oxygen gas, O_2), only carbon dioxide (CO_2) and water (H_2O) form. In general for **complete combustion**:

  hydrocarbon + oxygen → carbon dioxide + water

  Incomplete combustion occurs when there is a limited supply of air or the oxygen is poor. In general for incomplete combustion:

  hydrocarbon + oxygen → carbon dioxide + carbon monoxide + carbon + water

  Complete combustion releases more energy than incomplete combustion, and is preferred. Incomplete combustion also creates carbon monoxide (CO, a poisonous gas) and more soot (carbon). Most carbon monoxide comes from vehicle exhausts, burning coal and oil, and smouldering leaves. Unburned hydrocarbons may be carcinogenic (lead to cancers) and are also involved in photochemical smog.

3 Physical and chemical changes that release energy are called **exothermic**. Energy is often released as heat but other forms, such as light and sound, can occur. Here are some examples.

- **Condensation of water** (physical change)

  gaseous water molecules → liquid water molecules + heat
- **Reacting a base with an acid** (chemical change). A salt and water are formed, and the final mixture is warmer than the reactants.
- **Combustion of wood** (chemical change). In this reaction some of the chemical energy stored in the wood is released as heat. This reaction can be used to warm a room (wood-fired heater) or for cooking (wood stove or barbecue).

Physical and chemical changes that absorb energy are called **endothermic**. The surroundings cool down as some energy is absorbed by the reactants. The following are some examples.

- **Melting ice** (physical change)

  solid water + energy → liquid water
- **Electrolysis** of water (chemical change). In this process electrical energy is passed into the water through electrodes. This energy breaks down the chemical bonds in the water producing hydrogen and oxygen.
- **Thermal decomposition** of compounds (chemical change). For example, on heating mercury oxide decomposes into mercury and oxygen.

Checklist

Can you: ✓

1. *Describe what fuels are and the process of combustion?* ☐
2. *Explain pollution of the atmosphere, including complete and incomplete combustion?* ☐
3. *Explain the difference between exothermic and endothermic reactions, and give examples of each?* ☐

COMBUSTION

Chemical reactions

REVISION TEST

1 True or false?

a Wood is an example of a fossil fuel. (1 mark)
b Hydrocarbons are organic compounds consisting of carbon and hydrogen. (1 mark)
c Combustion occurs when a hydrocarbon compound reacts with oxygen. (1 mark)
d Air pollution can harm the health or comfort of humans and other animals. (1 mark)
e Incomplete combustion occurs when the supply of air is limited. (1 mark)
f When ice melts energy is released. (1 mark)
g In the electrolysis of water, carbon dioxide and hydrogen are produced. (1 mark)
h LPG is an example of a gaseous fuel. (1 mark)
i Nitrogen dioxide and nitrogen monoxide can produce components of photochemical smogs. (1 mark)
j Hydrocarbons are not used as fuels. (1 mark)
k An exothermic chemical reaction absorbs energy and requires energy for the reaction to occur. (1 mark)
l Photosynthesis is an endothermic chemical reaction. (1 mark)

2 The diagram below shows complete and incomplete combustion for methane in natural gas.

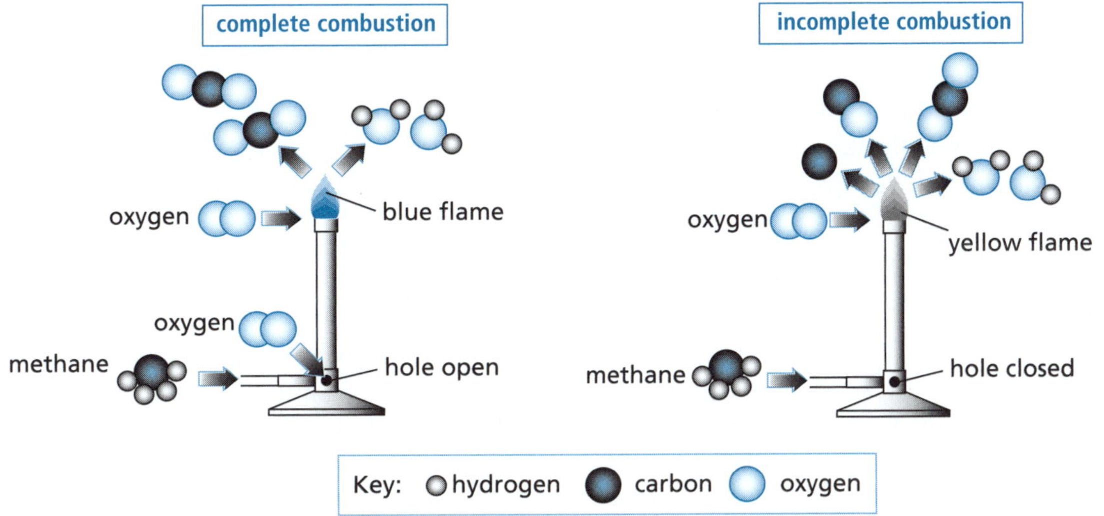

a How do you know from the model of the methane gas going into the Bunsen burner that it is a hydrocarbon? (1 mark)
b Explain why, when the hole is open, complete combustion occurs. *Hint 1* (2 marks)
c Explain the difference between complete and incomplete combustion. (2 marks)
d According to the diagrams, incomplete combustion may produce which carbon-containing gases? (1 mark)
e When using the Bunsen burner to heat something, the hole should be open and not closed. Give two reasons for this. (2 marks)
f What indications are there that the reactions are exothermic? (2 marks)

(cont.)

3 Explain these statements.

a There is less heat released in incomplete combustion compared to complete combustion. (3 marks)

b It is important that for the greatest efficiency, any combustion system, such as a gas heater or furnace, must have excellent ventilation. (2 marks)

c Faulty gas appliances have led to tragic deaths. *Hint 2* (2 marks)

4 The formation of acid rain has a number of terrible effects on the natural and built environment. List three of these. (3 marks)

5 The following diagram shows the contribution to Australia's energy use by primary fuel type.

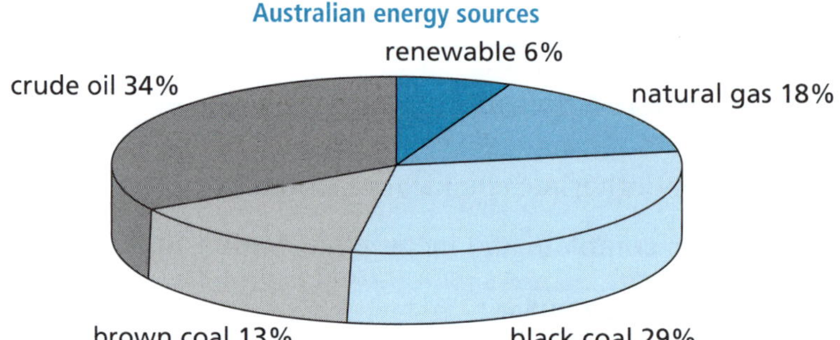

a Which fuel is the largest source of Australia's energy use? (1 mark)

b How do you think this largest source is used? *Hint 3* (1 mark)

c True or false? Energy from coal and crude oil comprised around three-quarters of Australia's total energy use. (1 mark)

d Give one example of a renewable energy source. (1 mark)

6 Read the following information and then answer the questions.

Aeroplanes could soon be flying on straw or fuel extracted from sawdust as airlines search for cleaner alternatives to aviation fuel (kerosene). Agricultural and forest wastes, and general non-food biomass, could lead to a new biofuel to be used in a 50-50 blend with kerosene. A researcher stated: 'We know how to set up the basic production line but now need to move towards turning that into large-scale industrial use. We need to translate what is done in laboratories to the real environment, improving efficiency and minimising costs.' The process involves using enzymes to break down industrial or farm wastes into sugars, mixing the result with microorganisms such as yeast and fermenting it. The fats and lipids obtained are transformed into hydrocarbons using a hydrogen treatment, which then has properties similar to fossil fuels. The project aims to have a 2 million tonne output of aviation biofuels by 2020.

a What are biofuels? (1 mark)

b Why is it important that alternatives to kerosene fuel be developed? *Hint 4* (2 marks)

c State two reasons for sourcing alternative fuels from agricultural and forest wastes. (2 marks)

d Explain why it is necessary to translate what is done in the laboratory to the real environment. (2 marks) e Fill in the missing steps (marked A to E) in the process of transforming plant wastes to hydrocarbons. (2 marks)

Industrial or farm wastes $\xrightarrow{A}$ B $\xrightarrow{C}$ D $\xrightarrow{E}$ hydrocarbon

7 Explain the following statement.

Biofuels should help decrease the airline industry's carbon footprint while using renewable energy resources. *Hint 5* (2 marks)

8 Explain the following statement.
The use of biofuels as an additive to petroleum-based fuels can also result in cleaner burning. (2 marks)

9 In the thermite reaction iron(III) oxide and aluminium react to produce molten iron and aluminium oxide. The products emerge as liquids. Its first commercial application was to weld (join together metal pieces by heating the surfaces) rail tracks.

a Balance the thermite reaction equation.
$Fe_2O_3 + Al \rightarrow Al_2O_3 + Fe$ (2 marks)

b What indications are there that the thermite reaction is strongly exothermic? (2 marks)

c The thermite reaction cannot be smothered, will burn well while wet and won't be easily extinguished with water. Why? (2 marks)

10 Propane is a fuel used in engines, barbeques, portable stoves and residential central heating. The chemical equation for the complete combustion of propane is:

$$C_3H_8 + 5O_2 \rightarrow 3CO_2 + 4H_2O$$

Typically, the equation for the incomplete combustion of propane in oxygen is:

$$2C_3H_8 + 7O_2 \rightarrow 2C + 2CO + 8H_2O + 2CO_2$$

a How do you know propane is a hydrocarbon? (1 mark)

b What additional products are formed during incomplete combustion that are not formed in complete combustion? (2 marks)

c For each molecule of propane that reacts, how many oxygen molecules react during:

i complete combustion (1 mark)

ii incomplete combustion. (1 mark)

d What does the answer to the previous question tell you about the proportion of oxygen needed for complete combustion and incomplete combustion? (1 mark)

Hint 1: What can enter when the hole is open, and how will this affect the reaction?
Hint 2: Any appliance can develop faults over time which can affect its optimal operation.
Hint 3: There are a number of products obtained from this: one of them is very important for motion.
Hint 4: What are some features of fossil fuels?
Hint 5: A carbon footprint is the total of all emissions of carbon dioxide from the activities of an individual or organisation. Usually a carbon footprint is calculated over the period of a year.

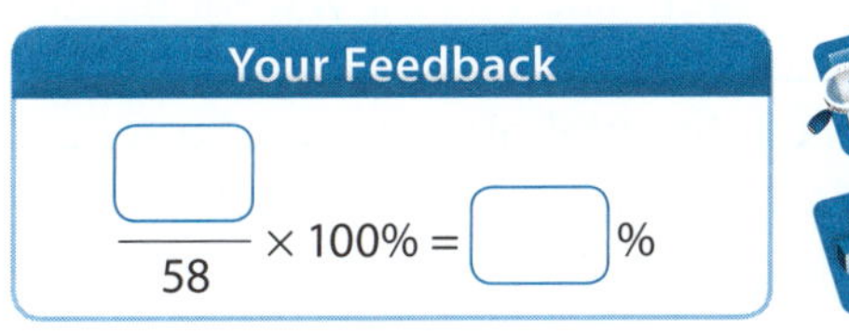

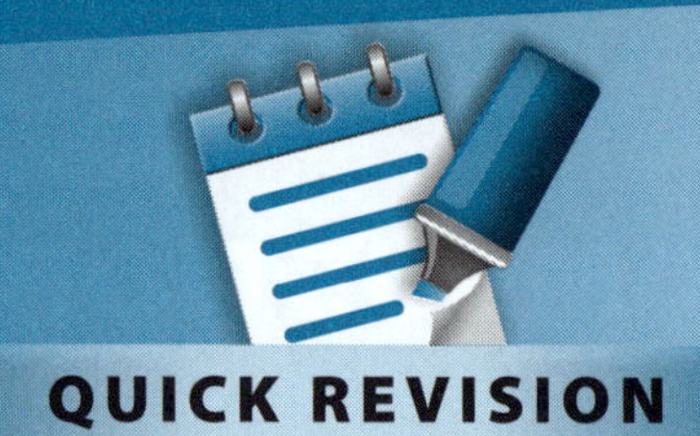

TECTONIC PLATES

QUICK REVISION

1 The internal structure of the Earth was not known until scientists analysed patterns of __________ waves as they were refracted inside the Earth. These studies showed that the Earth has __________ major layers. The surface layer is the __________ which varies in thickness. Under continental masses the crust is much __________ than under the oceans. The average thickness is about __________ km. The crust is composed of many different types of rocks including sedimentary, __________ and metamorphic rocks. As you move downwards from the surface the crust gets considerably __________ reaching about 1000 °C at its base. Below the crust is a very thick layer called the __________. The mantle represents about __________% of the volume of the Earth. The upper part of the mantle is rigid like the crust and together they are called the __________. Below the lithosphere is a region of the mantle called the __________. The rocks here are __________ than those above, and it also contains magma. Convection currents circulate in the asthenosphere. The __________ of the mantle is quite hot, reaching about 2500 °C. Deeper down are the final two layers. These are the __________ core and the inner core. They are both composed of __________ and nickel but in the outer core this material is __________. The inner core is solid.

2 When you look at a map of the Earth it is quite apparent that the west coast of Africa and the east coast of __________ America fit together like a jigsaw puzzle. Scientists wondered whether these two continents were once __________ in the distant past. Other continents also showed similar jigsaw fits. Eventually a theory developed called the theory of __________ drift. This theory proposed that the continents were once joined to form a __________ that was called Pangaea. Over time, this continent started to break up due to internal forces in the Earth. Initially it formed two large continents: __________ in the north and Gondwana in the south. Australia, Antarctica, __________ and South America were part of Gondwana. Eventually Gondwana and Laurasia started to __________ apart to produce the continents we have today. Support for continental drift eventually came from the theory of __________ tectonics. __________ measurements have confirmed that the continents are moving. Radioactive dating of __________ and fossils on continental margins also supports this theory.

3 The plate tectonics theory states that the Earth's crust is divided into large __________ of rock. These plates move over the asthenosphere below due to convection currents as well as __________. Tectonic plates are not all the same size. They also move in different __________ at different speeds. Australia lies on the __________ Plate which is currently moving in a north-easterly direction at 67 mm per year. This plate is colliding with the __________ Plate on the east and north-east and with the Eurasian Plate in the __________ and north-west. The Australian continent is __________ near the edge of the Indo-Australian Plate and therefore Australia does not experience __________ earthquakes.

Plates interact in a variety of ways. In the centre of the Atlantic Ocean is a large ridge on the sea floor. This __________ ridge is volcanically active and magma wells up to the surface creating new __________. The new crust forces the older crust sideways and so causes sea-floor __________. When this spreading happens in cracks in continental masses then __________ valleys result. The Himalayan Mountains are getting taller every year. This is caused by the __________ between the Eurasian and Indo-Australian Plates. When an oceanic plate collides with a continental plate a __________ zone is created. At a subduction zone the oceanic crust __________ into the mantle below the continental crust. Considerable frictional __________ leads to volcanic action and earthquake activity. The San Andreas Fault in California is an

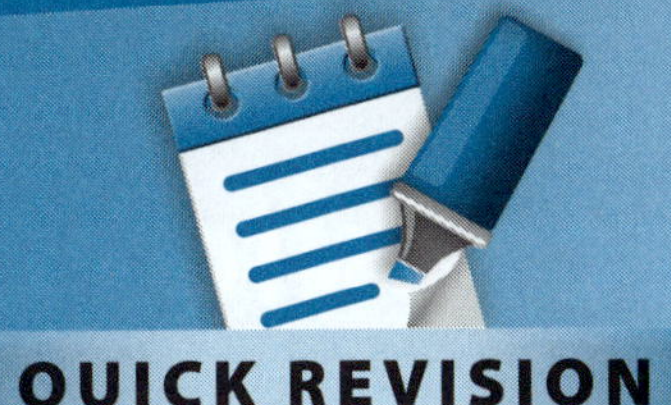

example of a __________ fault. Transform faults occur where two plates __________ past each other in the same or opposite directions. The following diagram shows a subduction zone where an oceanic plate collides with a continental plate.

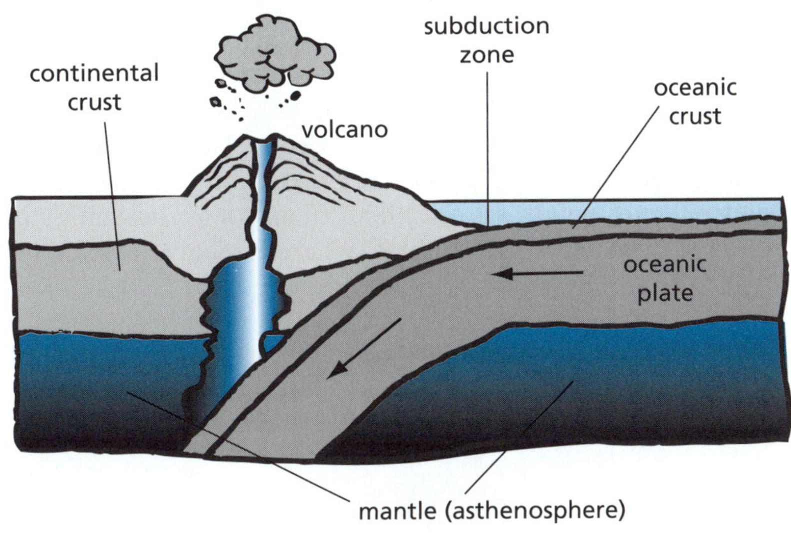

Answers **1** earthquake; four; crust; thicker; 20; igneous; hotter; mantle; 80; lithosphere; asthenosphere; weaker; base; outer; iron; molten **2** South; joined; continental; supercontinent; Laurasia; Africa; split (move); plate; Satellite; rocks **3** plates; gravity; directions; Indo-Australian; Pacific; north; not; large; mid-ocean; crust; spreading; rift; collision; subduction; dives (moves); heating; transform; slide

TECTONIC PLATES

Plate tectonics

REVISION SUMMARIES

1 The furthest humans have dug underground is around 3.9 km, at a gold mine in South Africa, so information about what lies inside the Earth has to be obtained by other means. By studying **earthquake waves** scientists have determined that the inside of the Earth consists of four main layers. The outermost layer is called the **crust**. The crust is thinner under the oceans and thickest below the continents. Below the crust is the **mantle** which occupies over 80% of the Earth's volume. Together, the crust and upper part of the mantle are called the lithosphere. The upper part of the mantle, below the lithosphere, is a weak, deformable layer called the **asthenosphere**. Below the mantle is a molten layer called the **outer core**. The innermost layer is called the **inner core**. New measurements propose that the Earth's inner core is far hotter than previous experiments suggested, putting it at around 6000 °C; as hot as the Sun's surface. Temperatures of the outer core range from 4400 °C to 6100 °C where it meets the solid inner core.

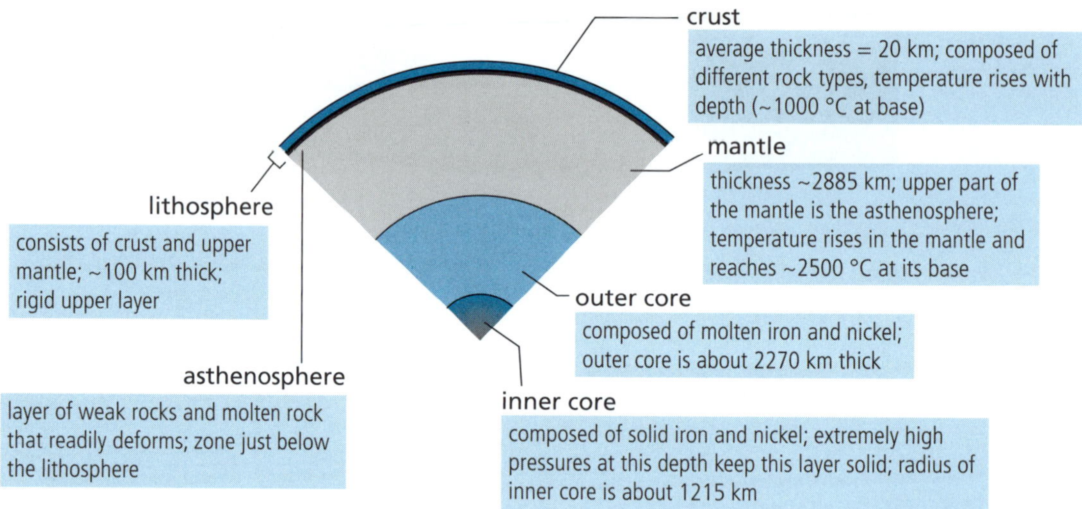

2 The continents of the Earth have not always been in their current locations. Over long periods of Earth's history, the continents have moved in response to forces within the asthenosphere as well as gravity. Before 225 million years ago the existing continents were all part of a giant supercontinent called **Pangaea**. This supercontinent eventually broke up into two large continents called **Laurasia** and **Gondwana**, which in turn broke up into the continents we see today. Australia was part of Gondwana. Together with India and Antarctica it broke away from Gondwana about 135 million years ago. The continents are still moving today but, at the rate of only a few centimetres each year, no difference is noticed during any person's lifetime. However, scientists are able to measure this movement using sophisticated instruments and **satellites**.

Considerable evidence has been collected to support this theory of **continental drift**.

- The east coast of South America fits like a jigsaw piece into the west coast of Africa. This is also true of other continental margins as shown in the diagram on the next page.
- Radioactive dating of rocks along the continental margins of South America and Africa show they are of the same age. This is true of other continental margins. Fossil dating along continental margins also confirms this theory.
- Laser beam measurements made from satellites show the current continents are moving relative to each other. Australia, for example, is moving at a rate of 67 mm per year in a north-easterly direction.

TECTONIC PLATES

Plate tectonics

Pangaea

Laurasia and Gondwana

3 Continental drift can be explained by the theory of **plate tectonics**. This theory is based on evidence that the Earth's crust is made up of a number of slabs of rock called tectonic plates. Tectonic refers to the forces which lead to movement of the Earth's crust. The tectonic plates are moving in different directions and at different speeds. They move in response to the combined effects of **convection currents** in the hot, partly molten region of the mantle (called the asthenosphere) that lies below the plates, and gravitational forces that help to pull heavy plate edges downwards. Australia and India are located on the Indo-Australian Plate as shown in the following diagram.

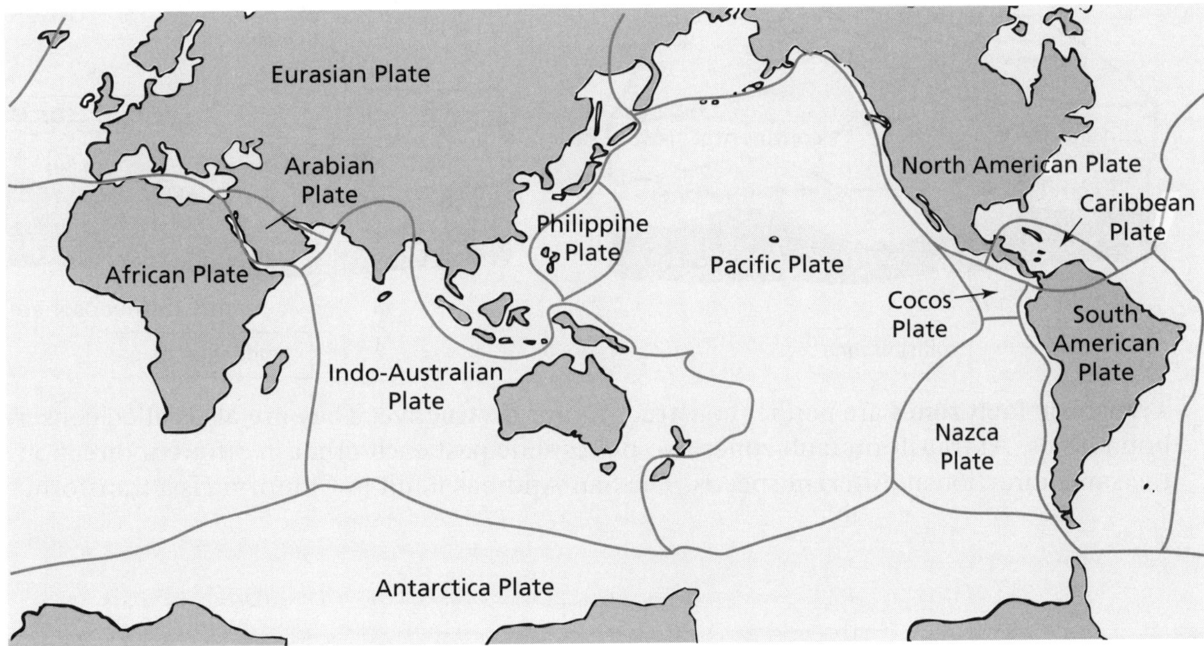

Plates can interact in three main ways: constructive plate boundaries, destructive plate boundaries and transform fault zones.

Examples of constructive (or divergent) plate boundaries include the following.

- **Mid-ocean ridges.** Magma rises to the surface under oceanic crust and creates new oceanic crust at ridges on the sea floor. The sea floor spreads and creates a spreading zone. The Mid-Atlantic Ridge is an example.

(cont.)

TECTONIC PLATES *(continued)*

Plate tectonics

- **Rift valleys.** These form in a similar way to mid-ocean ridges but involve rising magma under continental masses. New continental crust forms along fault lines and creates a plain between mountains on either side. The East African Rift valley is an example.

Examples of destructive (or convergent) plate boundaries include the following.

- **Collision zones.** Mountain ranges form when the edges of two continental plates collide. These zones are also called orogenic belts. The Himalayas have formed in this way.
- **Subduction zones.** When an oceanic plate collides with a continental plate, the oceanic plate moves down and underneath the continental plate. Mountain building, vulcanism and earthquakes result from such an interaction. The Andes Mountains of South America are formed in this way. There is also a subduction zone where the Pacific Plate moves below the Indo-Australian Plate to the east and north-east of New Zealand.

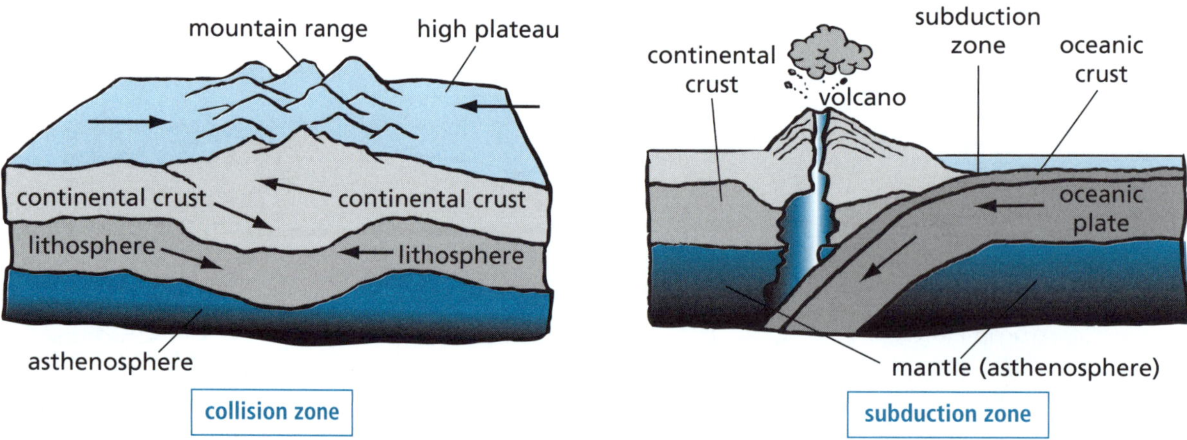

collision zone

subduction zone

Transform fault zones are neither constructive nor destructive. They are also called conservative boundaries. At transform fault zones two plates slide past each other in different directions or in the same direction at different speeds. The San Andreas Fault in California is a transform fault.

Checklist

Can you:

1. *Describe the layered structure of the Earth?* ☐
2. *Describe the evidence for continental drift?* ☐
3. *Explain how the theory of plate tectonics can be used to distinguish between constructive and destructive plate boundaries?* ☐

TECTONIC PLATES

Plate tectonics

REVISION TEST

1 The following diagram shows an example of an interaction between tectonic plates.

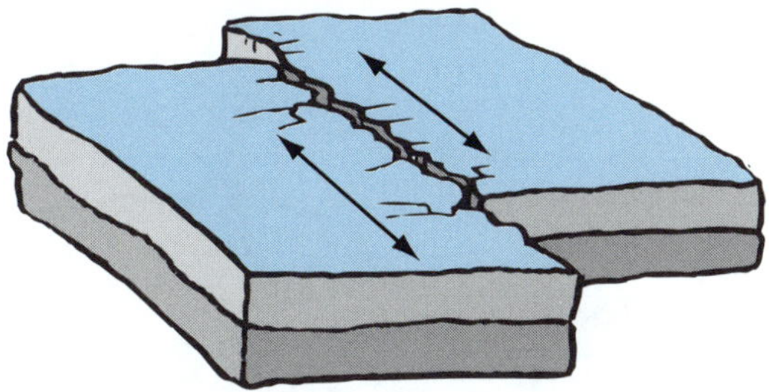

a Name this type of plate interaction. (1 mark)

b If the interaction occurs along the Californian coast of the United States, name the two plates that are interacting. (2 marks)

2 True or false?

a The lithosphere is composed of the crust and the whole mantle. (1 mark)

b Gondwana is the name of the giant supercontinent that broke up into two smaller continents called Laurasia and Pangaea. (1 mark)

c Mid-ocean ridges are classified as constructive plate boundaries. (1 mark)

d The asthenosphere is a weak layer of material below the lithosphere in the upper mantle. (1 mark)

e The inner core of the Earth is composed of molten iron and nickel. (1 mark)

3 Similar fossil plants and animals have been found on the east coast of South America and the west coast of Africa. Why is this important evidence for the theory of plate tectonics? *Hint 1* (1 mark)

4 Fossilised corals and seashells can be found in rock layers high on the Andes Mountains along the west coast of South America. Explain how this is possible. *Hint 2* (1 mark)

5 The following diagram shows an example of a plate interaction.

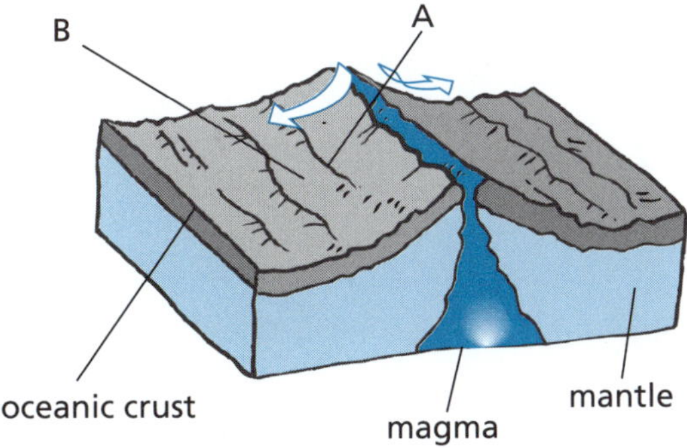

a Name this type of plate interaction. (1 mark)

b Name this process and what it is forming. (1 mark)

c Compare the age of the oceanic crust at A and B. Explain. (2 marks)

(cont.)

6 The following diagram shows another plate interaction.

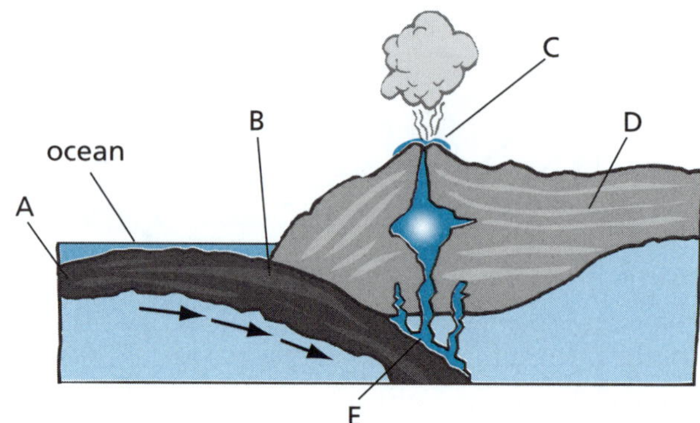

a Identify the structures that are labelled A to E using the appropriate term from the following list: continental plate; subduction zone; frictional heating; oceanic crust; volcano. (5 marks)

b Identify two locations from the following list where subduction zones occur. *Hint 3* (2 marks)

- east of New Zealand's North Island
- Himalaya Mountains
- west coast of Africa
- west coast of South America

c Are subduction zones divergent or convergent boundaries? (1 mark)

7 The words lithosphere and asthenosphere are derived from Greek words: *lithos* meaning 'stone', *sphaira* meaning 'ball' and *astheneia* meaning 'weak'. Explain why these layers of the Earth's structure were so named. (2 marks)

Hint 1: How does the plate tectonic theory explain why continents moved apart?

Hint 2: In what environment do coral and shell fossils normally form?

Hint 3: Examine the plate diagram of the world and determine where oceanic plates and continental plates come together destructively.

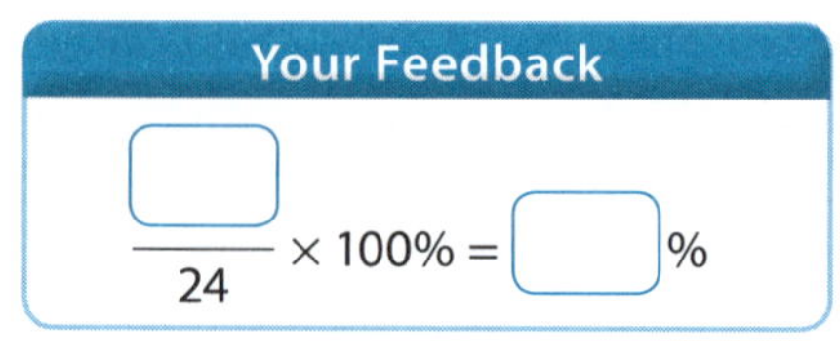

VOLCANOES

1 When molten rock (or __________) emerges through cracks in the Earth's crust it becomes __________ and can form volcanoes. The magma originates in the upper __________. At these depths the magma is under great pressure and extremely hot, and it may __________ some of the chambers it flows through. The magma finds its way to the surface through new or existing __________ in the rocks in the crust. The emerging lava may __________ quietly out of these fissures or it may be ejected violently along with gases, __________ and rocks. Some volcanoes erupt from the sea floor. The Hawaiian Islands formed in this way.

When you watch a volcano __________ you can see smoke emerging from the vent. This smoke is composed of __________ and dust. Along with the smoke are invisible gases which include carbon dioxide, hydrogen __________ and sulfur dioxide. Large amounts of carbon dioxide can lead to asphyxiation of people living nearby as it is colourless and __________ people do not realised they are breathing it in. Hydrogen sulfide and sulfur dioxide, however, have an odour and this __________ people that they need to escape. Sulfur dioxide can also dissolve in rainwater, causing the rain to become __________. If large pieces of lava are thrown up into the air they are called __________ bombs. The outside of these bombs hardens into rock, but they are still __________ inside.

2 There are hundreds of __________ volcanoes around the world. The term *active* does not mean that they continuously __________ but that they release lava and gases regularly. Other volcanoes are intermittent and still others are __________ or quiet. It is difficult to predict when these volcanoes will erupt again. If a volcano does not erupt over hundreds or thousands of years then it is said to be __________. Australia has a number of extinct volcanoes that were active millions of years ago.

There are a number of different __________ of volcanoes. The tallest and most steep-sided are called __________ volcanoes. Their sides consist of __________ layers of ash and lava. The lava in these volcanoes is typically rich in light __________ such as silica, quite stiff and does __________ flow readily. Volcanoes with lava rich in dark minerals are called __________ volcanoes. This lava is highly __________ and flows long distances from the vent before solidification. These volcanoes are very __________ with gently sloping sides. Cinder cones are quite __________ volcanoes that are exclusively made of pyroclastic fragments (__________ and small fragments of rock) called cinders. Lava is not normally released from these cones.

3 Volcanoes are more common in some locations than others. The __________ of the Pacific Ocean is a volcanic zone that is often referred to as the __________ of Fire. Approximately three-quarters of all volcanoes are found here. This region contains __________ zones where oceanic crust moves __________ continental crust. The friction that results generates large amounts of heat which __________ rock to form magma. The volcanoes that form here are usually composite volcanoes, as the lava is rich in __________ and other light minerals. Mount Fuji in Japan is a composite volcano that was formed as a result of the subduction of the __________ and Philippines Plates with the Eurasian Plate. There is a deep ocean __________ at this subduction zone.

Volcanoes can also be formed at __________ zones of continental plates. Mount Etna and Mount Vesuvius in the Mediterranean area of Europe formed in this way. Wherever mid-ocean ridges occur, igneous activity is observed. Igneous rocks or minerals are formed by cooling and hardening of magma or lava. Magma emerges from fissures in the __________ floor and the

(cont.)

lava that results is quite fluid. Iceland in the North Atlantic is situated on a mid-ocean ridge and the islands are __________ from lava. The Hawaiian Islands, however, were formed from lava emerging from a __________ spot or isolated magma chamber in the mantle. As the Pacific Plate moves above this hot spot, new islands are formed.

4 Many features of the landscape are due to past igneous activity. Inside the Earth magma can exist in different sized __________ or it can move along fissures creating thinner, sheet-like structures. These structures may be exposed millions of years later due to __________ and erosion. Dykes and sills are common igneous structures. They both consist of sheets of __________ rock. Sills were formed when magma flowed __________ the bedding layers of sedimentary rock, while dykes resulted from magma cutting __________ the bedding layers. This magma often comes from very large magma chambers. Once these chambers cool and solidify a __________ is formed. When a volcano becomes extinct and erosion begins, the lava that solidified in the vent is exposed. This tall structure is called a __________. In the case of shield volcanoes, lava can flow large distances over the ground. As it cools and solidifies a lava __________ is created. Erosion of these large lava fields usually yields very rich __________ which are idcal for agriculture.

5 Active volcanoes have a significant effect on the environment. They emit a variety of poisonous __________ including sulfur dioxide and hydrogen sulfide. The carbon dioxide emitted adds to global warming as carbon dioxide is a __________ gas. Acidic gases such as sulfur dioxide can dissolve in __________ to produce acid rain. If this acid rain falls into rivers or lakes then they become acidified. The low pH that results causes the __________ of various aquatic organisms. Ash is also released from many volcanoes. In very violent eruptions the ash can be thrown high in the atmosphere and can cause __________ to the jet engines of aeroplanes. If hot ashes fall into bodies of water, such as lakes, then they can cause considerable __________ pollution. This can harm the organisms in the lake as warm water has __________ dissolved oxygen and carbon dioxide. Aquatic plants need carbon dioxide for __________ and oxygen is required for respiration in animals.

Answers **1** magma; lava; mantle; melt; fissures (cracks); flow; ashes (cinders); erupting; steam; sulfide; odourless; warns; acidic; volcanic; molten **2** active; erupt; dormant; extinct; types; composite; alternating; minerals; not; shield; mobile (fluid); wide; small; ashes **3** rim (edge); Ring; subduction; beneath (under); melts; silica; Pacific; trench; collision; ocean; formed; hot **4** chambers; weathering; igneous; between; across; batholith; plug; field; soils **5** gases; greenhouse; rain; death; damage; thermal; less; photosynthesis

VOLCANOES

Plate tectonics

1 **Volcanoes** are openings in the Earth's crust through which molten rocks, cinders and gases emerge. Volcanoes can be found on the land or on the sea floor. While molten rock is underground it is called **magma**, but when it emerges at the surface it is called **lava**. Magma some 30 to 70 km deep within the Earth's surface is very hot and under extreme pressure. It can even melt parts of the rock chamber it is in. Sooner or later it will reach a weak part of the Earth's crust and escape. A conduit or vent is formed to the surface where gases can blast out. Lava differs from magma in that the volatile components (such as gases or liquids) boil off when the molten material reaches the surface. This is because it is no longer under pressure. Another difference is that on the way to the surface, the lava has picked up (or lost) other chemicals. Some of this is rock that the magma has melted its way through to get to the surface. Nevertheless, the lava is still an important source of information about the upper mantle.

A volcano emits a large variety of products.

- **Smoke.** This is a gaseous mixture made up mainly of steam. It may appear dark because of the extremely fine dust it carries.
- **Gases.**These include hydrogen, hydrochloric acid, hydrofluoric acid, carbon dioxide, hydrogen sulfide and sulfur dioxide.
- **Pieces of rock.** These vary from ash and little stones to very large lumps.
- **Volcanic bombs.** These are huge pieces of molten lava with an outer surface that has hardened on contact with cooler air. Some volcanic bombs can be 5 m across.

However, most volcanoes don't throw lava high into the air. The lava just oozes out and flows down the sides of the volcano, eventually solidifying as it cools.

2 An erupting volcano is said to be **active**. There are hundreds of active volcanoes in the world today. Other volcanoes can be described as **intermittent**. These erupt at fairly regular intervals. Some volcanoes are quiet, but have not been dead long enough for us to know if they will erupt again. These are called **dormant**. When we know that a volcano will certainly not erupt again, it is referred to as **extinct**.

Volcanoes can be classified according to their shape and composition.

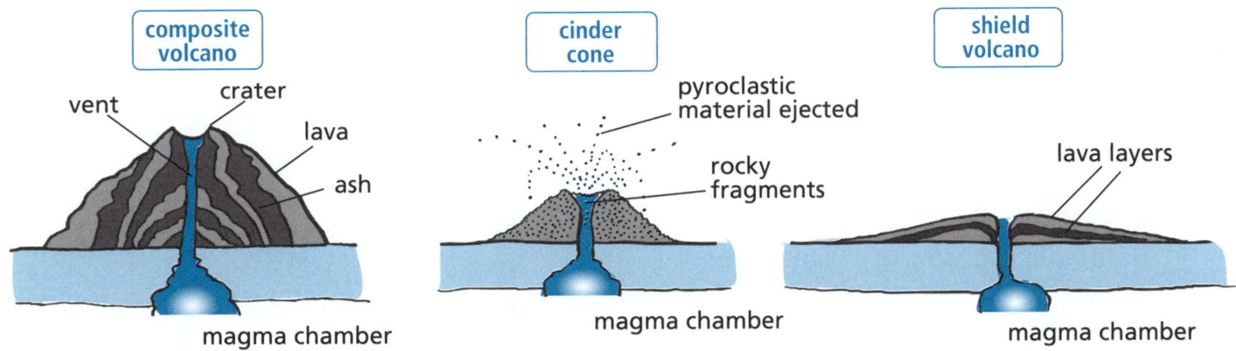

- **Composite volcanoes** are typically tall with symmetrical cones and steep sides. They consist of alternating layers of lava flows, cinders and volcanic ash. The lava that forms composite volcanoes is usually rich in silica minerals that make the lava stiff and slow moving.
- **Cinder cones** are composed of ashes and small fragments which are collectively called pyroclastic fragments or cinders. These volcanoes are much smaller than composite volcanoes.

(cont.)

- **Shield volcanoes** are formed from highly mobile lava flows. The lava from these volcanoes is quite low in silica and rich in dark minerals which ensure greater fluidity. These repeating flows fan out in all directions from a central summit vent and crater creating a broad, gently sloping cone.

3 Volcanoes can be found in a number of different locations around the world. Volcanic action occurs at the following locations.

- **Subduction zones.** This is where an oceanic plate dives beneath a continental plate. Such zones occur around the Pacific Rim. In this region an arc of volcanoes known as the Ring of Fire contains 75% of the Earth's volcanoes. The lava is rich in silica and does not flow quickly. Composite volcanoes tend to form. Along the west coast of North and South America these volcanoes are located on the continental mass. In eastern Asia the volcanoes are found on islands created near the deep ocean trench formed by the subduction zone. The islands of Japan were formed in this way.
- **Continental plate collision zones.** The volcanoes in the Mediterranean area of Europe and those in the Red Sea areas of the Middle East are the result of the African Plate colliding with the Eurasian Plate. Mount Etna in Italy is a well-known example of such a volcano.
- **Mid-ocean ridges.** Magma emerges in the crack between two oceanic plates as they move apart. The lava is rich in dark minerals and quite fluid and so shield volcanoes tend to form. Iceland was formed by such volcanic action.
- **Hot spots.** When cracks in the lithosphere form above isolated magma chambers hot spots are formed. The lava that emerges produces shield volcanoes, usually on the ocean floor. Over time these volcanoes grow larger and emerge above the ocean to form islands. The Hawaiian Islands were formed in this way. As the Pacific Plate moved over the hot spot, new islands were formed. Currently the hot spot is below the biggest island in the group.

4 There are various landforms associated with igneous activity. The weathering and erosion of the surface of the land exposes igneous structures that were previously underground. The general name **pluton** is applied to these underground structures. Some are small and some are very large. The most common of these features are as follows.

- **Dykes** form when magma cuts across sedimentary rock layers and cools to form sheets of igneous rock. These structures may be vertical or at an angle to the layers of sedimentary rock.
- **Sills** form when magma flows between layers of sedimentary rock and cool to form sheets of igneous rock.
- **Batholiths** are very large igneous masses that solidified in the Earth's crust. When exposed on the surface due to erosion, these structures typically have an area greater than 100 km^2.
- A **plug** is the solidified remains of magma that was still inside the vent of a volcano after it stopped erupting. It becomes exposed as the sides of the volcano erode.

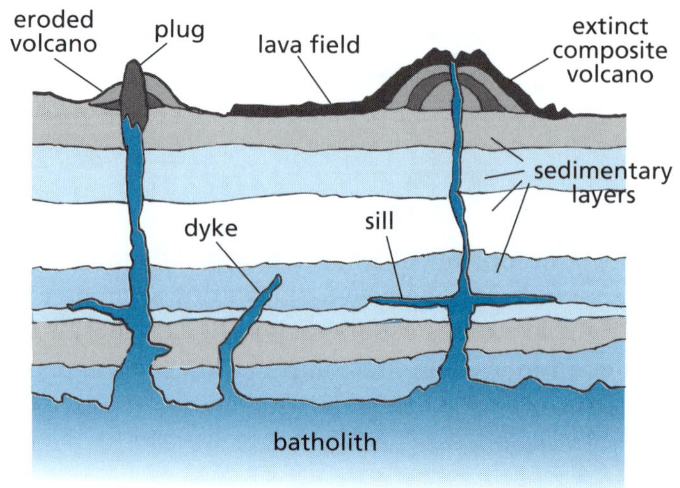

- **Lava fields** are large areas of lava that form on the surface when very fluid lava spreads out from a volcanic vent and solidifies. Lava fields are typically composed of basalt.

5 Volcanic activity has a large effect on the environment. Volcanoes affect the atmosphere, hydrosphere and biosphere.

- **Atmospheric effects.** Apart from water vapour, volcanoes emit carbon dioxide, carbon monoxide, sulfur dioxide, hydrogen sulfide and acidic vapours such as hydrochloric and hydrofluoric acid. Carbon dioxide has a significant effect on global warming. Solid ash particles are also produced in an eruption. In violent eruptions the ash cloud can rise to great heights and be carried around the world. This greatly affects the weather and it is also dangerous for jet aeroplanes as ash clouds can stall their engines.
- **Hydrosphere effects.** When acidic gases and vapours are dissolved in rain, the acid rain that is formed can fall into lakes and rivers, and acidify these bodies of water. When hot ashes fall into lakes and streams the temperature of the water rises. This leads to a decrease in dissolved oxygen and carbon dioxide concentrations in the water.
- **Biosphere effects.** As a result of acid rain being deposited in bodies of water, the pH of the water decreases, causing stress to aquatic plants and animals. Fish and their eggs can die if the pH decreases below 5.5. Acid rain can also damage the leaves of forest trees as well as materials in the built environment such as steel and marble. Aquatic plants rely on dissolved carbon dioxide for photosynthesis and aquatic organisms require oxygen for respiration. When the temperature of the water increases due to volcanic action, the concentration of these gases decreases. In these conditions aquatic plants do not photosynthesise efficiently and aquatic animals cannot respire efficiently, and many die.

  Sulfur dioxide that is emitted from volcanoes has serious health effects on humans. Conditions such as breathing difficulties, headaches, eye damage and allergies can result.

Checklist
Can you:

1 *Describe the different types of materials emitted from volcanoes?* ☐
2 *Name the common types of volcanoes and describe their composition?* ☐
3 *Identify the common types of locations where volcanoes are found?* ☐
4 *Name common landforms associated with igneous activity?* ☐
5 *Identify the effects that volcanic eruptions can have on the environment?* ☐

1 The following diagram shows two igneous structures labelled X and Y.

a Name these two types of structures. (2 marks)

b Structure X is exposed and stands higher than the surrounding soil on the land surface. Explain why this structure has not eroded as quickly as the conglomerate. *Hint 1* (2 marks)

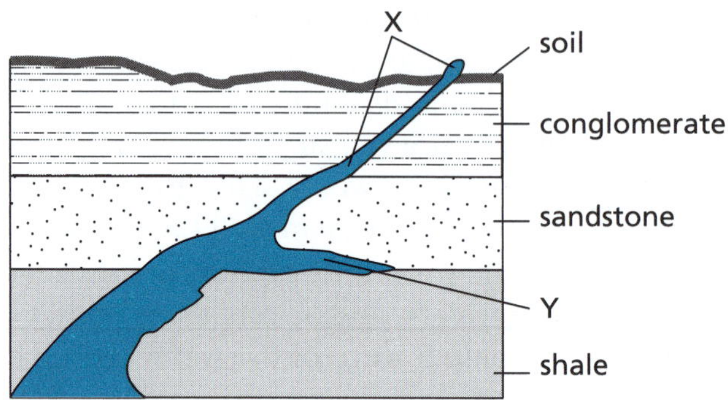

2 Igneous rocks have different sized crystals or none at all, depending on the speed at which heat is lost and crystallisation occurs: the slower the crystallisation, the larger the crystals. The following diagram shows some labelled points along an igneous intrusion.

a At which point (W, X or Y) would the igneous rock have:

i the largest crystals? (1 mark)

ii the smallest crystals? (1 mark)

b Tuff is a layer of igneous origin. It is composed of cinders and ash that have been deposited on the ground following a volcanic eruption. The igneous intrusion is a much later event. What evidence is there in the diagram that the igneous intrusion is a much later event?
Hint 2 (1 mark)

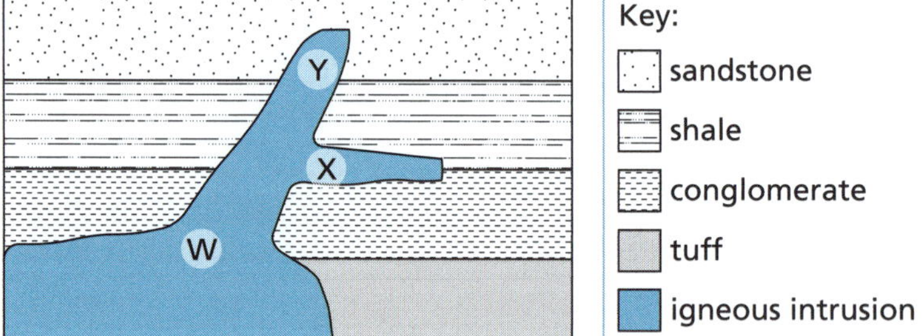

3 True or false?

a Volcanic activity can lead to the production of acid rain. (1 mark)

b Volcanoes can emit carbon monoxide and carbon dioxide during eruptions. (1 mark)

c Sills form when magma cuts across sedimentary layers. (1 mark)

d The Ring of Fire is associated with subduction zones in the Atlantic Ocean. (1 mark)

e A composite volcano is very wide and not very tall. (1 mark)

4 Use the code letters to match the terms in column 1 with the terms in column 2. (5 marks)

Column 1	Column 2
A magma	F volcano that could erupt again
B dormant	G pyroclastic fragments
C extinct	H molten rock underground
D cinder cone	I hot spot
E Hawaiian Islands	J volcano that will not erupt again

5 The following diagram shows a chain of volcanic islands and sea mounts. Explain how this chain of volcanic islands and sea mounts formed. (3 marks)

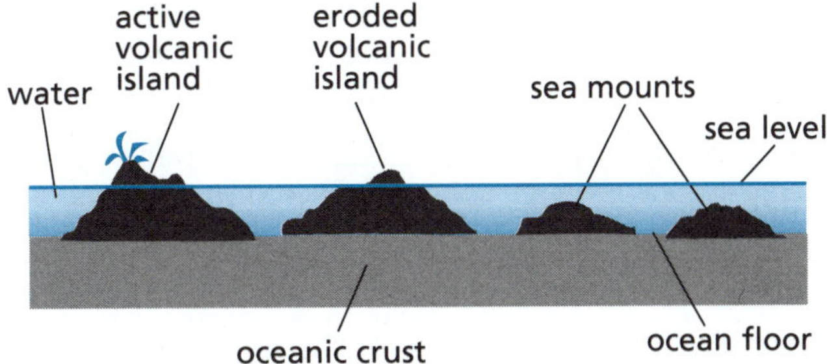

6 The diagram to the right shows a volcanic landform labelled X.

a Name this type of volcanic landform. (1 mark)

b Explain how this structure forms. (2 marks)

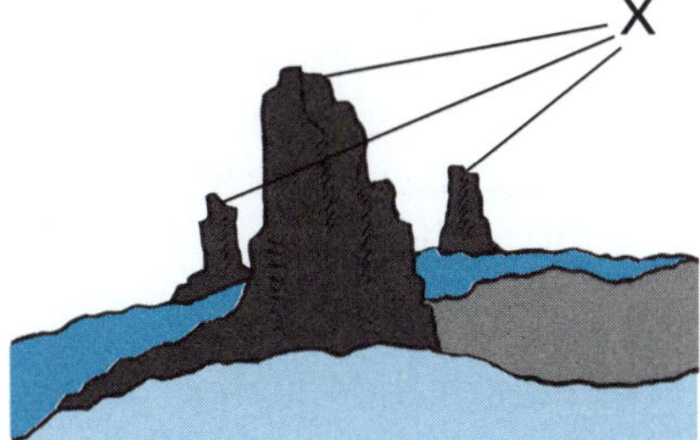

7 A volcanic eruption occurs near a large freshwater lake. This eruption continues for many months and then ceases. A team of geologists investigate the lake and make the following observations.

- The water is moderately acidic.
- There are many dead fish and freshwater invertebrates.
- The aquatic plants in the lake have died.

Explain the likely cause of each of these observations. *Hint 3* (3 marks)

Hint 1: The igneous structures are made of interlocking crystals which help to resist agents of erosion.

Hint 2: Magma intrudes into and often melts existing geological structures.

Hint 3: Hot ashes will heat the water and change its acidity.

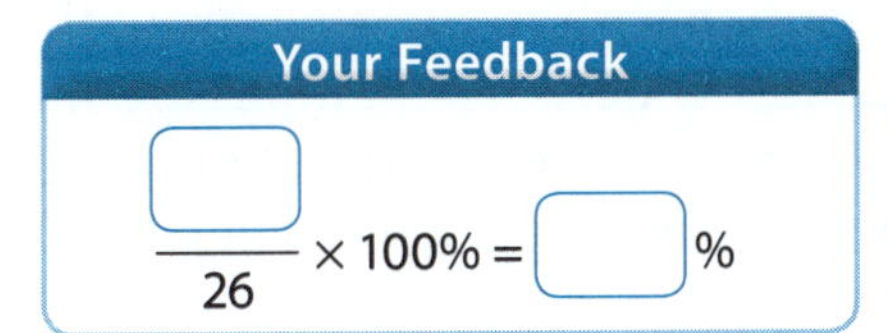

PAGES 236–237

PAGE 252

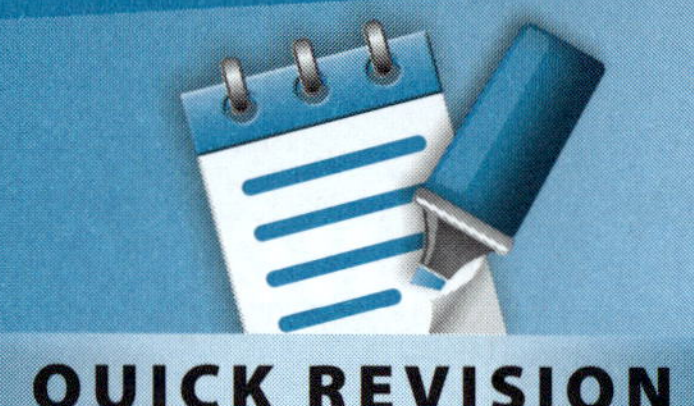

EARTHQUAKES

Plate tectonics

QUICK REVISION

1 Over many __________ rocks are increasingly deformed and twisted out of __________ by forces within the Earth. Eventually the rock yields in a wrenching break (__________), and the two sides of the fault slip to a new position to relieve the accumulated strain. An earthquake occurs when an enormous amount of __________ is released as a result of this breaking and movement.

The __________ of an earthquake is the point where the rocks first fracture and move during an earthquake. The vibrating seismic (__________) waves travel away from the focus of the earthquake in all __________. These waves carry energy and can be so powerful that they may reach all parts of the globe. Directly above the focus on the Earth's __________ is the epicentre. Shallow-focus earthquakes (less than 60 km below the surface) are much more dangerous than __________-focus earthquakes. Shallow-focus earthquakes release 75% of all the __________ produced by earthquakes each year. Deep-focus earthquakes occur over 290 km below the __________. These earthquakes occur in island arcs or in deep ocean trenches where one plate is __________ over another.

Seismic waves are defined by the type of __________ of a particle in the path of the wave.

- P waves travel through __________ and solids, in all directions. These compression waves move the material particles that transmit the energy parallel to the __________ of propagation. P waves are the __________ waves and travel at around 6.0 km/s in the crust.
- S waves are transverse shear waves that travel through solids, but not through __________ or gases. They cause rock particles to __________ at right angles to the direction of movement. S waves do not travel as fast as P waves and have a speed of around 3.5 km/s in the crust. S waves do not penetrate the molten outer __________ of the Earth.
- L waves only occur at the __________ of the Earth. It is these waves that do the most __________ during an earthquake, especially at distances far from the epicentre. They move the surface of the Earth up and down. The speed of surface waves varies with their wavelength but is always __________ than P and S waves.

2 No part of the Earth's surface is __________ from earthquakes, but some regions experience them more frequently than others. They are most common at __________ plate boundaries, where different plates meet. An earthquake is a sudden movement of the Earth's crust, which originates naturally at or __________ the surface. There are two main causes.

- Earthquakes are commonly associated with explosive __________ eruptions; they either precede (come before) or accompany eruptions.
- Earthquakes can be triggered by tectonic plate activity at the __________ margins and faults. Most worldwide earthquakes are of this sort, as the constant __________ of the plates causes faults to slip.

3 A __________ is a device that scientists use to measure earthquakes. Its purpose is to accurately record the motion of the __________ during an earthquake. Good seismographs are isolated and connected to bedrock. With today's technology, digital seismographs record ground shaking over a large __________ (band) of frequencies and amplitudes. These seismographs can sense ground motion with frequencies from thousands of seconds to less than a hundredth of a __________. Older seismographs are limited by the amount of __________ possible between the mass, or pendulum, and its housing. Modern seismographs work by measuring the amount of electrical energy needed to keep the weight __________ in the housing in the presence of

strong ground shaking. The Global Seismographic Network consists of over 150 seismographs in 60 countries and monitors earthquakes all around the world.

The most common method of measuring the __________ of an earthquake is the Richter scale. This rates the amount of __________ an earthquake releases, and requires information gathered by seismographs. Richter scale ratings are determined soon after an __________, when __________ (scientists who study earthquakes) can compare the data from different seismograph stations. The following photo shows earthquake waves being recorded on a seismograph.

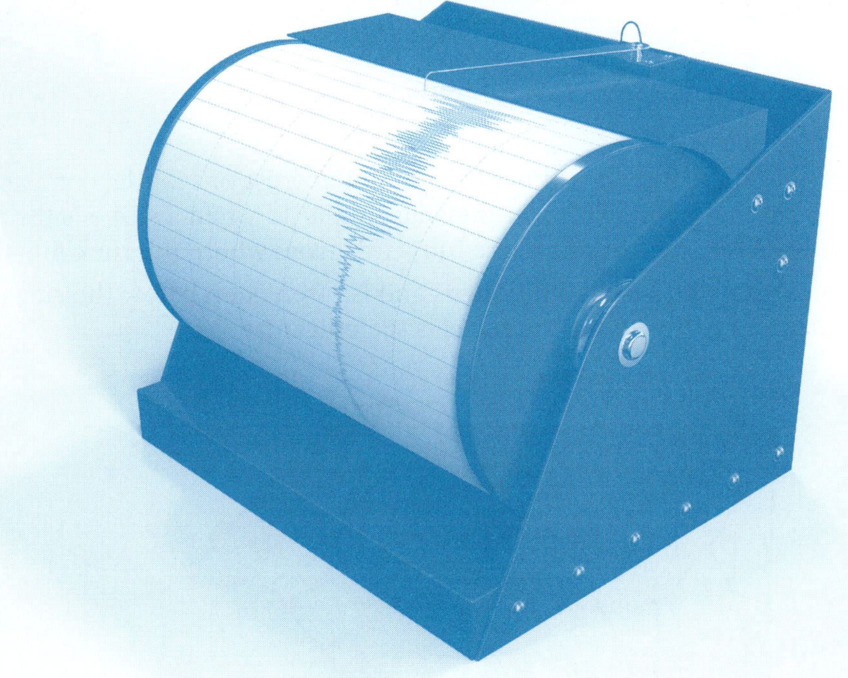

Answers **1** centuries (decades, millennia); shape; fault; energy; focus; earthquake; directions; surface; deep; energy; surface; slipping (moving, sliding); motion; fluids (liquids); direction; fastest; liquids; oscillate (vibrate, move); core; surface; damage; slower **2** free (absent); tectonic; below; volcanic; plate; movement (motion) **3** seismograph; ground; range; second; movement; centred (stationary, motionless); magnitude (size); energy; earthquake; seismologists

EARTHQUAKES

Plate tectonics

1 Within the Earth's crust strains can build up over a long period of time. These strains can **deform** and **fold** layers of rocks. Eventually a crack or fracture (fault) occurs and movement takes place to release this tension. **Earthquakes** result from this sudden release of energy.

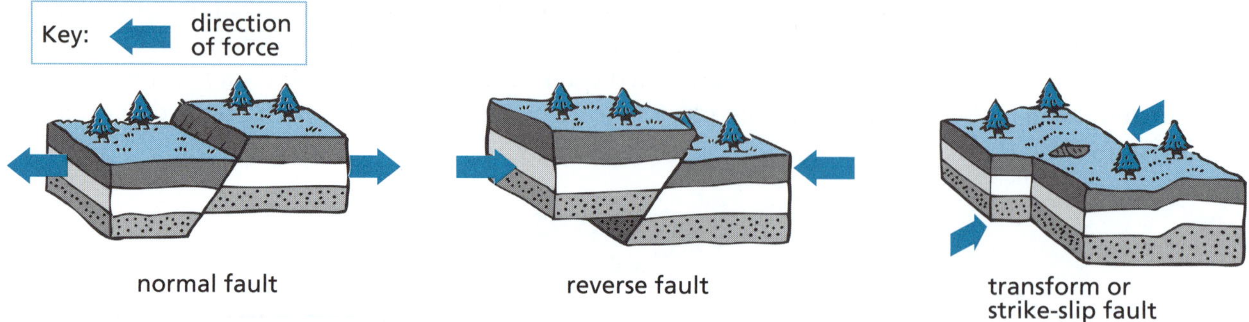

Once this sudden crack and movement occurs within the crust or mantle, concentric **shock waves** move out from that point. This is somewhat similar to when a stone is thrown into still water and a series of concentric waves move outwards from where the rock hits the water.

The origin of the earthquake is called the **focus** and is often deep below the surface. The point on the Earth's surface directly above the focus is the **epicentre**.

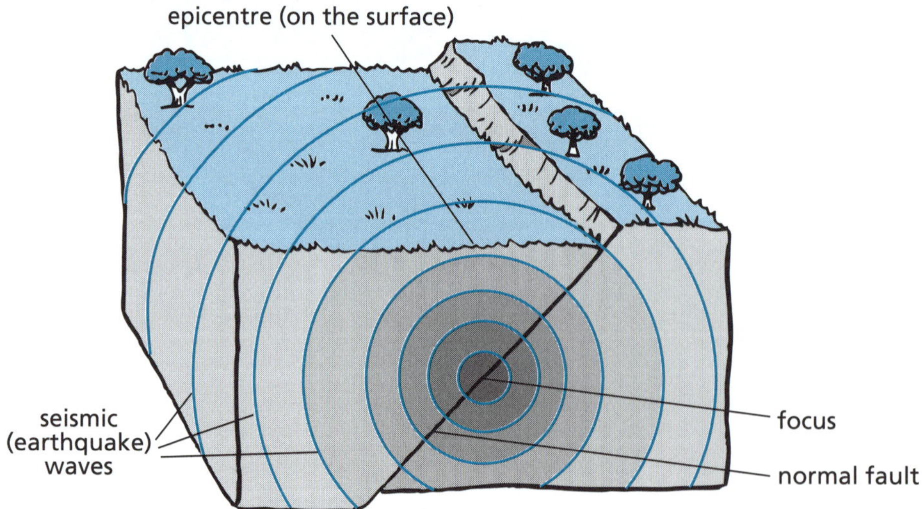

Earthquake waves move outwards from the focus, and can travel both horizontally and vertically. There are three different types of waves.

- **P waves** (primary waves) are similar to sound waves. They are high-frequency, short-wavelength, **longitudinal** (or compression) waves that can pass through both solids and liquids. The ground moves forwards and backwards as it is compressed and decompressed.
- **S waves** (secondary waves) travel more slowly than P waves and so they arrive after the P waves. Like P waves they are high-frequency, short-wavelength waves, but instead of being longitudinal they are **transverse**. Their speeds depend upon the density of the rocks through which they are moving but they cannot move through liquids.
- **L waves** (surface waves) are low-frequency transverse vibrations with a long wavelength. They are formed near the epicentre and only travel through the surface part of the crust.

They account for most of the damage caused by earthquakes. Their motions are similar to that of sea waves, rising and falling as visible waves move across the land.

2 Earthquakes can occur everywhere. When plotted on a world map, most earthquake locations look like narrow bands meandering through the continents and oceans. These zones closely align with **plate boundaries**. Earthquakes happen where these plates crush into or slide past each other. Earthquakes occur not only where **plates collide** but also where two plates move away from each other, in **spreading zones** such as the Mid-Atlantic Ridge where Europe is moving away from North America.

Unlike Australia, which lies in the middle of a tectonic plate, New Zealand experiences many earthquakes because it lies along a plate boundary.

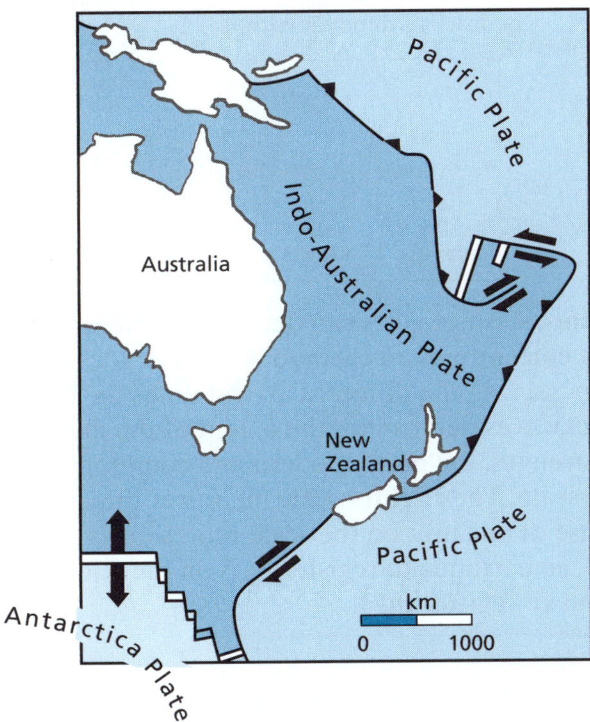

While most earthquakes occur along tectonic plate boundaries, a small number do occur along faults in the normally stable interior of plates, such as reactivated fault systems (where there is movement along formerly inactive faults). Stresses are set up where plates lock against each other at the boundaries. Those stresses may transmit across the plate and under Australia, for example, popping up somewhere in the country as an occasional small earthquake. Volcanic activity can also trigger earthquakes due to the build up of pressure in the Earth's surface and movement of molten material.

3 A **seismograph** is an instrument that records the shaking of the Earth's surface caused by **seismic** (earthquake) waves. Most modern seismographs are electronic, and basically consist of a drum with a roll of paper, a bar or spring with a hinge at one or both ends, a weight and a pen. When an earthquake occurs the seismograph, which is anchored to the ground, moves except for the weight with the pen attached. This stays still because of inertia. As the drum and paper

(cont.)

shake next to the pen, the pen makes squiggly lines on the constantly rotating drum and paper, creating a record of the earthquake.

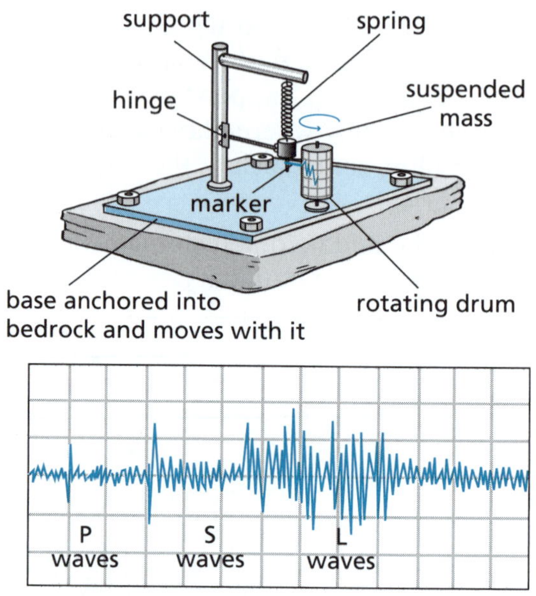

By studying **seismograms**, seismologists can determine where the earthquake occurred and how strong it was. The epicentre of an earthquake can be located by using three different **seismic stations**, and measuring the different arrival times for P and S waves. The strength, or magnitude, of the shockwaves determines the extent of the damage caused. Two main scales exist for defining the strength, the more modern and commonly used Richter scale and the older and less used Mercalli scale. The **Richter scale** measures the amount of energy released by an earthquake. An increase of one unit on the scale represents a 31.6 times increase in the energy released. For example, an earthquake registering 5 on the Richter scale releases 31.6 times more energy than an earthquake registering 4.

The Richter scale	
Earthquake	**Richter magnitude**
minor	0.0–3.9
light-moderate	4.0–5.9
strong	6.0–6.9
major	7.0–7.9
great	≥ 8.0

Checklist

Can you:

1. *Describe the causes of earthquakes?* ☐
2. *Explain why earthquakes mainly occur in particular zones around the world?* ☐
3. *Describe how earthquakes are measured using seismographs and the Richter scale?* ☐

EARTHQUAKES

Plate tectonics

REVISION TEST

1 True or false?

a Folding of rocks is most likely to take place when rocks are under tension. (1 mark)

b The twisting of rock without breaking is called faulting. (1 mark)

c Rocks are most likely to break near the Earth's surface. (1 mark)

d Tension in rocks causes reverse faults. (1 mark)

e A fault is a break in a rock layer which allows the bordering rocks to slide past each other. (1 mark)

f The place where slippage first occurs is called the earthquake's epicentre. (1 mark)

g Most earthquakes occur along or near the edges of tectonic plates. (1 mark)

h Surface waves are seismic waves that travel the fastest. (1 mark)

i P waves cause rock particles to move together and apart in the same direction as the wave is moving. (1 mark)

j In order to determine how far away from a seismograph station an earthquake occurred, scientists find out the difference in arrival times between P and S waves. (1 mark)

2 Examine the diagram below and identify each of the types of faults. (3 marks)

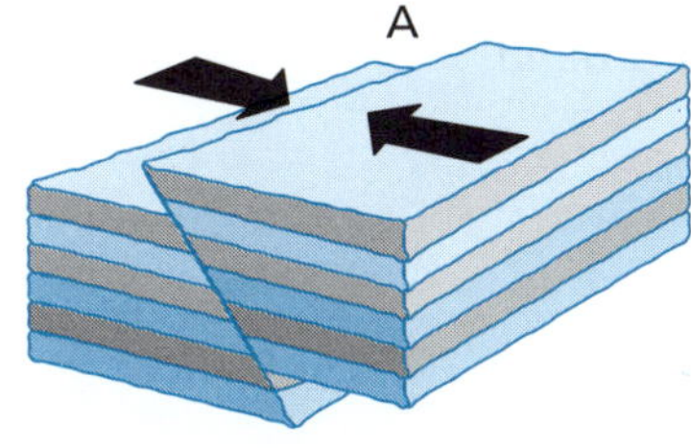

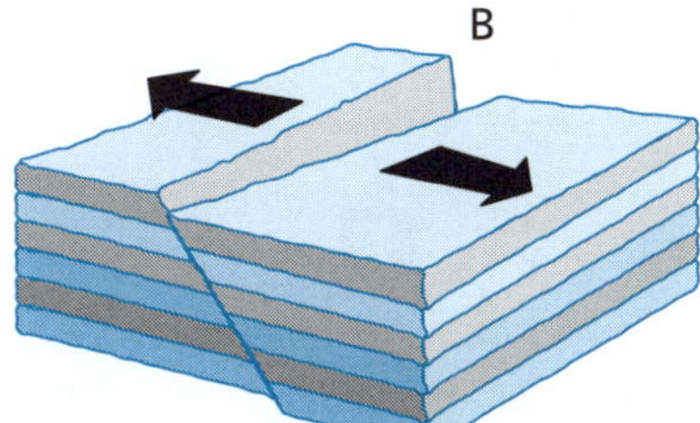

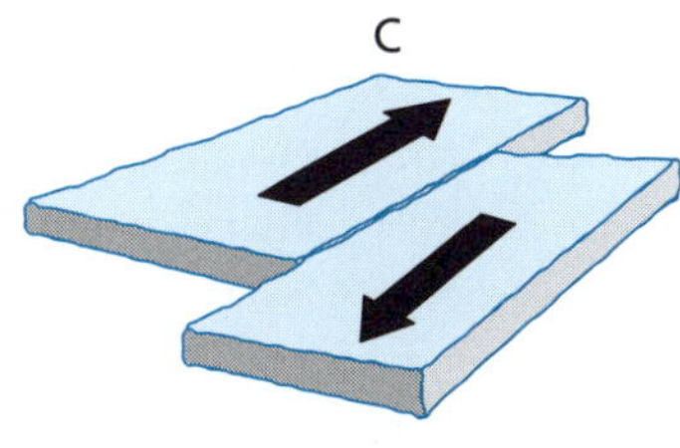

3 The map shows the locations of major major earthquakes around the world. Account for their distribution.
Hint 1 (2 marks)

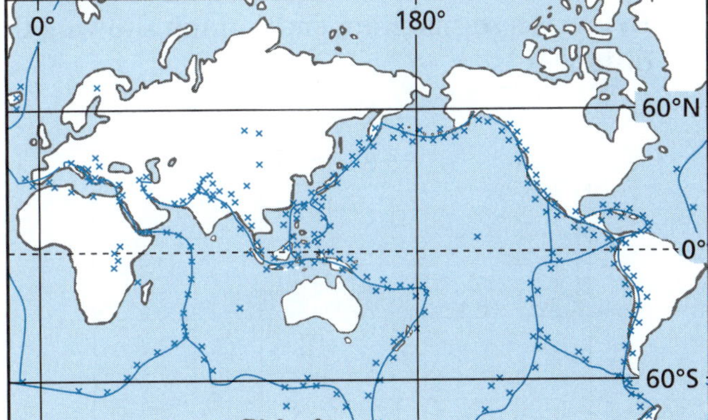

4 It is estimated that several million earthquakes occur in the world each year. Many go unnoticed.

a Suggest two reasons for this. (2 marks)

b The table gives the frequency of occurrence of earthquakes of different sizes around the world. (Magnitude is based on the Richter scale.)

Magnitude	2.0–2.9	3.0–3.9	4.0–4.9	5.0–5.9	6.0–6.9	7.0–7.9	≥ 8.0
Annual average	1 300 000	130 000	13 000	1300	130	14	1

State two conclusions that can be made from this table. (2 marks)

(cont.)

c As more seismographs are installed around the world more earthquakes will be detected. For which magnitude earthquakes would this *least* affect? (1 mark)

5 The following table shows the changes in ground motion (displacement) and energy that correspond to changes in Richter magnitude.

Change in magnitude	0.1	0.3	0.5	1.0
Displacement change	1.3 times	2.0 times	3.2 times	10.0 times
Energy change	1.4 times	3.0 times	5.5 times	32.0 times

For example, compared to an earthquake of magnitude 5.5, a 6.5 magnitude earthquake produces 10 times more ground motion, and 32 times more energy.

What is the change in ground motion (displacement) and change in energy between two earthquakes, one with magnitude 5.8 and the other with magnitude 6.1? *Hint 2* (2 marks)

6 The diagram below shows a different type of seismograph to the one shown in the Revision Summaries section. Use this diagram to describe how a seismograph works. (4 marks)

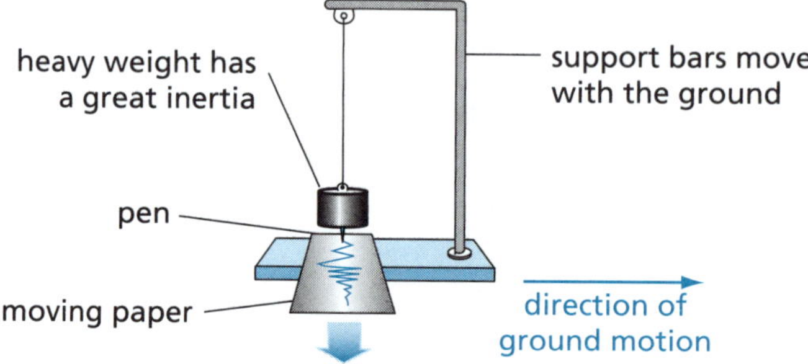

7 Below is a seismogram and a graph showing the time it takes for P and S waves to cover certain distances.

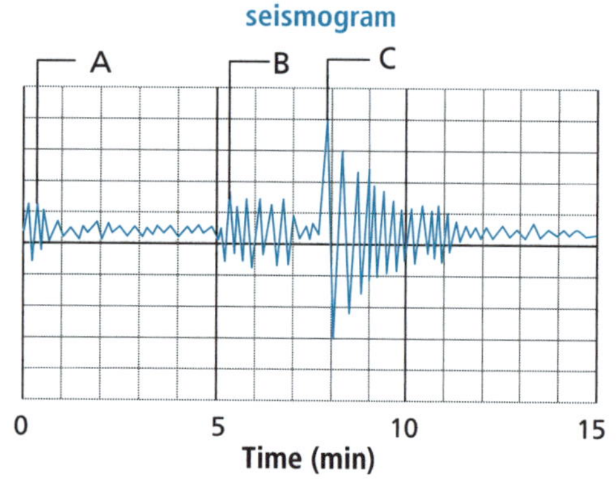

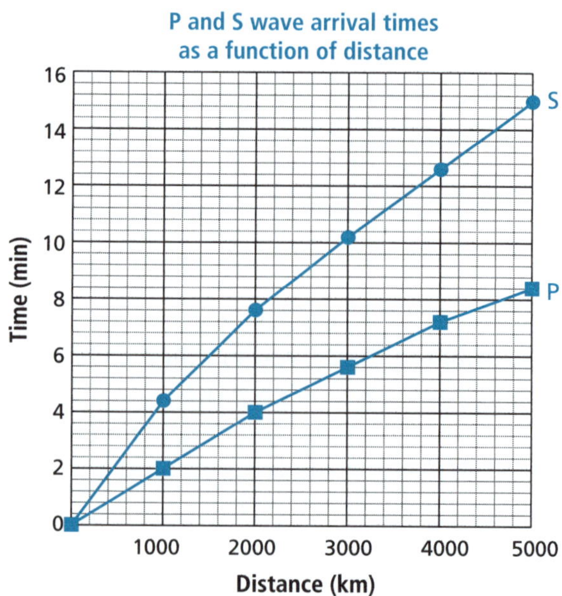

a Identify the waves labelled A, B and C. *Hint 3* (3 marks)

b Use the seismogram time scale to determine the time difference between the arrival of the P and S waves. Answer to the nearest whole minute. *Hint 4* (1 mark)

c Use this time difference and the time–distance graph to determine how far away the earthquake occurred from this seismograph station. *Hint 5* (1 mark)

d In order to locate the position of an earthquake three seismic stations, located in Sydney, Adelaide and Perth, recorded seismograms.

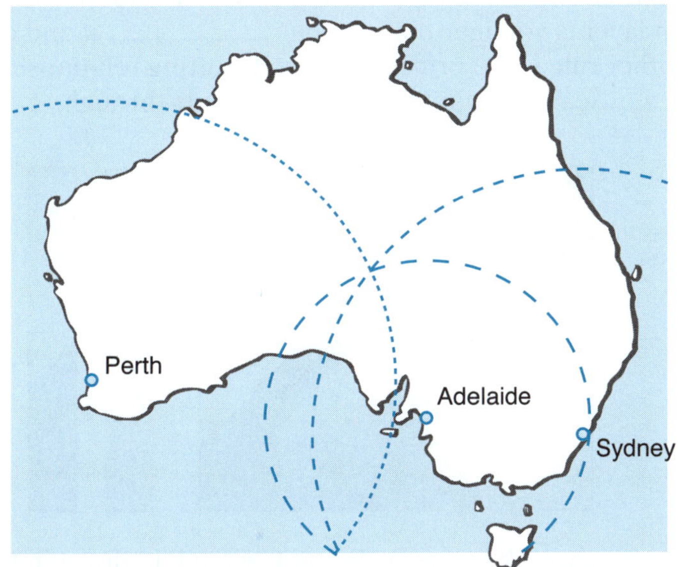

i What do the circles around each city represent? (1 mark)

ii Explain how the stations calculated how far away the earthquake occurred. (3 marks)

iii Explain why it takes at least three different seismograph stations to locate the epicentre of the earthquake. (3 marks)

Hint 1: Compare the location of the earthquake locations with the map of tectonic plates earlier in this chapter.

Hint 2: Determine the change in magnitude.

Hint 3: The order of arrival, and recording, will depend on their speed.

Hint 4: Use the 1-minute scale provided on the seismogram.

Hint 5: The further away the earthquake, the greater is the time difference between the arrival of the two waves. Use a piece of paper to record the length between 0 minutes and the time difference you found in part b. Now use this length between the P wave curve and S wave curve to determine how far away the earthquake took place.

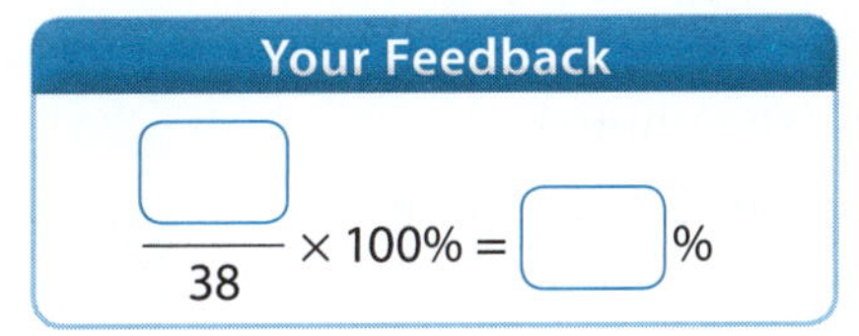

GEOLOGICAL HISTORY OF AUSTRALIA

Plate tectonics

QUICK REVISION

1 Stratigraphy is the study of __________, or layers. Specifically, stratigraphy refers to the study of the composition, relative positions and distribution of rock strata in order to determine their __________ history and the relative ages of the layers. Stratigraphy can tell us a great deal about the processes affecting the deposition of __________ and soils, and the condition of sites and artefacts. The law of superposition states that in any undisturbed sequence of rocks deposited in layers, the __________ layer is on top and the oldest is on the __________. In other words, each layer is younger than the one __________ it and older than the one __________ it. Another rule is the principle of cross-cutting relationships which states that a fault or intrusion is __________ than the rocks that it cuts through.

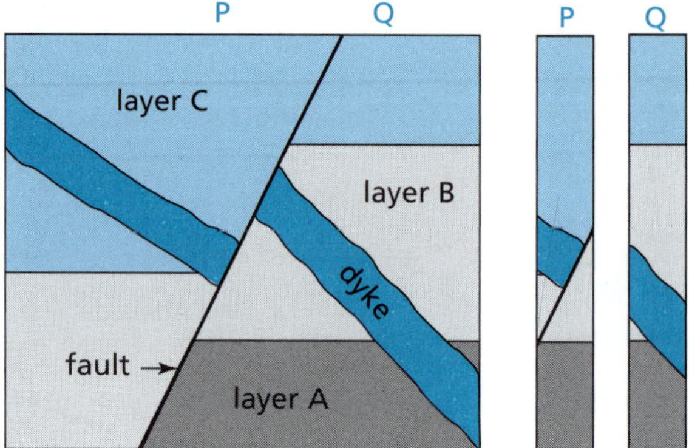

In the diagram sedimentary rock layers A, B and C were laid down in that __________. Then the __________ (a volcanic feature) cut through these layers. At a later time a __________ occurred and the two blocks shifted. A geologist taking stratigraphic columns at P and Q would see there is a __________ in P, but it's not present in Q, and has enough information to infer what lies between __________ and Q.

2 Australia is an ancient land and one of the __________ continents. Originally, Australia was joined to a southern continent called __________, before it separated and drifted across the Earth's surface, experiencing a variety of different climatic changes and geological events. In the long distant past there were __________ eruptions and earthquakes that helped shape the face of Australia, but in the recent past the continent has undergone a long period of geological __________ with little volcanic or tectonic activity. Accordingly, Australia has been open to the forces of __________ and erosion, which are processes that break apart and move rocks and soil around. As a result, Australia is an old, flat land (see photo).

This long period of stability, and Australia's isolation from other continents, allowed __________ plants and animals to evolve on the continent. The lack of any large __________ animals allowed monotremes and __________ to gain dominance. Marsupials and monotremes also existed on other continents, but only in Australia and New __________ did they out-compete placental __________ and flourish.

3 Within Australia, the last eruption of an active __________ was about 5000 years ago in the area around Mount __________ in South Australia. Mount Gambier is now dormant.

Earthquakes can strike anywhere and at any time as tectonic plates __________ and grind against each other. But it's hard to predict when and where a major __________ will hit. While Australia is relatively safe, being in the middle of a __________ plate, we do get roughly one small earthquake a day, and stronger ones every few years or so. As these are minor tremors and often occur in __________ areas, most people don't notice. However, if the quake takes place under a __________, especially if shallow, it can cause catastrophic damage. One of Australia's worst earthquakes occurred in Newcastle, New South Wales, in 1989 (__________ 5.6). Although it was not the strongest earthquake, its location in a highly populated area meant 13 people were killed and 160 others were hospitalised. It caused an estimated $4 billion of damage to the city and surroundings. Christchurch in New Zealand has suffered devastating effects due to earthquakes as shown in the photo below. Besides obvious fault lines, folding is another way that rock layers respond to stress. They may crumple or crinkle sideways, without fracturing, like wrinkles in a rug.

Answers 1 strata; geological; rocks; youngest; bottom; beneath (below); above; younger; order (sequence); dyke; fault; fault; P 2 oldest; Gondwana; stability; weathering; volcanic; unique (distinctive); placental (carnivorous); marsupials; Guinea; mammals 3 volcano; Gambier; collide (mash); earthquake; tectonic; remote (isolated, uninhabited); city (town); magnitude

GEOLOGICAL HISTORY OF AUSTRALIA

Plate tectonics

1 **Stratigraphy** is a branch of geology which studies rock layers and layering (strata). Stratigraphy deals principally with sedimentary rocks but may also include layered igneous rocks, such as those ensuing from successive lava flows. The law (or principle) of **superposition** is based on observations and states:

Sedimentary layers are deposited in a time sequence, with the oldest on the bottom and the youngest on the top.

Consider the two cross-sections of rock layers taken some 30 km apart. In each case the rocks higher up were laid down after those lower down and so are younger. By lining up similar layers an **ordered sequence** is obtained. But, of course, it doesn't tell you when the layers were deposited, just that some may be around the same age, or younger or older, than other layers.

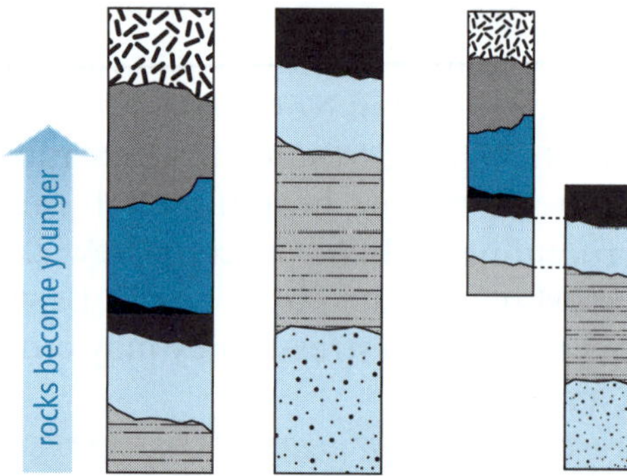

Absolute dating of a rock or rock layer is obtaining an actual date. The chief method for this is radioactive dating. **Relative dating** is basically placing rocks in age sequence. This is most useful in a section of rock layers (or stratigraphic column), where the rock at the bottom is older and a rock further up the column is younger. A number of such stratigraphic columns from different areas in the same region allow geologists to **reconstruct** its history.

Consider the following cross-section of a landform.

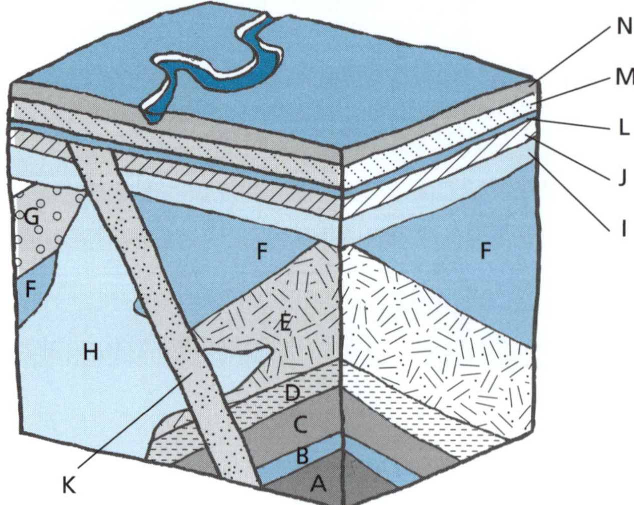

A brief history of this area follows.

- Layers A to G laid down (under water) in that (alphabetical) order.

- An igneous intrusion, H, cuts through these layers. (H is possibly formed as magma melts its way through the layers.)
- Rocks A to H are now tilted by tectonic forces, uplifted and then eroded to form a flat surface.
- Layers I and J are laid down (under water). (There may have been more layers here, but none survive.)
- Dyke K cuts through rocks A to J. (If a rock is cut by another rock, then it is older than the rock which cut it.)
- Uplift followed by further erosion flattens the landscape.
- Layers L, M and N (and possibly more layers) are laid down under water).
- More uplift is followed by further erosion, then a stream traverses the landscape.

2 Australia sits on the **Indo-Australian Plate** and once was joined with Antarctica as part of the southern supercontinent **Gondwana**. Around 96 million years ago Australia separated from Antarctica and began to drift northwards. When the last ice age ended, about 12 000 years ago, rising sea levels produced Bass Strait and separated Tasmania from the mainland. Then, around 10 000 and 8500 years ago New Guinea and Australia were separated.

In this geologically and climatically stable land unique plants and animals developed. The main reasons for this diversity were as follows.

- **Temperatures** remained fairly constant allowing a variety of plants and animals to evolve into specific ecological niches.
- Few outside species could colonise this isolated land allowing **native organisms** to develop without hindrance.
- Australian rocks and soils remained relatively unchanged and **infertile**. Flora and fauna had to overcome the unique challenges imposed by this environment in order to survive.

The age of rocks and fossils can be determined using **radioactive dating**. Among the best known techniques are radiocarbon dating (half-life 5730 years), potassium–argon dating (half-life 1.3 billion years) and uranium–lead dating (half-life 4.5 billion years). These methods have provided considerable information about the ages of fossils and rocks. For example, we know that marsupials evolved relatively recently, with the oldest fossil dated to around 55 million years ago.

3 While most earthquakes occur at plate boundaries, shallow earthquakes can occur in the relatively stable interior of continents. These are less common and do not follow simple identifiable patterns. Australia sits on the Indo-Australian Plate which is being pushed north and is crashing into the Eurasian, Philippine and Pacific Plates. This pushing builds up **compressive pressure** in the interior of the Indo-Australian Plate. This pressure is liberated during earthquakes. In Australia there are on average 80 earthquakes (magnitude ≥ 3.0) each year, while magnitude ≥ 5.5 quakes occur every two years. Potentially catastrophic earthquakes (magnitude ≥ 6.0) occur around every five years.

Adelaide has the highest number of earthquakes for any Australian capital city. This is because South Australia is being gradually compressed in an east-west direction by around 0.1 mm each year. Australia's largest recorded earthquake occurred in 1941 in the remote and largely unpopulated area of Meeberrie, Western Australia. It had a probable magnitude of 7.2.

(cont.)

GEOLOGICAL HISTORY OF AUSTRALIA *(continued)*

Plate tectonics

There are only two **active volcanos** in Australian territory; one on Heard Island in the southern Indian Ocean, and the other on McDonald Island nearby. On the Australian mainland, the last active volcano erupted about 5000 years ago around Mount Gambier in South Australia. This volcano is now **dormant**, rather than extinct, so there is the potential for another eruption.

The following photo shows road damage caused by an earthquake.

Checklist

Can you:

1 *Use stratigraphy and the law of superposition to interpret geological histories?* ☐

2 *Outline the steps in the formation of the Australian continent, from Gondwana to the present day, and include radioactive dating?* ☐

3 *Comment on the nature of earthquakes and vulcanism in Australia?* ☐

GEOLOGICAL HISTORY OF AUSTRALIA

Plate tectonics

REVISION TEST

1 The stratigraphic columns in the following diagram were taken in the same area but 10 km apart from each other.

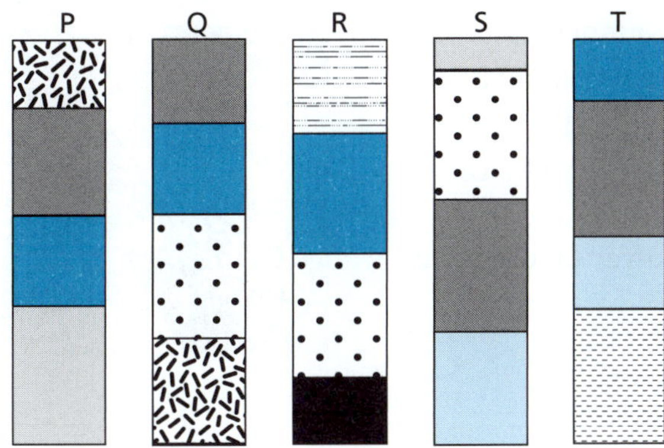

a By drawing lines between the columns, or otherwise, correlate (line up) the various layers. *Hint 1* (3 marks)

b Which column holds the:

i oldest rocks? (1 mark)

ii youngest rocks? (1 mark)

2 Rock layers were set down in an area, as shown by these stratigraphic columns.

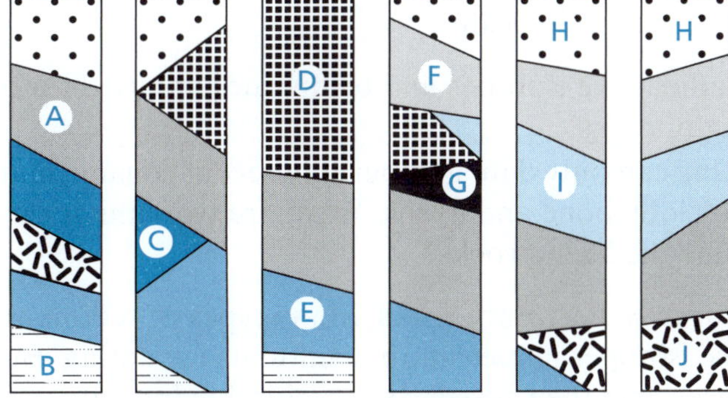

List the order of layers A to J from oldest to youngest. *Hint 2* (3 marks)

3 The diagram on the right shows a number of features, labelled A to E. List these features in order from oldest to youngest. (2 marks)

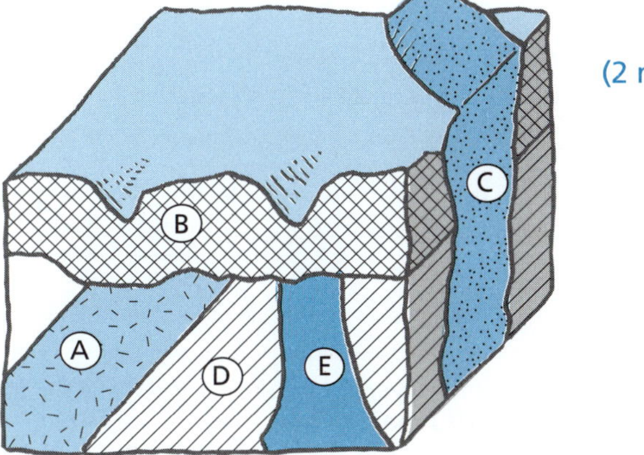

(cont.)

4 Consider the cross-section shown.

- **a** Which layer is the oldest? (1 mark)
- **b** Which layer is the youngest? (1 mark)
- **c** Which letter shows a volcanic dyke? (1 mark)
- **d** Which letter shows a fault? (1 mark)
- **e** Which occurred first: the dyke or the fault? (1 mark)
- **f** Did folding of the lower layers occur before or after the:
  - **i** dyke? (1 mark)
  - **ii** fault? (1 mark)
  - **iii** depositing of layer B? (1 mark)
- **g** Which letter shows rocks where fossils would *not* be found? *Hint 3* (1 mark)
- **h** What occurred after folding, but before B and A were laid down? (1 mark)
- **i** What is feature K most likely to be? (1 mark)

5 Australia is different from other continents, and even unique, both in its geology and its flora and fauna.

- **a** Suggest the two main reasons for the development of Australia's unique plant and animal life. (2 marks)
- **b** Why is Australia stable geologically? (2 marks)

6
- **a** Many different radioactive isotopes and techniques are used for dating. Explain the radioactive dating of rocks. (2 marks)
- **b** Radiocarbon dating can only date geologically recent organic materials, usually wood and charcoal, but also cloth, bone and horns. What are two limitations to using radiocarbon dating to determine the age of rocks? (2 marks)

7 Fossils typically occur in sedimentary rocks. Many types of sediment don't normally include the necessary radioactive isotopes in measurable amounts for dating. But scientists can accurately date sedimentary rocks, and their fossils, by considering igneous rocks around them. Suggest how this can be done. *Hint 4* (2 marks)

Hint 1: Look for similar layers.
Hint 2: Use the law of superposition.
Hint 3: In what kinds of rocks do you find fossils?
Hint 4: Use the law of superposition.

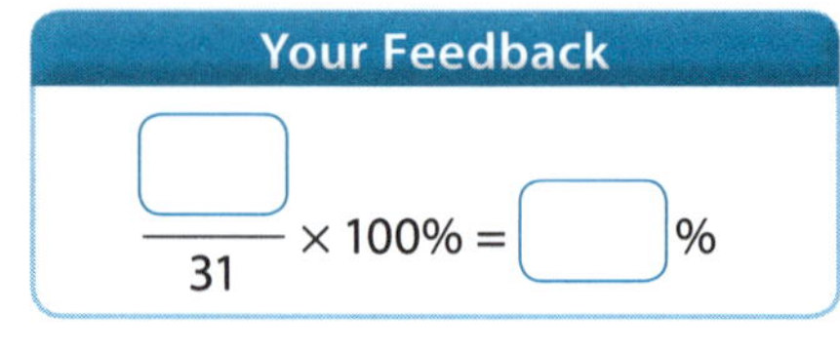

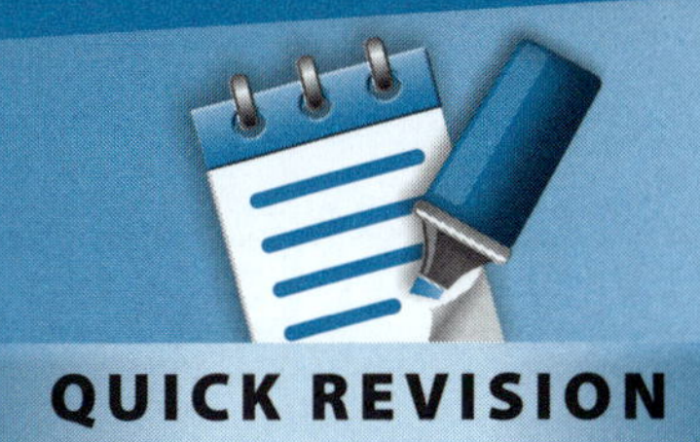

HEAT ENERGY

Energy on the move

1 When matter is heated it __________ and occupies a larger space. This property can be used to measure the __________ of an object using a thermometer. A typical thermometer consists of a small, thin-walled glass __________ of liquid connected to a thin __________ tube. As the liquid in the bulb is heated it expands and moves __________ the capillary tube. The tube is marked at regular intervals with a temperature scale, such as in Celsius degrees. The 0 °C line is marked on the scale. This is the temperature of a mixture of __________ and ice water, and is called the lower fixed point. The upper fixed point is 100 °C and represents the temperature at which water __________ and at which liquid water and water vapour coexist. Most school thermometers contain a dyed __________ and the scale is typically from –10 °C to +110 °C. Some thermometers contain __________. These thermometers are used for more accurate measurements because mercury expands more evenly on heating and has little attraction for the glass.

The particles of a hot body have __________ kinetic energy than the particles in a cold body. In a solid, the particles increase their __________ kinetic energy when they are heated but do not move from one place to another. In liquids and gases, the particles increase their kinetic energy by moving faster from one __________ to another, and their vibrational and __________ motions also increase on heating. The greater the __________ of a body the greater amount of heat it contains. A 2 kg block of steel at 60 °C contains more __________ energy that a 1 kg block of steel at the same temperature. Equal masses of different materials at the same temperature contain __________ amounts of heat energy. For example, 1 kg of water at 25 °C contains more heat energy than 1 kg of steel at the same temperature. Heat energy is measured in joules (J).

2 If a hot object is in contact with a __________ object, heat will flow from the higher temperature object to the lower temperature object. If you place a hot water bottle on a sore muscle, the heat travels from the __________ into the muscle. If you were to apply ice to a sprained muscle, the heat from the muscle would flow __________ the block of ice and melt it. Heat energy flows through matter by hot particles __________ with cooler particles and transferring some of their energy. This is called conduction. The particles in solids are __________ together, so conduction of heat occurs more readily in solids than in liquids or gases. Solids vary in their ability to conduct heat energy. Metals are the __________ heat conductors as they contain __________ electrons that can transfer heat more rapidly than the vibrating atoms. Silver and copper are examples of __________ heat conductors. Copper can be used on the bases of saucepans as it conducts heat quickly to the contents of the saucepan. If you stir a hot pot of soup with a metal spoon, the spoon soon becomes too __________ to handle.

Non-metals have no mobile electrons and so they are __________ heat conductors. They are called insulators. Compounds, such as plastics, are also insulators. Most saucepans have plastic handles to prevent __________ to your hands. A hot pot of soup can be stirred safely with a wooden spoon as the wood is a __________ heat conductor, so will not burn your skin. Air is a good insulator. In winter we wear thick clothes, often made of wool or synthetic fibres, as they trap __________ between the fibres and act as a barrier between our skin and the cold air outside.

3 Heat conduction in liquids and gases is quite __________. The exception is liquid mercury which is a __________ and therefore a good conductor. Liquids and gases mainly transfer heat

(cont.)

by a process called __________. If you have a convection heater in a room, heat is transferred __________ the room by a convection current. The air is heated by the heater and, as warm air is __________ dense than cold air, rises upwards towards the ceiling pushing the colder (more __________) air aside. This colder air sinks towards the floor and moves towards the heater where it is warmed. This continued displacement of cold air by warm air causes the air in the room to __________ up. Similar convection currents exist in the atmosphere and the oceans. The Sun's energy heats up the land surface __________ than the water surface. The air over the land becomes hotter and less dense, and so __________ upwards. This air is replaced by cooler, denser air from __________ the sea. The convection current that is established brings __________ breezes onto the land. Convection currents are also established in a beaker of water that is heated over a Bunsen burner. The heated glass __________ heat to the water at the base of the beaker. The warm water rises and displaces the cooler water at the top which __________ downwards. The circulation of the water in this way distributes the heat throughout the beaker.

4 The Earth receives large amounts of energy from the Sun. The energy produced by the Sun is called __________ energy and it consists of electromagnetic waves. These waves have the ability to travel through space as they do not need a __________ in which to travel. These waves are then __________ by the atmosphere, hydrosphere and the lithosphere. The absorbed radiation is converted into __________ energy. The highest energy waves are gamma and X-rays and these are absorbed by the atmosphere. Ultraviolet and infra-red rays are the main forms of radiant __________ that heat the planet. Objects vary in the extent to which they absorb radiant energy. Silvery coloured materials reflect radiant energy more than black materials which __________ radiant energy quite readily.

Radiant energy is also produced by fires and electric radiators. Bushfires produce considerable radiant energy, which travels out as waves at the __________ of light. These waves can cause trees to burn even though the flames from the fire have not reached them. The wires in the element of an electric radiator get hot as the electricity flows through them. Red-hot wires emit large amounts of __________ radiation which travels rapidly to heat the people and objects in the room. When the infra-red rays are absorbed they are converted to __________ energy.

Answers **1** expands; temperature; bulb; capillary; up; ice; boils; alcohol; mercury; greater; vibrational; place; rotational; mass; heat; different **2** colder; water; into; colliding; close; best; mobile; good; hot; poor; burns; air; poor **3** slow; metal; convection; around; less; dense; heat; faster; rises (moves); over; cool; conducts; moves **4** radiant; medium (substance); absorbed; heat; energy; absorb; speed; infra-red; heat

HEAT ENERGY

Energy on the move

REVISION SUMMARIES

1 Bulb thermometers measure the temperature of matter using the principle that liquids, such as mercury or alcohol, **expand** when heated. These thermometers contain liquid in a bulb that has a very thin glass wall to allow rapid **heat exchange**. Connected to the bulb is a fine capillary tube that the liquid expands into as it is heated. A temperature scale is printed on the capillary tube. In most school thermometers the temperature range on this scale is –10 °C to +110 °C. The **Celsius scale** of temperature is created by setting 0 °C as the temperature of melting ice and 100 °C as the boiling point of water. The scale is then divided into 100 equal divisions. In a mercury thermometer the meniscus is convex (∩) whereas in an alcohol thermometer the meniscus is concave (∪). The top of a mercury meniscus or bottom of the alcohol meniscus is used to measure the temperature.

The temperature of an object is a measure of the average **kinetic energy** of the particles in that body. In a gas, the molecules are in constant motion. As the gas is heated, the kinetic energy of the molecules increases. When a thermometer is inserted in the heated gas, the molecules of gas collide with the bulb and transfer some of their energy to the thermometer. This energy transfer heats the liquid in the bulb, which expands and moves up the capillary, registering a higher temperature.

Heat is a form of energy. The particles of a hot body have greater energy than the particles of a cold body. As matter is heated its particles move faster. **Molecular vibrations** and rotations increase and, if the sample is a liquid or gas, the particles move from one place to another with greater speed. In solids, the particles vibrate but do not move away from their positions in the solid crystal. The total heat content of a body depends on its mass as well as its temperature. This can be shown by a simple experiment. If samples of brass (A, B and C) of different mass are heated to 100 °C in boiling water and then each is added to 100 mL of cold water at 20 °C in different beakers, then the rise in temperature of the water is greatest for the largest mass of hot brass. The greater the mass of a body the greater its heat content at a fixed temperature. This experiment is summarised in the following diagram.

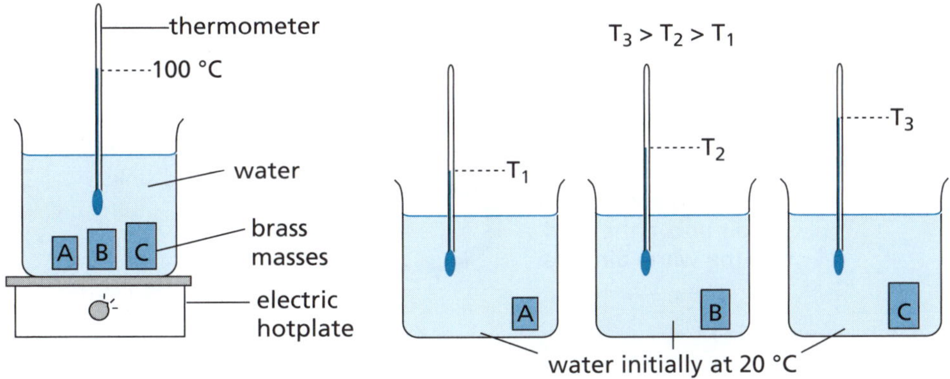

2 Heat energy can be **transferred** from one place to another. Heat always flows from a region of high temperature to a region of lower temperature. The transference of heat through matter is called **conduction**. Different materials conduct heat at different rates. **Metals**, for example, are good heat conductors and some metals are better heat conductors than others. **Non-metals** are poor heat conductors. Hot atoms vibrate faster than cooler atoms and transfer their energy by colliding with neighbouring atoms. In metals the heat is conducted by collisions between electrons as well as vibrating atoms, as electrons in metals are not locked into one position and

(cont.)

they are relatively free to move. In non-metals there are no mobile electrons, so heat conduction is due only to energy transfer from the vibrating atoms. Poor heat conductors are called **insulators**. Glass and plastic are insulators.

Gases and molecular liquids are also insulators. The air is a good thermal insulator. Blankets and woollen clothes, for example, trap air between their fibres, reducing the heat loss from a person so they stay warm. Metal saucepans have handles that are made of insulators. This prevents burns from the hot metal. The following diagram shows the results of an experiment on heat conduction on three metals and glass. In this experiment, silver is the best conductor of heat while glass, an insulator, is the poorest.

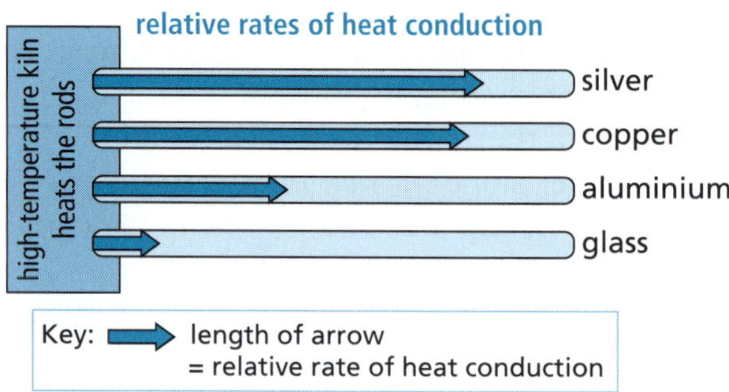

3 Another method of heat transfer is **convection**. This process allows heat transfer in liquids and gases. When matter is heated it expands. Its density decreases because the same mass of matter occupies a larger space. Warm air, for example, is **less dense** than cold air. If there are differences in air temperature from one place to another then convection currents are established. These convection currents transfer heat and matter from one place to another. The following diagram shows how a convection current is established between the air over the land and the sea. These convection currents in the atmosphere are responsible for cooling sea breezes on hot days.

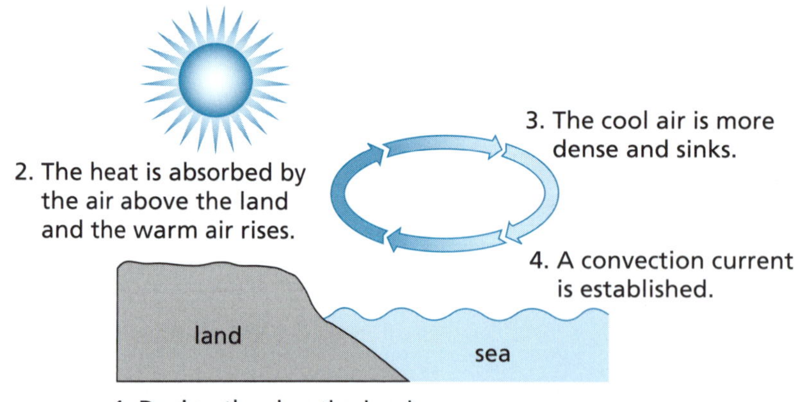

Convection currents also transfer heat in liquids. For example, warm ocean currents from the equatorial oceans move towards cold polar oceans and, in turn, colder currents move back. The cold currents move deep in the ocean. Warmer water is less dense, so the warmer currents are near the surface.

4 Heat transfer by conduction and convection involves the motion of particles of matter. A third method of heat transfer does not involve particle movement. **Radiant energy** is transferred by electromagnetic waves or **electromagnetic radiation**. This type of wave does not need a medium in which to travel. There are many different types of radiant energy. These include ultra-violet and infra-red radiation. The Sun emits vast amounts of radiant energy which travels to the Earth through space. This radiant energy is absorbed by the Earth and converted into heat energy. The atmosphere and the oceans absorb large amounts of this radiant energy. Life depends on the warming from this radiant energy. Objects of different colours absorb radiant energy to different extents. Silvery coloured objects **reflect** more radiant energy than they absorb, while black objects absorb large amounts of radiant energy. Hot black objects also cool down faster than hot silvery or white objects.

When you use an electric radiator in the winter, the hot wires emit infra-red radiation. The infra-red rays are absorbed by your skin and converted to heat energy. The electric radiator also heats air molecules near the hot wires and convection currents are formed that warm the air in the whole room. The following photo shows an electric radiator in use. Both visible light and infra-red radiation is produced.

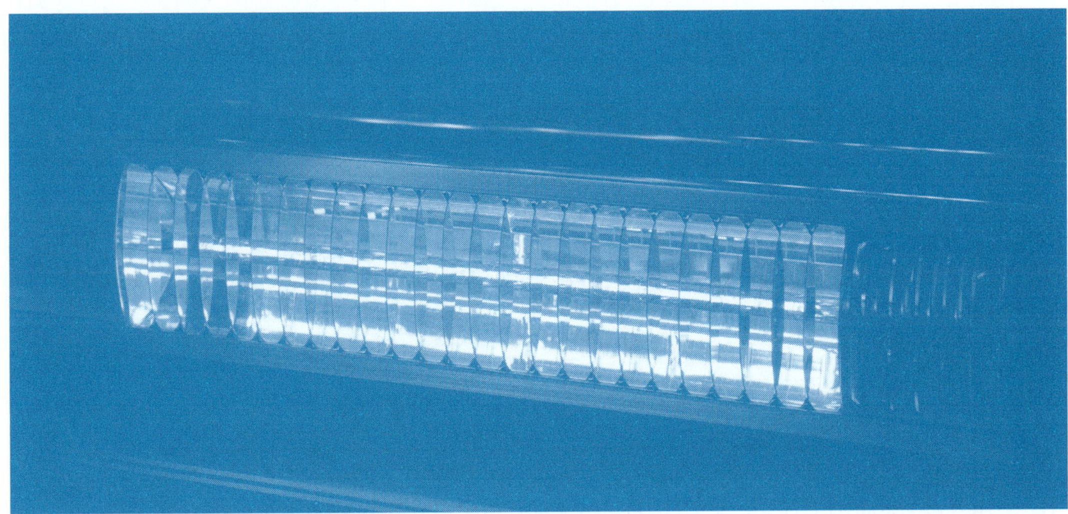

Checklist

Can you:

1 *Explain the difference between heat and temperature?* ☐
2 *Describe how heat energy is transferred by conduction and identify common examples of conductors and insulators?* ☐
3 *Explain how heat can be transferred by convection?* ☐
4 *Explain how heat is transferred by radiant energy?* ☐

HEAT ENERGY

REVISION TEST

1 Heat lamps are often used when you have sore back muscles as shown in the following diagram.

a The heat lamp emits radiant energy (X). Identify this type of radiant energy. (1 mark)

b Identify the type of energy that the radiant energy is converted to on the skin at Y. (1 mark)

c If the heat lamp is replaced by a hot water bottle placed in direct contact with the skin, describe how the heat is transferred to the sore back muscles. (1 mark)

2 Kites fly well near the coast. The following diagram shows a convection current which is formed along the coast.

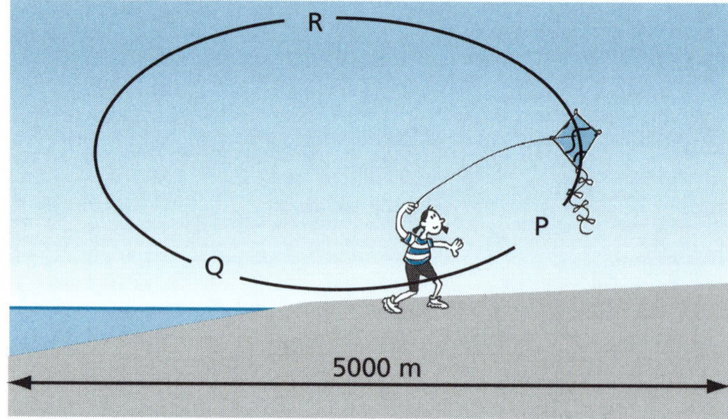

a In which direction (PQR or PRQ) does the convection current rotate? (1 mark)

b Explain why the convection current rotates in the direction selected in part a. (3 marks)

c Relate your answer in part a to kite flying. *Hint 1* (1 mark)

3 Complete the following table by first selecting the way that heat travels from the following list: (3 marks)

- travels as invisible rays
- particles move from one place to another
- particles vibrate and bump into their neighbours.

Then indicate if heat can travel through empty space. (3 marks)

	Conduction	Convection	Radiation
How heat travels			
Can heat travel through empty space?			

4 The following diagram shows the results of an experiment to compare the relative heat conductivity of four rods made of different materials. As the heat moves along the rod it will melt the paraffin grease and allow small pieces of chalk to fall onto the bench. Initially there are five chalk pieces stuck to each rod with the paraffin grease.

a Rank the four solids (A, B, C and D) in order of decreasing heat conductivity. (1 mark)

b Which rod is most likely to be an insulator? (1 mark)

c If the best conductor in the experiment is copper, explain how heat is conducted through copper. *Hint 2* (2 marks)

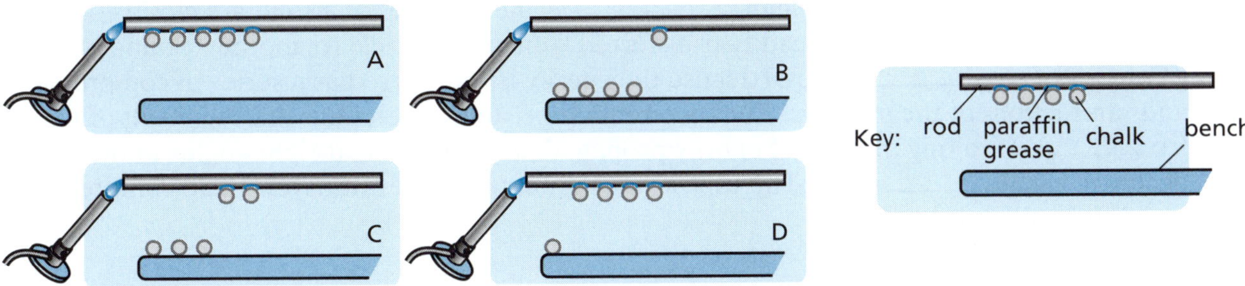

5 While out in the sports field on a summer day, should you wear bright, shiny clothes, or dull, dark ones? Should the same choice be made on a cold day in winter? (2 marks)

6 A thermos flask is often used to store hot or cold liquids over long periods of time. Two of the features of the thermos flask are as follows.

- There is a vacuum between the double glass walls.
- The glass walls have silvered surfaces.

Explain why these features help to reduce the rate of heat transfer. *Hint 3* (2 marks)

7 The following diagram shows part of the scale of an alcohol-in-glass thermometer.

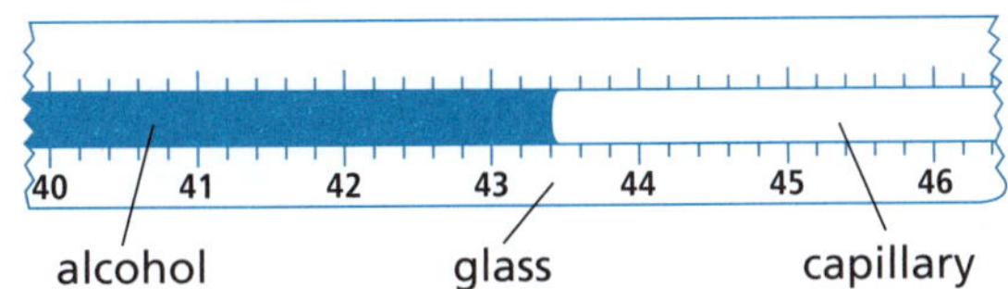

a What name is given to the curved surface of the alcohol? (1 mark)

b What is the temperature shown on the thermometer? (1 mark)

Hint 1: In kite flying air currents must be able to lift the kite upwards.
Hint 2: Copper contains atoms and mobile electrons.
Hint 3: Silver-coloured objects are good reflectors.

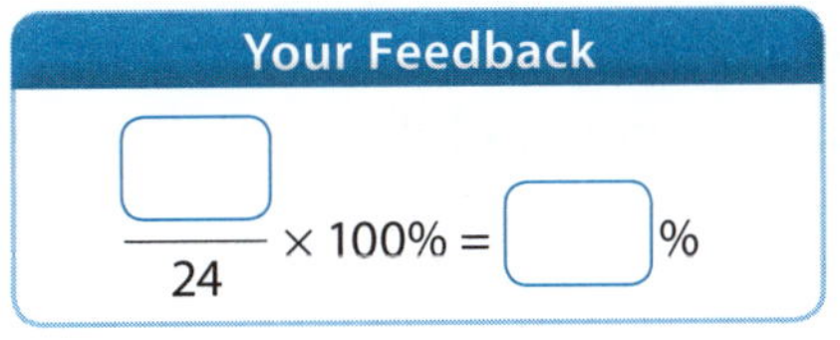

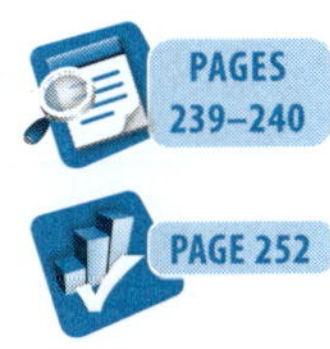

PAGES 239–240
PAGE 252

SOUND ENERGY

Energy on the move

QUICK REVISION

1 The waves on the ocean are caused by __________ blowing on the water surface. The __________ energy of the wind is transferred to water molecules which oscillate up and down. Because of the attraction of water molecules for each other this energy is transmitted to other water molecules nearby. Waves form as this __________ is transferred. The waves are classified as transverse waves because the oscillations are at __________ angles to the direction of energy propagation.

Sound waves, along with water waves, are classified as __________ waves as they need a medium through which to travel. Sound waves can travel through solids, liquids and gases. Sound waves travel fastest through __________ as the particles in a solid are closer together and the wave energy is more readily transferred from one particle to another. Sound waves are examples of __________ waves, because the energy is transferred as a series of compressions and rarefactions of the medium. These compressions and rarefactions are the result of to-and-fro __________ along the axis of energy propagation rather than at right angles to the axis as occurs in a __________ wave. The following diagram compares transverse and compressional (longitudinal) waves.

2 Some sound waves have longer wavelengths than others. The wavelength of a compression wave is the __________ between the centres of adjacent compressions. Sound waves also vary in their frequency. The frequency of a wave is the __________ of waves that pass a fixed point each second, and is measured in hertz (Hz). The speed or velocity of a sound wave depends on the __________ and the temperature. Sound waves travel __________ as the temperature increases. The wave equation ($v = f\lambda$) relates the velocity of a wave to its __________ and wavelength.

Sound waves __________ off surfaces. Some surfaces, like bare walls and floors, reflect a __________ amount of the sound energy. Rooms containing curtains and carpeting reflect __________ sound energy because large amounts of sound energy are absorbed. Music studios have walls that __________ most of the sound energy so as to avoid echoes. Sound reflection

SOUND ENERGY

Energy on the move

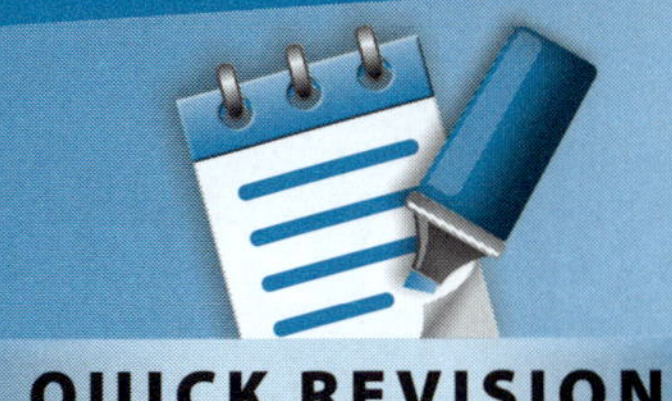

QUICK REVISION

can be of use in a variety of ways. Boats may use __________ soundings to determine the depth of water under the boat or to detect schools of fish. This technology is called __________. In medicine, ultrasounds can be used to image unborn babies to determine if there are any medical problems.

3 Musical instruments can be played loudly or softly. The loudness of the sound is a measure of the __________ of the sound wave. When a piano is played the keys can be struck with great force to produce a __________ note as the piano wire vibrates and produces a transverse wave with __________ amplitude. The amplitude of a transverse wave in a guitar or piano wire is the __________ of the wave in the wire when measured from the median position of the wire. In a sound wave in air, however, the amplitude of the vibration refers to the __________ that the air molecules are displaced from their median positions. Quiet sounds are ones of lower energy and __________ amplitudes than loud sounds.

The pitch of a sound is different to its loudness. The pitch is related to the __________ of the sound wave. Sopranos sing with a much __________ pitch than bass singers as their vocal cords are thinner and more taut. A high pitched sound is caused by waves of higher frequency. Consider tuning forks that are used to tune various musical instruments. The tuning fork labelled 256 Hz produces a __________ pitched sound than the fork labelled 512 Hz.

4 Very loud sounds can cause __________ to our hearing. Machinery in factories can cause noise pollution, and workers in such factories must have their hearing protected. The loudness of a sound is measured in units called __________ (dB). The zero on this noise scale is set at the limit of __________ hearing and noises greater than 90 dB are known to cause hearing damage. As the energy of a sound wave __________ with distance the noise becomes less. Even at a distance of 30 m from a jet engine the noise level is still 140 dB.

5 Our ears are designed to capture sound waves and relay the information to the brain. Sounds that enter the ear __________ cause the ear drum to __________. On the inside of the ear drum are small bones called __________ that transmit the sound energy into a structure called the cochlea. Inside the cochlea is fluid and fine __________ that transmit the energy to the __________ nerve which converts the sound wave into an __________ signal that travels to the brain. Injury to any of these structures causes degrees of deafness. Loud music, for example, can damage the fine hairs in the __________ and lead to deafness over extended periods of exposure.

Cochlear implants have been developed by Australian scientists. These help deaf people to hear, but they still require a working auditory __________. A microphone and __________ are worn behind the ear. The sounds that it detects and transmits are detected by the implant inside the head. The implanted receiver __________ the signals and causes the cochlea to be stimulated by implanted electrodes. The signal is then passed onto the auditory nerve and then the __________.

Answers **1** winds; kinetic; energy; right; mechanical; solids; compression; oscillations; transverse **2** distance; number; medium; faster; frequency; reflect; large; less; absorb; echo; sonar **3** energy; loud; large; height; distance; smaller; frequency; higher; lower **4** damage; decibels; human; decreases **5** canal; vibrate; ossicles; hairs; auditory; electrical; cochlea; nerve; transmitter; decodes; brain

SOUND ENERGY

Energy on the move

REVISION SUMMARIES

1 Energy can be transferred from one place to another in a variety of ways. One way is through **mechanical waves**. These waves require a **medium** (such as a solid, liquid or gas) in which to move. Water waves are mechanical waves that are formed when wind pushes on the surface of the water and transfers **kinetic energy** to the water particles. The water particles start to oscillate up and down and some of this energy is transmitted to neighbouring water molecules. This creates waves which move outwards across the water surface. Water waves are called **transverse waves** as the water particles oscillate at right angles to the direction of movement of the wave.

Sound is another example of a mechanical wave. When you play a guitar, the strings vibrate as they are struck. This vibration is transmitted to the surrounding air molecules which begin to vibrate. Initially the air molecules bunch up. This is called a **compression**. Then the air molecules spread out again. This is called a **rarefaction**. As they collide with other air molecules a sound wave is produced which travels outwards as a **compression wave**. Unlike a transverse wave, the particles in a compression wave vibrate to-and-fro along the same axis as the direction of the wave motion. Your ears detect the sound wave when the air particles strike your ear drum. The following diagram shows the production of compressions and rarefactions in air as a tuning fork vibrates.

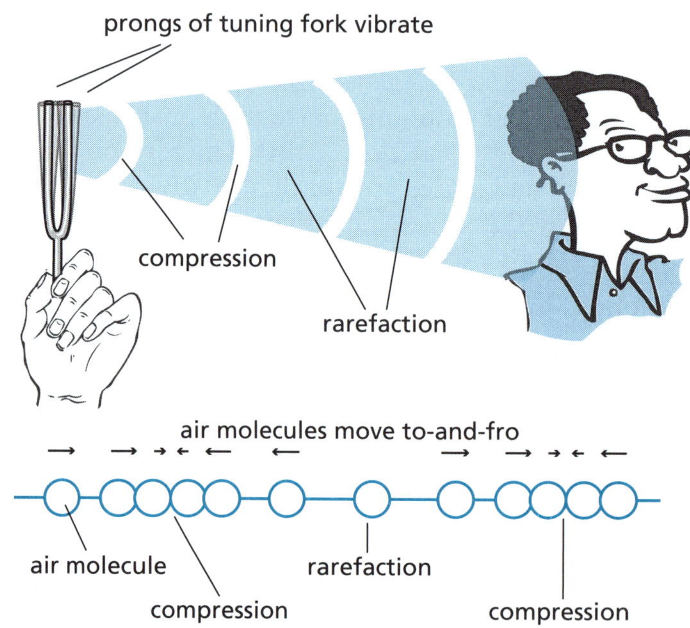

2 Sound waves have various properties which can be measured.

- **Wavelength** (λ, the Greek letter lambda). The wavelength of a sound wave is the distance between the centres of two adjacent compressions, or two adjacent rarefactions, or just two adjacent points undergoing identical motion. The unit of wavelength is the metre (m).
- **Frequency** (f). The frequency of a sound wave is the number of waves that pass a fixed point in one second. The unit of frequency is the hertz (Hz).
- **Velocity** (v). The velocity (or speed) of a sound wave is the distance travelled in one second. The unit of velocity is metres per second (m/s).

These three properties are related via an equation called the **wave equation**:

$$v = f\lambda$$

The wave equation shows us that the wavelength of a sound wave depends on both its speed and its frequency. Decreasing the frequency of a sound will increase the wavelength at constant velocity.

The speed of sound varies with temperature and the medium through which the sound wave travels. The speed of sound increases as the temperature increases. In dry air sound travels at 331 m/s at 0 °C and at 20 °C its speed is 343 m/s. Sound waves also travel much faster in solids than in liquids and they travel faster in liquids than in gases. In water at 20 °C, sound waves travel at 1484 m/s. Sound travels through granite rock at 6000 m/s at 20 °C. Sound waves travel faster in solids because the particles are very close together and vibrational energy is transferred rapidly.

Sound waves can **reflect** off various surfaces, producing an echo. An echo is undesirable in a music auditorium so the walls are made from sound absorbent materials, or are hung with heavy drapes. **Sonar** location, however, makes practical use of echoes. Boats can send sound waves downwards into the water to determine the depth or to locate fish. Bats use the same principles to avoid flying into objects at night. Bats use ultrasonic waves that are not detected by the human ear. Dolphins and some whales use a similar process in echolocation, where they use echoes to locate and identify objects. Ultrasound is also used during a human pregnancy to check the progress of a developing foetus.

3 Sound waves can vary in their **loudness** and their **pitch**. The loudness of a sound is related to the amount of **energy** of the sound waves. Loud sounds have more energy than soft or quiet sounds. If you strum guitar strings with great force the sound produced is loud. Gentle plucking of the strings produces a quiet sound. With wind instruments, such as trumpets and pipe organs, the energy input determines the loudness of the sound. If air is blown through the tubes with great force then a loud sound is produced. In these cases high-density **compressions** are formed inside the tube. Quiet sounds are produced by low-density compression waves moving down the tube. The loudness of a sound wave is related to the **amplitude** of vibration of air molecules. The amplitude of a compression wave refers to the distance that an air molecule is displaced from its median position. The greater the amplitude of the wave the greater is the loudness of the sound. This principle is shown in the following diagram.

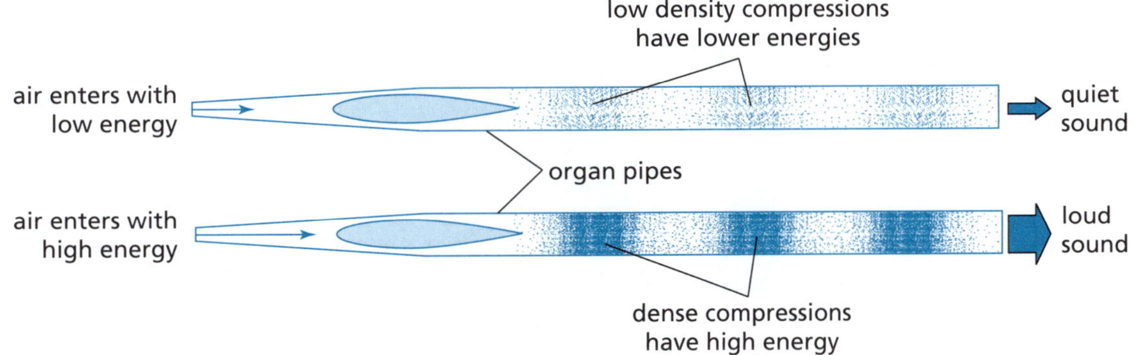

The **pitch** of a sound wave is related to its **frequency**. The higher the frequency, the higher is the pitch of the sound. On a guitar some strings are thicker than others. The thinner strings are used to play higher pitched notes. The pitch can also be increased by tightening the strings. All the instruments in an orchestra have to be in tune with one another. This is usually done to the note A on a piano, oboe or tuning fork.

(cont.)

High-pitched sound waves have more waves passing per second past a fixed point. In air at a fixed temperature, the high-pitched waves will have a shorter wavelength than lower pitched waves. This principle is shown in the diagram where different tuning forks are vibrating.

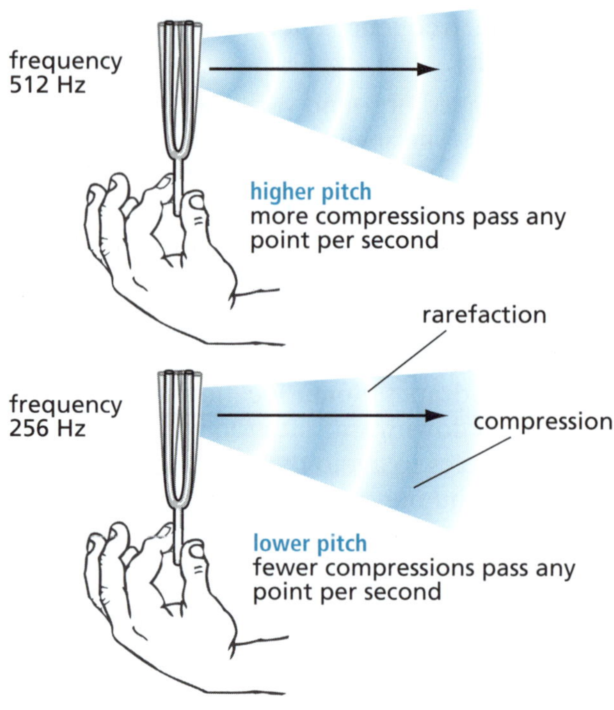

4 **Noise pollution** is a problem in many workplaces. Many machines produce loud sounds during their operation. Jet aircraft and some music concerts also produce very loud sounds. These sounds can damage our hearing unless protective measures are taken. Different levels of permanent **deafness** can result. The loudness of a sound is measured in a unit called the decibel (dB). On the decibel scale, a sound that can only just be heard has a loudness of 0 dB. Sounds with a loudness greater than 90 dB can cause hearing damage with repeated exposures. Loudness decreases with distance from a sound source, so the loudness is often quoted at a specific distance. For example, a whisper at 1 m has a loudness of about 25 dB and a jet engine at 30 m has a loudness of 140 dB. Listening to loud music through ear phones can cause hearing damage because the sound energy is completely directed into the ear canal.

5 The ear allows us to hear sound waves. The structure of the ear is shown in the following diagram.

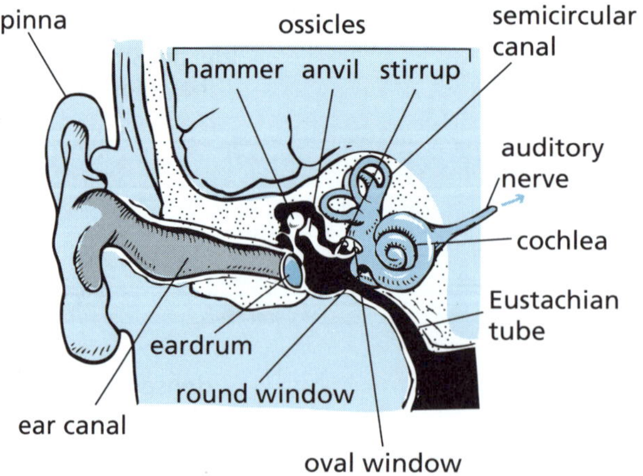

Sound waves are collected by the pinna and funnelled down the **ear canal** towards the **eardrum** which then starts to vibrate. There are three small bones (or **ossicles**) connected to the eardrum which move in response to the eardrum vibrations. This transfers sound to the oval window of the **cochlea**. The cochlea contains fluid and very fine hair cells that move in response to the sound waves. This movement carries the sound to the **auditory nerve** which is connected to

the brain. Damage to any of these structures can lead to partial or complete deafness. If the ossicles are damaged then conductive deafness results, as sound vibrations cannot reach the oval window. Loud noises can permanently damage the fine hair cells in the cochlea and prevent sounds reaching the auditory nerve.

Cochlear implants were the first implanted electronic devices developed to help deaf people to hear. They were developed by Australian scientist Graeme Clark and his team in Melbourne in 1978. Further developments in electronics led to great improvements in these implants over the following 20 years. These implants use a microphone (worn behind the ear) to collect sound waves. The sound is converted into an electrical wave which is then converted by an external electronic speech microprocessor into encoded waves. These waves are transmitted to an implanted receiver that decodes the signals and stimulates the cochlea through implanted electrodes. The cochlea then passes the signal to the auditory nerve.

The following illustration shows the typical features of a cochlear implant.

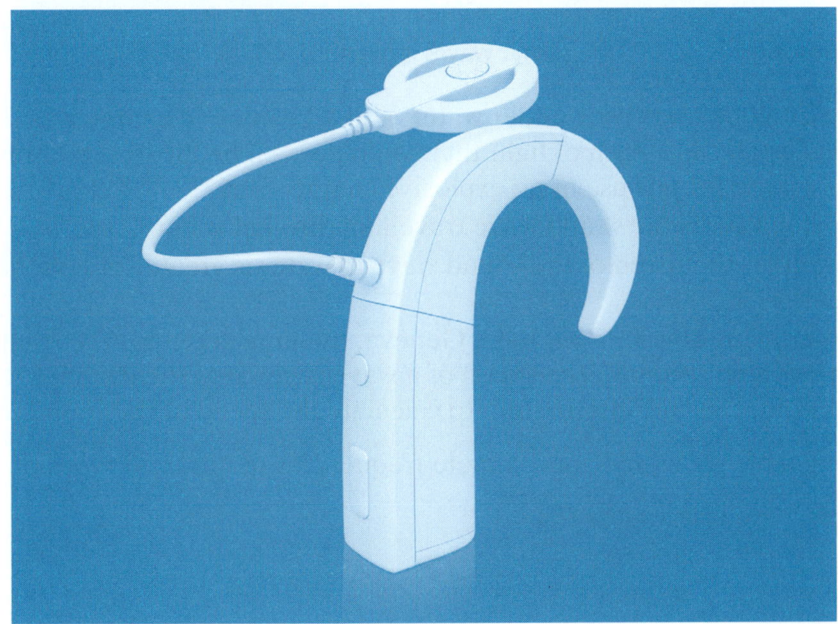

Checklist

Can you:

1. *Explain how sound waves are transmitted through matter?* ☐
2. *Explain the key properties of sound waves (frequency, speed and wavelength) and explain how sound wave reflection can be useful?* ☐
3. *Distinguish between the loudness and pitch of a sound wave?* ☐
4. *Explain the damage to our hearing caused by loud sounds?* ☐
5. *Explain how our ear allows us to hear and how cochlear implants can help deaf people to hear?* ☐

SOUND ENERGY

Energy on the move

REVISION TEST

1 The following diagram shows an experiment that was first performed in Switzerland in 1826, to measure the speed of sound in water on Lake Geneva. A sound wave was created by ringing a bell under water. At the same time, the observer in a boat 13.5 km away started a clock. The time for the sound wave to reach the observer was recorded.

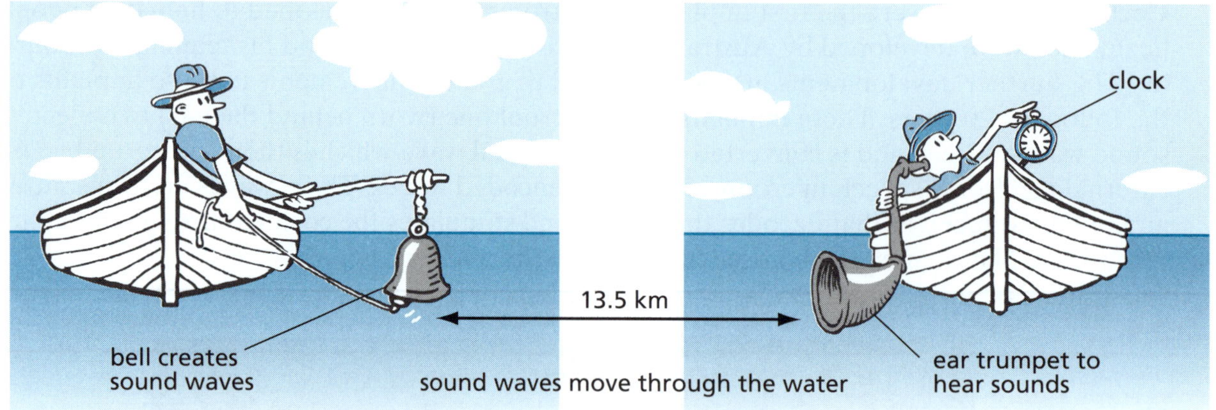

- **a** At the time the bell was rung under water, a small explosive charge was set off on the same boat which produced a flash of light that could be seen by the observer on the second boat. Suggest a reason for this use of an explosive charge. *Hint 1* (1 mark)
- **b** What type of wave is the sound wave travelling through water? (1 mark)
- **c** Predict whether the speed of the sound wave in water is greater or less than the speed of the sound in air. (1 mark)
- **d** In the experiment described on Lake Geneva, the time taken for the sound wave to reach the observer was nine seconds. The speed of a wave is equal to the distance it travels divided by the time taken. Use this information to calculate the speed of the sound in water. (2 marks)

2 The following table shows the approximate frequencies of different types of singing voices.

Singing voice	Frequency range (Hz)
alto	200–600
bass	80–400
soprano	250–1350
tenor	170–500

- **a** Rearrange the data in the table so it is in order of increasing frequency. (1 mark)
- **b** The pitch of a note depends on the thickness of the vocal cords as well as the tension of the cords. Which of the tabulated singing voices would have the thickest vocal cords? *Hint 2* (1 mark)

3 True or false?

- **a** The ear ossicles are composed of two bones. (1 mark)
- **b** The cochlea of the ear transmits sound waves to the auditory nerve. (1 mark)
- **c** Damage to the fine hairs in the cochlea can lead to deafness. (1 mark)
- **d** The cochlear implant can help people hear even if the auditory nerve does not work. (1 mark)
- **e** The eardrum vibrates when sound waves travel down the ear canal. (1 mark)

4 The following photo shows a tuning fork being used to tune a piano.

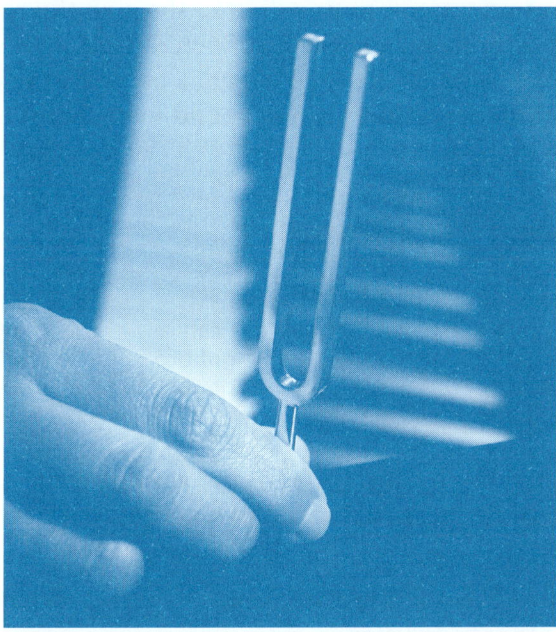

A 256 Hz tuning fork was struck and a sound wave was produced. If the speed of the sound wave in air was 350 m/s, use the wave equation to calculate the wavelength of the sound wave.
(2 marks)

5 The following diagram shows an experiment involving sounds produced by a radio. The radio was turned on and set on a music station. The radio was placed under a glass bell jar. The air inside the jar could be slowly removed using a vacuum pump.

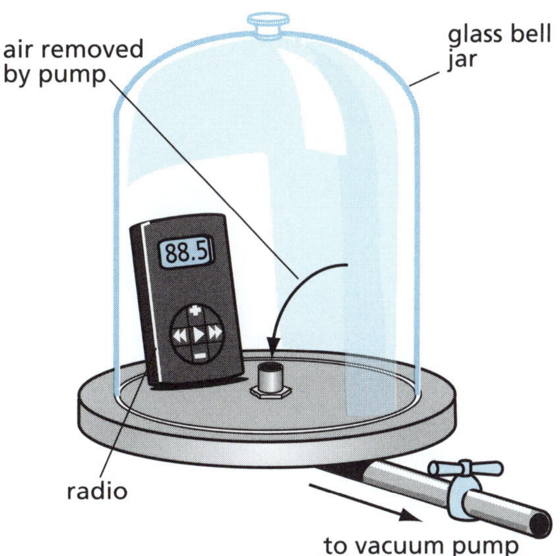

a As the air is removed from the bell jar the loudness of the sound produced by the radio decreases until no sound can be heard. Explain why this occurs. (1 mark)

b How does this experiment illustrate that sound waves are mechanical waves? (1 mark)

(cont.)

6 The following diagram shows a seismic survey carried out by geologists. An explosion is set off at the surface and the sound waves are transmitted through the rocks below. When the sound waves meet a new rock layer, some energy is reflected back to the surface. Reflections of these sound waves are detected by receivers. The distance between the explosion and the receivers can be varied. Information about the speed of sound in different rock types is also provided.

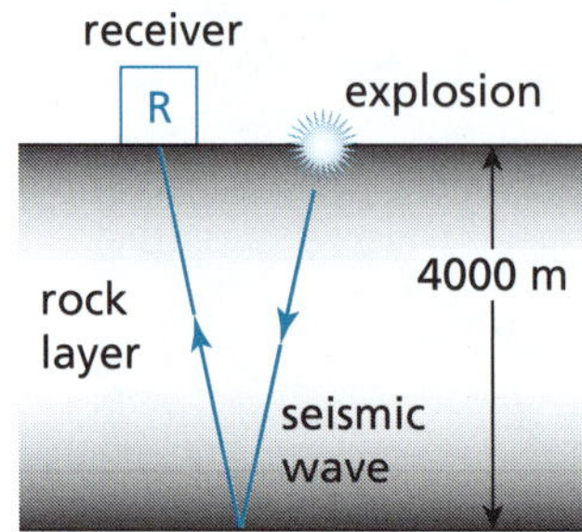

Rock	Speed of sound (m/s)
granite	6000
gabbro	7000
peridotite	8000

In a series of experiments the shortest time for the seismic wave to be detected following the explosion was 1.14 s. Use the information in the diagram to determine what type of rock layer is present. *Hint 3* (3 marks)

7 The following table shows the noise levels inside a suburban house throughout a day.

Time	5.00	6.00	7.00	8.00	9.00	10.00	11.00	12.00	13.00	14.00	15.00	16.00	17.00	18.00
Noise level (dB)	30	35	45	50	65	70	70	55	50	60	75	75	50	60

a Plot this data as a line graph and draw the line of best fit. (5 marks)

b i When was the quietest time in the house? (1 mark)

ii When was the noisiest time? (1 mark)

iii What could have produced the increase in noise in the middle to late afternoon? (1 mark)

c Are the noise levels registered in the house at a level that could cause hearing damage? (1 mark)

Hint 1: Compare the speed of sound to the speed of light.

Hint 2: Thick strings on a guitar play notes that are low in pitch.

Hint 3: The shortest time will occur if the seismic waves travel vertically down and up.

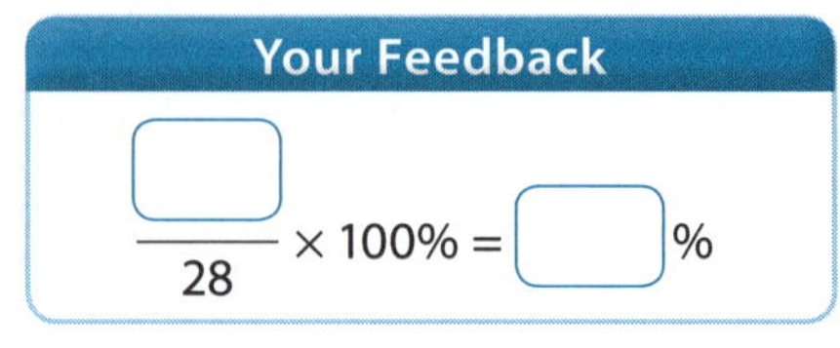

LIGHT ENERGY AND COMMUNICATIONS

Energy on the move

QUICK REVISION

1 Electromagnetic radiation includes visible ____________, radio waves, gamma rays and X-____________, where electric and ____________ fields vary simultaneously. Electromagnetic radiation carries ____________ that travels from its source and spreads out as it goes. The electromagnetic spectrum refers to the range of this type of radiation. Hotter, more energetic objects and events create ____________ energy radiation than cooler objects. Only extremely hot objects or particles moving very fast can create high-energy radiation like X-rays and ____________ rays. The following diagram shows the various radiation types.

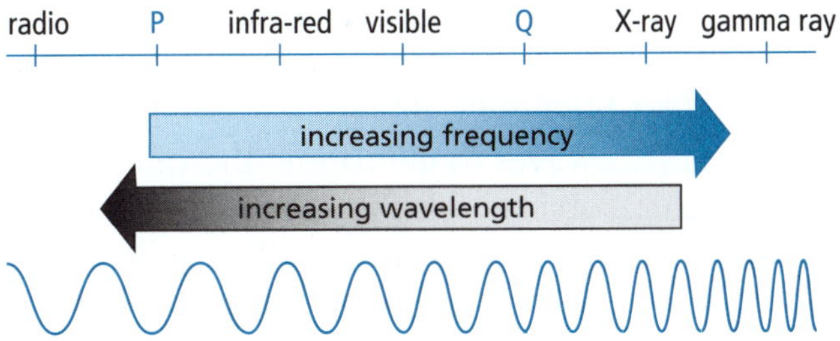

Two labels are missing—P is ____________ and Q is ____________. The greater the frequency, the ____________ is the energy that the wave carries. The ____________ spectrum—which we can detect with our eyes—makes up only a small portion of the electromagnetic spectrum. The colours of light (red, ____________, yellow, green, ____________ and violet) have wavelengths between 700 and 400 nm (1 nm = 10^{-9} m = 1 ____________ of a millimetre).

Electromagnetic radiation doesn't need a medium to travel through, so can travel through a ____________. Electromagnetic radiation crosses the vast emptiness of space to reach us from the Sun. In a vacuum all electromagnetic radiation travels at 300 000 km/s. This is called the ____________ of light.

2 Many objects are ____________ and do not allow light to travelling through them. A brick wall is opaque. But it does ____________ light. This is how we see it is there. Some materials allow light to pass through it, and are said to be ____________. Clear plastic film and a glass window are examples. Even though light passes through them, there still is some reflection—this is how we can see that these materials are there. There are other materials that allow light to pass through but scatter it so that people and objects on the opposite side are not clearly ____________. Frosted window glass is an example of this ____________ material.

Refraction is the ____________ of light as it passes from one material to another with a different density. Refraction occurs because in denser materials light travels slightly ____________ than in less dense materials.

3 Each day our eyes take in a lot of information about our ____________: shapes, colours, movements, and so on. They send this information to the ____________ for processing so we are aware of what's going on. The eye controls the amount of light that enters it and focuses the light rays onto the ____________ at the back of the eye. Embedded in the retina are millions of light-sensitive ____________: rods and cones. Rods are good for monochrome (black/white) vision in ____________ light, while cones are used for ____________ and to detect fine detail. Cones are packed into a part of the ____________ called the fovea, where our eyes focus most of the light. When light strikes either the rods or the cones of the retina, it is changed into an electric signal that is dispatched to the brain through the ____________ nerve. The brain then ____________ the electrical signals as the images we see.

(cont.)

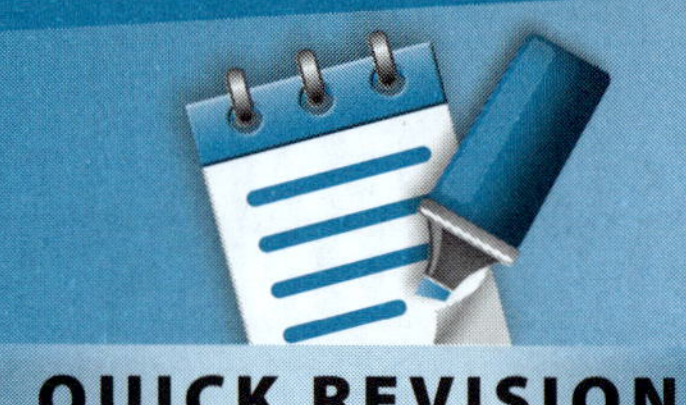

LIGHT ENERGY AND COMMUNICATIONS *(continued)*

Energy on the move

QUICK REVISION

4 Electromagnetic radiation and advances in ____________ have allowed many devices to be commonly used in communications. Mobile phones, also called ____________ phones, can be used to communicate from anywhere there is a ____________ to anywhere in the world almost instantly. A city is divided into cells of ____________ around 25 km^2. When using a mobile phone you use microwaves to connect to a ____________ station within that cell and it relays the call to wherever you want it to go. If you are moving, the ____________ can be transferred from one cell to the next seamlessly.

Modern communication ____________ are located in certain orbits above the Earth's surface. They can ____________ messages from the ground to another distant point on the ground very quickly. Not only can they relay telephone signals, but also radio and television signals.

Radar detectors use ____________ waves to determine the range, altitude, direction or speed of both moving and ____________ objects. This includes aeroplanes, rainfall, hurricanes and ____________ on the road. This technology is also used in radio astronomy. A radio dish, or ____________, sends out pulses of radio waves that bounce off any ____________ in their path. Some of the wave's energy is returned to the dish and detected. The military has developed stealth ____________ which avoid detection by employing several features that interfere with ____________. While no aircraft is totally hidden from radar, ____________ aircraft make it hard for conventional radar to be spotted or tracked successfully.

The following photo shows a radar dish.

Answers **1** light; rays; magnetic; energy; higher; gamma; microwaves; ultraviolet; greater; visible; orange; blue; millionth; vacuum; speed **2** opaque; reflect; transparent; visible (seen); translucent; bending; slower **3** surroundings (environment); brain; retina; cells; poor; colour; retina; optic; interprets (deduces) **4** technology; cell; signal; area; base; signal (call, conversation); satellites; relay; radio; fixed (stationary); cars (vehicles); antenna; object; aircraft; radar; stealth

LIGHT ENERGY AND COMMUNICATIONS

Energy on the move

REVISION SUMMARIES

1 Light, microwaves, X-rays, and television and radio transmissions are all examples of **electromagnetic waves**. They consist of the same kind of wavy disturbance that repeats itself over and over. Features of electromagnetic waves include the following.

- They do not require a medium (material) for their propagation.
- They are **transverse** waves.
- They involve oscillating **electric** and **magnetic** fields.
- They travel at their highest velocity (300 000 km/s, the speed of light) in a vacuum. Through other materials they travel slightly more slowly.

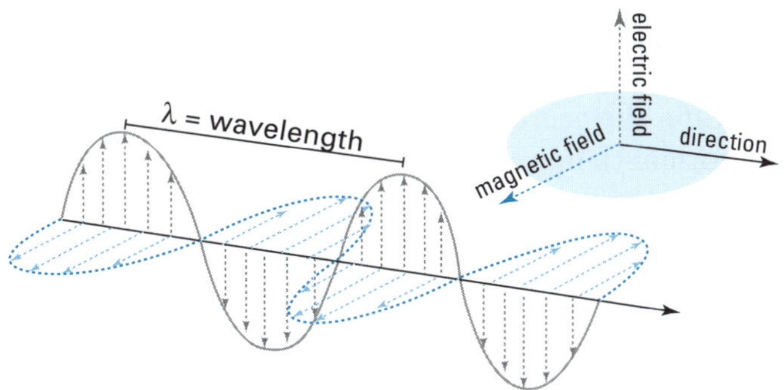

All segments of the electromagnetic spectrum are governed by the same rules; however, their different wavelengths and varying energies allow them to have diverse effects on matter. Visible light (the light we can see) makes up only a small part of the electromagnetic spectrum.

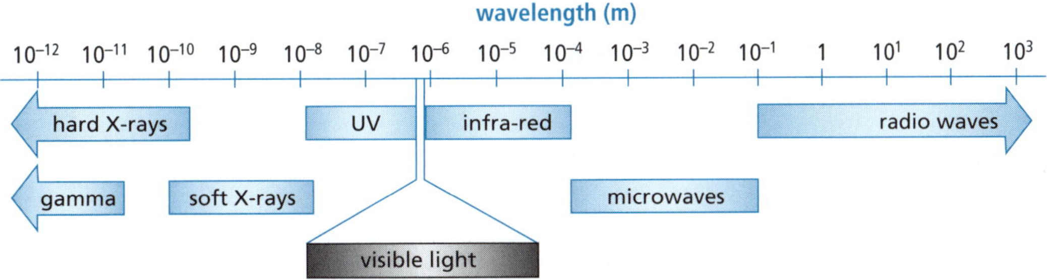

The higher the frequency, the greater is the energy the electromagnetic wave carries. Hotter objects emit higher frequency (more energetic) waves. Electromagnetic waves can affect us differently. While UV and gamma rays can cause cancers, infra-red rays can cause skin burns. Radio waves are used for **communications**. The messages can be encoded in the wave by varying the amplitude (AM) or by varying the frequency (FM).

2 When light strikes an object it may be:

- **transmitted**, that is, pass through the object with only a little loss of energy
- partially transmitted, with some light **reflected** and some **scattered**
- almost completely **absorbed** so that no light emerges. Absorption by a black body is almost complete.

When light waves fall on any material, they are reflected, absorbed or **refracted**. Substances that allow almost all the light falling upon them, such as glass and air, to pass through with very little change in direction are **transparent**. There is no known substance that is perfectly transparent.

(cont.)

Even a clear window will reflect some light. Without this reflection we would not see that the glass is there. Bathroom windows are often **translucent**. They allow light in, but the rays are scattered.

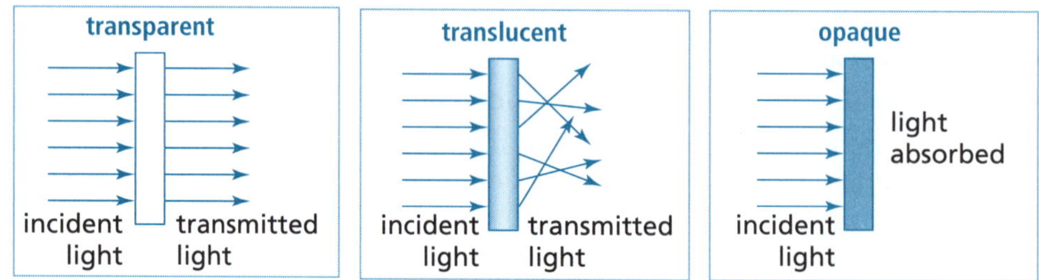

When light bounces off a surface, such as a mirror, it is reflected. The angle the incident ray makes with the normal (a line at right angles to the surface) is the same as the angle the reflected ray makes with the normal. In other words:

angle of incidence = angle of reflection

As light passes from one medium (such as air) into another (such as glass) its speed changes, and so the light ray is bent (refracted). It is slower in more dense material.

- If the ray passes from a less dense to a more dense medium, it is refracted towards the normal. The angle of incidence (*i*) is greater than the angle of refraction (*r*).
- If the ray passes from a more dense to a less dense medium, it is refracted away from the normal. The angle of incidence (*i*) is less than the angle of refraction (*r*).

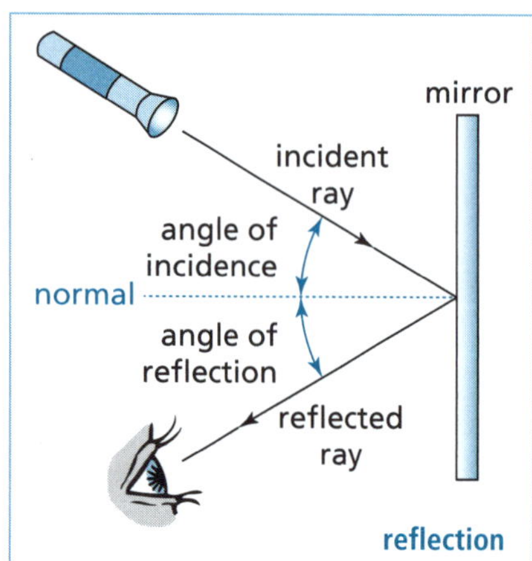

reflection

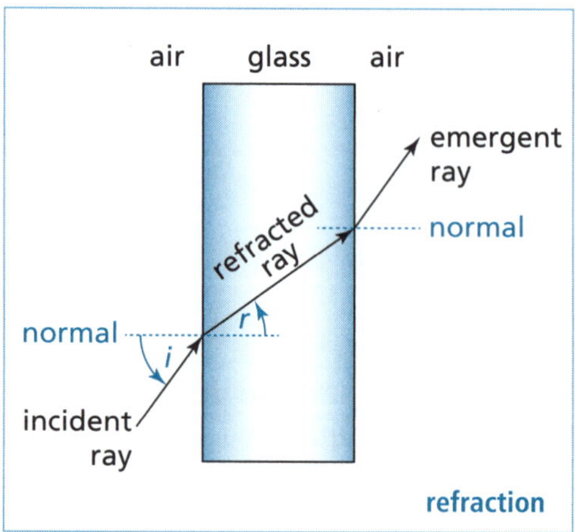

refraction

The change in angle of the light ray is the same when it enters and leaves the glass. If the sides of the glass are parallel, then the emergent ray is parallel to the incident ray.

3 The human eye is the organ we use to detect light. The transparent **cornea** allows light to enter the eye. The amount of light entering is controlled by the **iris**. This light passes through the transparent aqueous humor and vitreous humor and is focused by the **lens** onto the **retina** at the back of the eye. The retina consists of two types of light-sensitive cells.

- **Cones** are sensitive to colour and bright light; they cannot detect very dim light.
- **Rods** are especially sensitive to dim light and all wavelengths of the visible spectrum; they cannot detect colour.

The **optic nerve** carries the impulses formed by the retina to the brain. The structure of the eye is shown in the following diagram.

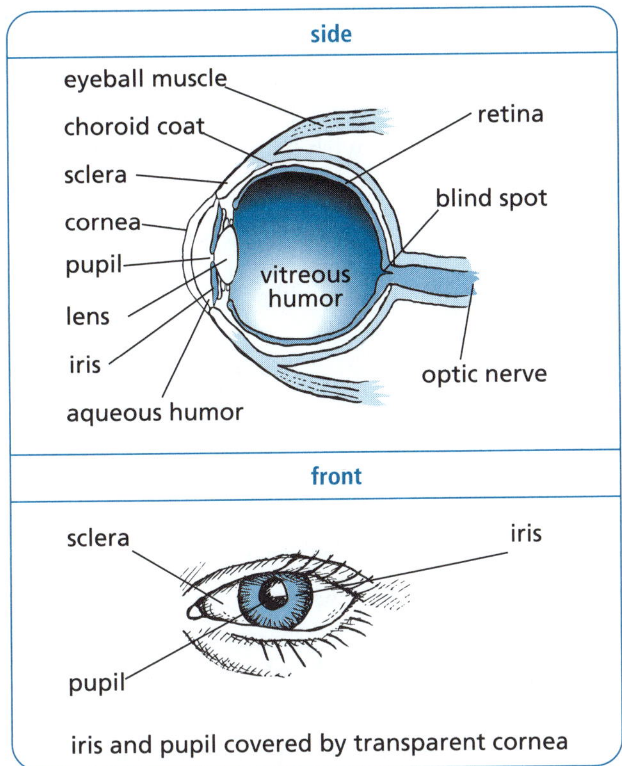

iris and pupil covered by transparent cornea

Blindness is a very devastating condition that impacts on people's capacity to lead independent lives. With advances in technology a **bionic eye** that can bring back sight to people with certain forms of vision injury is now being developed. It works by inserting electrodes into the retina that send electrical impulses to nerve cells in the eye. These impulses are then passed down along the optic nerve to the vision processing centres of the brain, where they are interpreted as an image. In early prototypes, researchers can control the information sent to the eye, allowing them to study how the brain reacts. With this feedback they are developing a vision processor so they can assemble images using flashes of light. The next step is to test various levels of electrical stimulation.

4 Electronics has made possible many different forms of communications that are based on electromagnetic waves. Computers, mobile phones, satellites, fibre optics, radar, radio and television are all examples. An event occurring on the other side of the world can now be instantly relayed to anywhere on the globe. Businesses and society today could not function without this technology. Mobile phones, for example, allow communication while people move from place to place. The city is divided into smaller areas called cells. In each cell there are mobile phone towers that transfer phone calls via microwaves from one cell to another. Radar is an important technology that uses radio waves. Radar dishes are used at airports to monitor the positions of aircraft. The radio waves reflect off the aircraft and return to the radar dish to create images of airport positions on a monitor.

(cont.)

An **optical fibre** is flexible, transparent and a little thicker than a human hair. It is typically made of glass (silica) or plastic, and coated with a polymer as shown in the diagram below. It transmits light between the two ends of the fibre. Optical fibres are extensively used in fibre-optic communications, especially over long-distances. They allow transmission over greater distances and at higher bandwidths (the speed of data transfer) than other types of communication, and transmit multiple signals simultaneously. Fibres are preferred to metal wires because signals travel along them with less loss of signal and are also unaffected by **electromagnetic interference**.

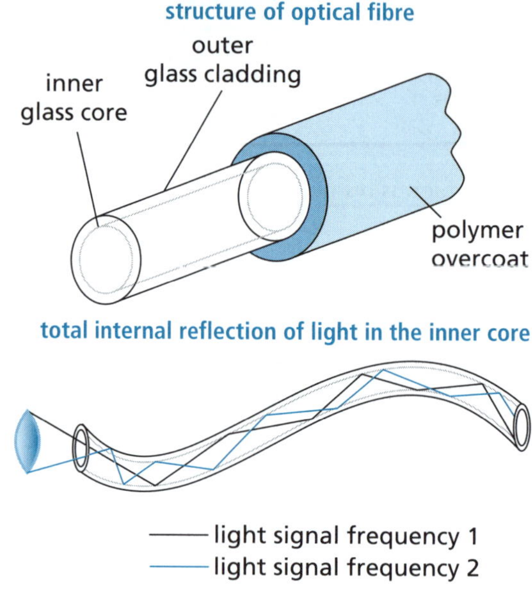

The light in a fibre-optic cable travels through the core by constantly bouncing from the cladding covering it. This is called **total internal reflection**. Since the cladding does not absorb any light from the core, the light wave can travel great distances before needing to be boosted (amplified).

Checklist

Can you:

1. *Describe features of the electromagnetic spectrum?* ☐
2. *Account for the transmission, reflection, refraction and absorption of light?* ☐
3. *Outline how the human eye and bionic eye function?* ☐
4. *Briefly outline the impact of electromagnetic waves in communications such as television, fibre optics, mobile phones and satellites?* ☐

LIGHT ENERGY AND COMMUNICATIONS

Energy on the move

REVISION TEST

1 Write the order of the following forms of electromagnetic radiation from highest frequency to lowest frequency: visible light, radio waves, gamma rays, ultraviolet, X-rays, microwaves, infra-red. (2 marks)

2 True or false?

a Sound waves are a type of electromagnetic radiation. (1 mark)

b Electromagnetic radiation consists of two major energy fields. (1 mark)

c The hotter the piece of matter is the higher is its emission of short wavelength energy. (1 mark)

d Microwaves are used to transmit mobile telephone messages. (1 mark)

e We can see frequencies in a large part of the electromagnetic spectrum. (1 mark)

f The speed of electromagnetic radiation in a vacuum depends on its frequency. (1 mark)

g X-rays are a type of electromagnetic radiation that can be used to check bones for damage. (1 mark)

h The main danger of infra-red light is that it causes skin burns. (1 mark)

i Light signals are used in fibre optics. (1 mark)

j Light rays do not bend as they pass from one transparent material to another. (1 mark)

3 Radio signals can be transmitted by AM (amplitude modulation) and FM (frequency modulation).

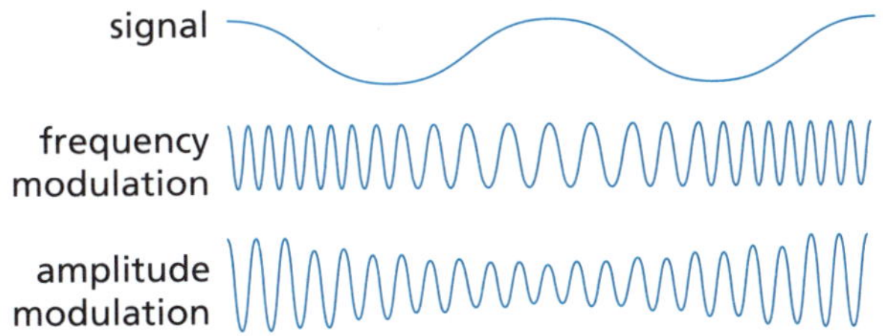

Describe the difference between these two modes of transmission. *Hint 1* (4 marks)

4 The diagram below shows the dispersion of white light by a glass prism.

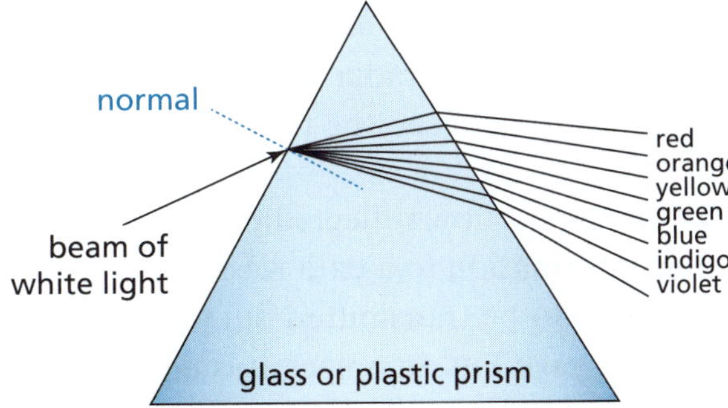

a As white light enters the prism, is it refracted towards or away from the normal? Explain this refraction. (2 marks)

b When the light emerges out the other side of the prism, all the colours of the rainbow are seen. What does this tell you about the composition of white light? *Hint 2* (1 mark)

(cont.)

c Comment on the following statement.
Colours of the visible light spectrum that have shorter wavelengths will deviate more from their original path than colours with longer wavelengths. (2 marks)

d A process called dispersion can be used to explain rainbows. Each individual droplet of water acts as a tiny prism that bends each component wavelength of the visible light to a different extent. Each coloured component of the light is reflected back to your eye. For simplicity, only the red and violet components of the dispersed light are shown.

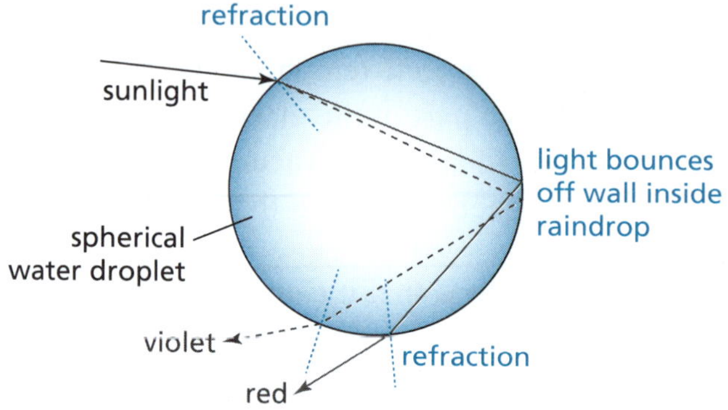

Use this diagram to explain why you see the colours of the rainbow. (5 marks)

5 The following diagram shows how radar works.

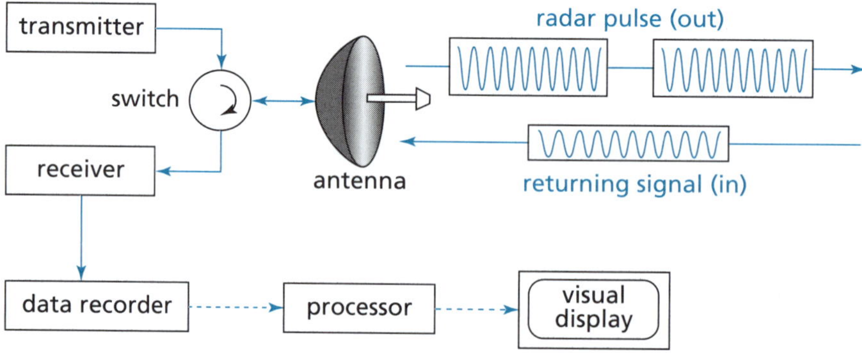

The following are the steps showing how radar works, but they are out of sequence.

1 The switch may toggle control between the transmitter and the receiver up to 1000 times each second.
2 Some of the outgoing waves are now reflected to the antenna.
3 The receiver sends this information to a data recorder for storage.
4 The switch directs the pulse to be transmitted out of the antenna.
5 When an electromagnetic wave hits an object, it is partly reflected away from the surface and partly refracted into the surface.
6 The transmitter sends a high power electromagnetic pulse to a switch.
7 Useful information is calculated such as the range of a distant object, the type of surface that reflected it, and how fast the object is moving.
8 The returned signal carries information such as the time taken for it to return, how strong the returned signal is and any change in its frequency.

9 The amount of reflection and refraction depends on the surface and the material the wave was travelling through.

10 Once the signals are received the switch then transfers control back to the transmitter to send another signal.

11 The data can then be interpreted into something useful by the processor and be displayed.

12 As soon as the antenna completes transmitting the pulse, the switch changes control to the receiver, allowing the antenna to receive echoed signals.

What is the correct sequence explaining how radar works? *Hint 3* (6 marks)

6 Match the following words about the eye with their descriptions: cornea; sclera; iris; aqueous humor; lens; retina; optic nerve; cones; vitreous humor. *Hint 4*

a Coloured part of the eye between cornea and lens controlling how much light enters the eye. (1 mark)

b Tough skin covering the outside of the eyeball; also called the 'white of the eye'. (1 mark)

c Cells in the eye that allow you to see colours. (1 mark)

d Clear watery liquid that assists the cornea to maintain its rounded shape. (1 mark)

e Images from the retina are converted to electrical signals and carried by this to the brain. (1 mark)

f A clear, see-through layer covering the front of the eye that doesn't contain blood vessels. (1 mark)

g Changes shape to focus light onto the retina; clear and convex in shape. (1 mark)

h The clear gel filling most of the inside of the eye. (1 mark)

i Where image formation in the eye takes place. (1 mark)

Hint 1: The answer lies in comparing the frequency and amplitude of each wave. Indicate what property has changed and what remains the same.
Hint 2: Think about whether or not visible white light consists of waves of only one wavelength.
Hint 3: It might help to write each of these statements on strips of paper first, order then in a logical sequence, and then stick them into your book.
Hint 4: Refer to the diagram of the eye earlier in this section.

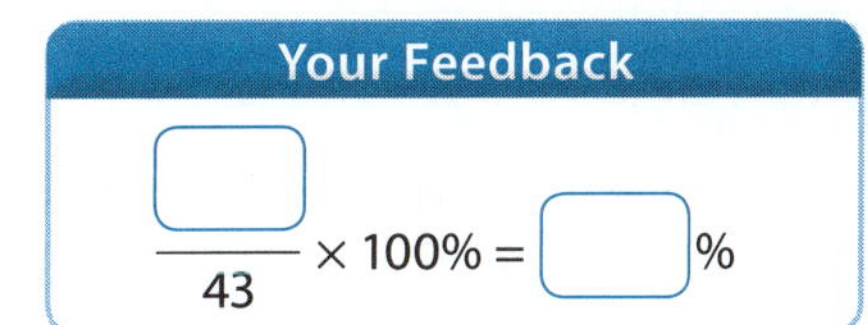

ELECTRICAL ENERGY

Energy on the move

QUICK REVISION

1 Electricity is one of the most common ____________ sources available. Electricity is the movement of ____________. These are tiny charged particles found in atoms. Electrons flow in closed loops called ____________. The path must be complete before electrons can move. If you break the loop, the circuit is incomplete and the electrons ____________ flowing. This is the function of a ____________, so you can turn electrical devices on and off.

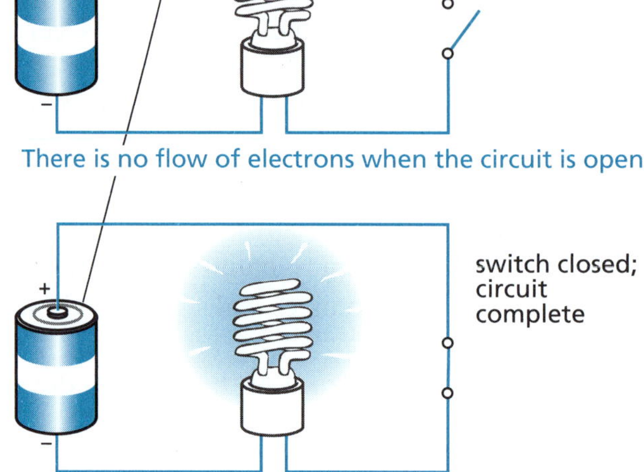

When the switch is closed, the circuit is complete and electrons can flow.

Conductors are made of materials that allow electricity to flow through ____________. The atoms of these materials, such as metals and graphite, contain ____________ that can move freely. Insulators have atoms where electrons are not easily freed, preventing or blocking the flow of ____________. Examples of insulators include glass, wood, plastic and rubber. The rubber or plastic on an electrical cord provides an ____________ for the wires, forcing the electricity to travel on the copper ____________. This also prevents people from touching a 'live' wire and being ____________.

2 A ____________ is a flow of electrons. While electrons can be made to flow in one direction, the ____________ current is defined as flowing in the opposite direction. In order to make this current flow a force is needed. This force is called ____________. Just as if you were to pick up one end of a desk covered in marbles, making them all ____________ in one direction, providing a voltage, or potential ____________, across a circuit will make the free electrons flow in one direction. In any electrical flow, electrons experience ____________, and their flow is reduced. Conductors have very ____________ resistance, although it is not zero. Resistance is measured in ohms. Among metals copper is one of the better conductors so it is used for electrical wiring. Power lines typically have a resistance of a few ohms per kilometre. Other materials exhibit a lot of resistance. They are called ____________ or non-conductors. Current is measured using a piece of equipment called an ____________; voltage is measured using a ____________.

3 Many devices in an electrical circuit exhibit resistance. These appliances can be connected in one of two ways: series or parallel.

ELECTRICAL ENERGY

Energy on the move

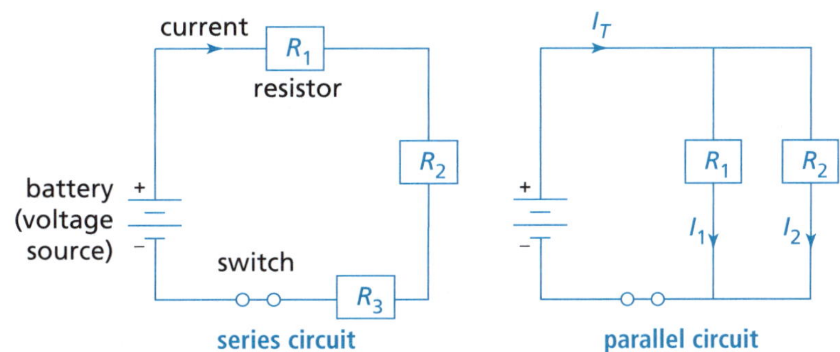

In the series circuit, when the switch is __________, the current flows through all three resistors, R_1, R_2 and R_3. In the __________ circuit, the total current, I_T, splits and some of it travels through R_1, while the remainder travels through R_2. The __________ current $I_T = I_1 + I_2$. (Current is often represented by the letter I.)

4 Lights and power points around the home are connected in __________. Should an appliance fail, there is a fuse or __________ breaker at the fuse box, usually located around the side of the home or in some convenient wall space. A circuit breaker is an automatically operated electrical __________ intended to protect an electrical circuit from damage caused by __________ (excessive current) or short __________. The essential function of a circuit breaker is to interrupt the connection to a faulty __________, and immediately __________ electrical flow. Older homes may still have __________ wires which melt when a certain current is reached. Unlike a fuse, which operates once and then must be __________, a circuit breaker can be reset to restart regular function. The following photo shows a typical fuse which contains a wire inside a glass tube. If the current is too high, the fuse wire melts.

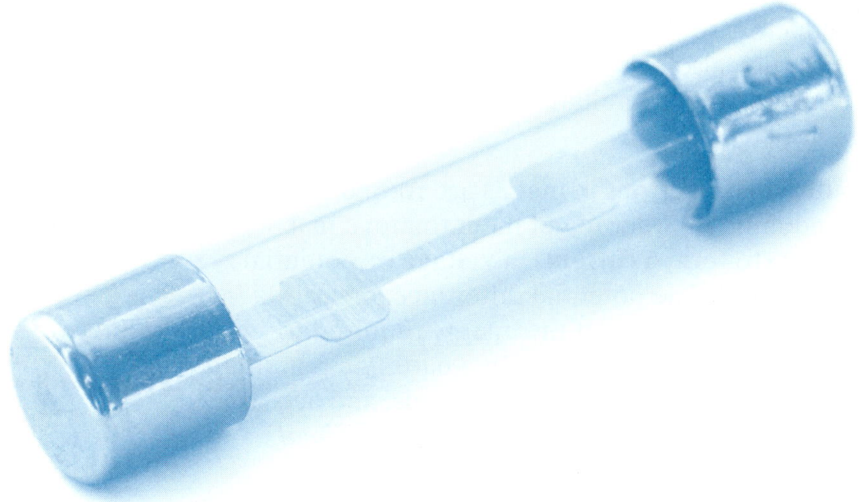

Answers **1** energy (power); electrons; circuits; stop (cease); switch; easily (effortlessly); electrons; electricity; insulator; wires (metal); electrocuted **2** current; conventional; voltage; move (slide, roll); difference; resistance; little (small, tiny, negligible); insulators; ammeter; voltmeter **3** closed; parallel; total **4** parallel; circuit; switch; overloading; circuit; circumstance (condition, situation); terminate (end, stop); fuse; replaced (renewed)

ELECTRICAL ENERGY

Energy on the move

REVISION SUMMARIES

1 **Electricity** is the presence and flow of an electric charge, which carries energy. Since the 1800s humans have been able to harness electricity and store the electrical charge for later use. Electricity is one of the cheapest and most convenient energy sources available. There are several different sources of electrical energy.

- Chemical energy is generated from chemical reactions in a **battery**. These drive your mobile phone, calculator, MP3 player, camera, torch and laptop. Many modern batteries are rechargeable, making it far cheaper than having to buy batteries all the time.
- Mechanical energy is generated when moving water turns a generator at a **hydroelectricity** plant. This electrical energy is transferred by wires to your home.
- Radiant energy is energy generated in the Sun and captured and stored in solar electric cells. Many people have installed solar panels on their roofs. Solar energy is useful for remote locations.
- Thermal energy is generated when steam turns generators in electrical power plants. These plants burn oil or coal, or use other material (such as radioactive nuclear fuel), to produce the steam that generates the electricity.

Metals form giant crystalline structures where electrons in the outer shells of the metal atoms are free to move away from those atoms. These electrons are called **mobile electrons**. This makes metals good conductors of electricity and heat, since these free electrons can carry a charge or heat energy through the metal. Some metals are better conductors than others as they allow electrons to move through more freely.

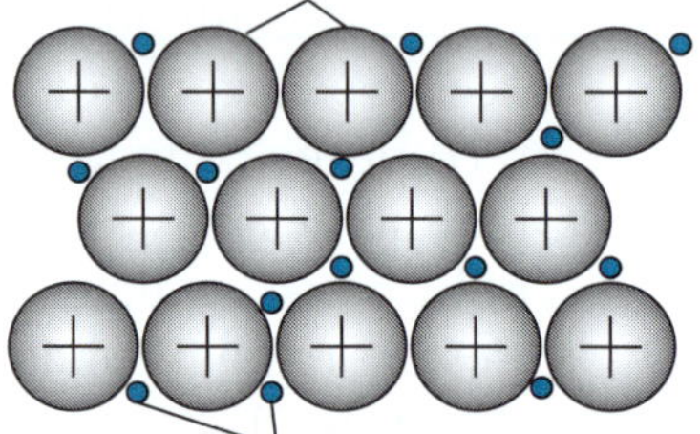

Insulators, or non-conductors, don't have free electrons. These materials can prevent the flow of electrical current. Electrical insulators are used to coat, protect or support electrical conductors so that the electrical current flows through the conductor. They also help prevent electrical shock or sparks. When you turn on a light switch your fingers are only millimetres from a live wire but are protected from it by the non-conducting plastic.

2 While the outermost electrons in metals are free to move, they do so randomly. In order to get them to move in a certain direction (to form a **current**), a **voltage** or potential difference is required. The potential difference, measured in volts (symbol V), is a measure of the energy required to move an electric charge along a path. In a battery a potential difference is maintained between the terminals, and acts as a pump to move electrons around. The term *conventional current* refers to the direction in which positive charge appears to move.

conventional current
flow of electrons
potential difference
battery

The current, measured in amps (symbol A), refers to the movement of the charge. In **direct current** (DC), the current travels in one direction out of the battery. With an **alternating current** (AC) the current changes direction repeatedly within an electrical system. AC is usually used to power homes, buildings, factories, schools and offices. In Australia, the

power supplied to homes is AC with 230 to 240 V, and a frequency 50 Hz. (This just means the current changes direction 50 times each second.)

Electrical resistance describes the forces that oppose the flow of an electron current in a conductor. All materials contain some resistance to an electron flow, including good conductors, although this resistance is low. Even the best conductors have some resistance per unit length, so the longer the wire the larger is this total resistance. In transporting electricity from the power station to your home, many kilometres away, some energy is dissipated (converted to heat) along the way.

Some materials are designed to have high resistance. It is often useful to add components, called **resistors**, to an electrical circuit to hamper the flow of electricity and so guard the components in the circuit. Resistance is measured in ohms (symbol Ω).

3 In a **series circuit** the current flows continuously around the circuit. A light globe (also known as a light bulb) acts as a resistor. In the diagram, the current flows through one globe after another making them both glow. In a series circuit if one globe burns out, the current can no longer pass through it, and so no current flows through the circuit. Both globes go out.

In a **parallel circuit** the current splits and passes at the same time through each globe. With a parallel circuit, if one of the globes blows and current can't pass through it, the current can keep on flowing through the remainder of the circuit.

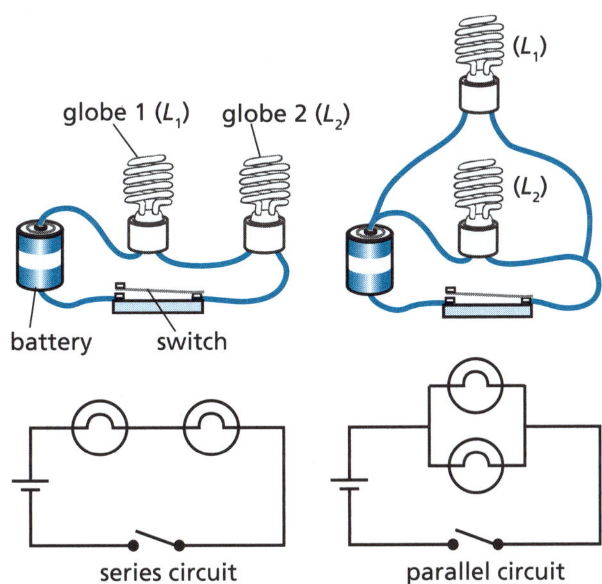

Current is measured using an **ammeter**; voltage is measured with a **voltmeter**. In order to measure the current through, and the voltage across, a resistor these meters are connected as shown.

The ammeter is connected in series with the resistor, while the voltmeter is connected in parallel with it. As the resistance of the resistor increases, the current decreases, but the voltage drop across it increases. A greater resistance impedes the flow of current (so it is lower), while using more energy for that current to make it across the resistor (hence greater voltage).

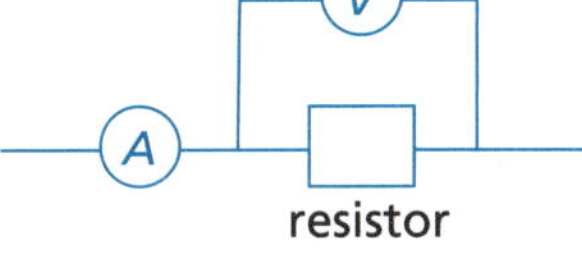

(cont.)

ELECTRICAL ENERGY *(continued)*

Energy on the move

4 In the home electrical **power points** and lights are connected in parallel. This is so if any lights go out, or switches get turned off, current still flows to the remaining parts of the circuit.

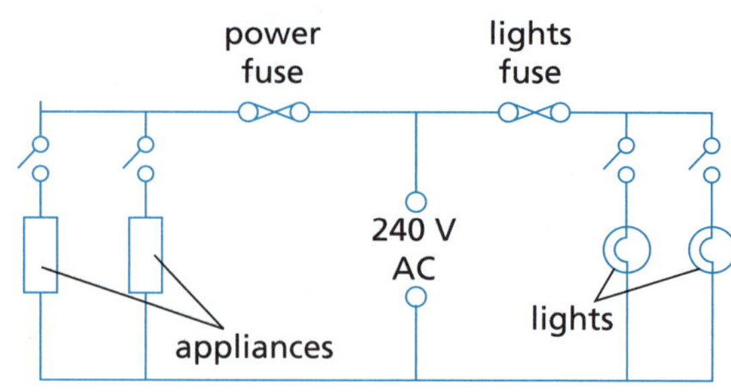

Power plugs have two or three distinct wired contacts. The two slanting pins are the live and neutral and carry the current back and forth. Many power plugs also have a third, earth, contact to connect to ground should the insulation or the device become dangerous. Power points (wall sockets) always have three pins. Should an appliance be faulty, the current is cut off by fuses or circuit breakers in the fuse box on the outside of your home. Circuit breakers can be reset whereas fuses melt and have to be replaced.

Checklist

Can you:

1 *Describe what electrical energy is and distinguish between conductors and insulators?* ☐
2 *Define current, voltage and resistance?* ☐
3 *Draw simple electric circuits showing series and parallel arrangements?* ☐
4 *Describe some features of circuits in the home?* ☐

ELECTRICAL ENERGY

Energy on the move

REVISION TEST

1 Which diagram shows a circuit in which a current will flow? (1 mark)

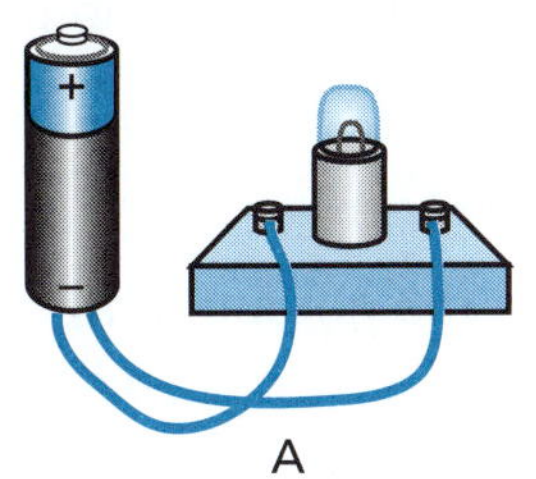

A

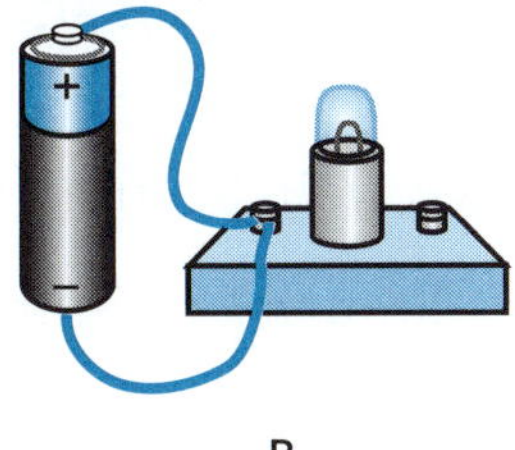

B

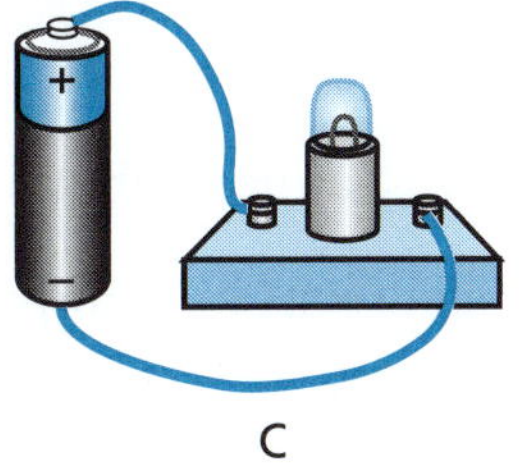

C

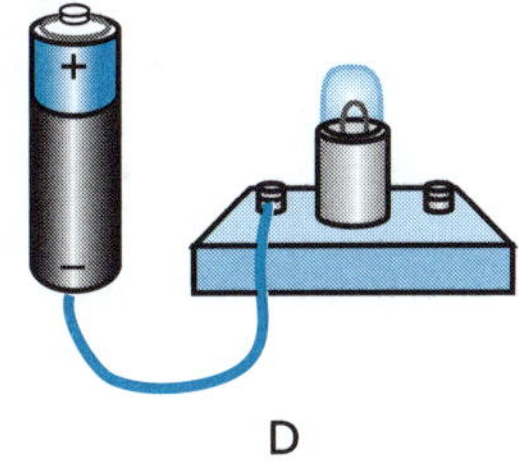

D

2 True or false?

a Circuits need to be complete for a current to flow. (1 mark)

b Materials that allow electricity to pass through are called insulators. (1 mark)

c Inserting a switch in a complete circuit allows the current to remain constant. (1 mark)

d The circuit below shows two globes in series and one in parallel.  (1 mark)

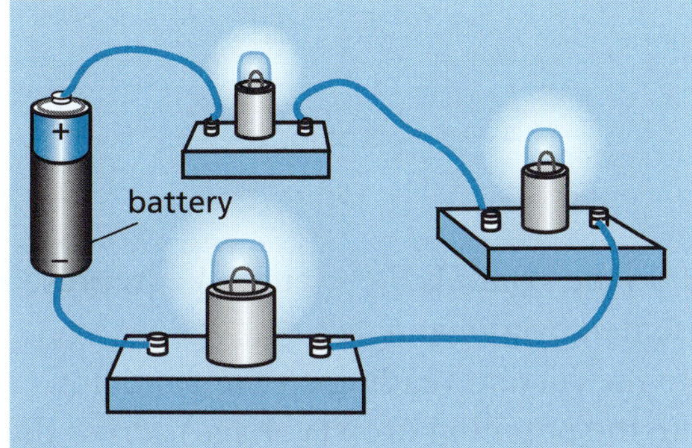

e If one of the globes in the circuit above blows, then all lights go out. (1 mark)

f If a fourth globe is added in series to the circuit shown, the brightness of the other globes remains the same. *Hint 1* (1 mark)

g Parallel circuits are useful because all other components keep working even when one fails. (1 mark)

h The following diagram shows the correct connection for an ammeter and voltmeter. (1 mark)

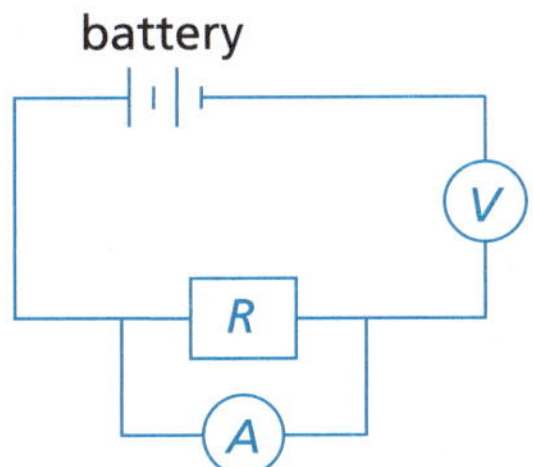

i A globe glows brighter when it is powered by two batteries rather than by one because the flow of electricity in the circuit is greater. (1 mark)

j Metals are good electrical conductors because some electrons have the ability to move freely and easily from atom to atom. (1 mark)

(cont.)

3 Ohm's Law was named after the Bavarian mathematician and physicist Georg Ohm who linked current, voltage and resistance together. He came to this result after much experimentation. Ohm's Law states that:

$R = \frac{V}{I}$ or, in words, resistance (ohms) = voltage (volts)/current (amperes).

In an electrical circuit a current of 3 amps passes through a resistor where the potential drop across it is 18 volts. Calculate the resistance. (2 marks)

4 The three diagrams represent part of an electrical circuit. In order to answer some of these questions you will need to use Ohm's Law ($V = IR$).

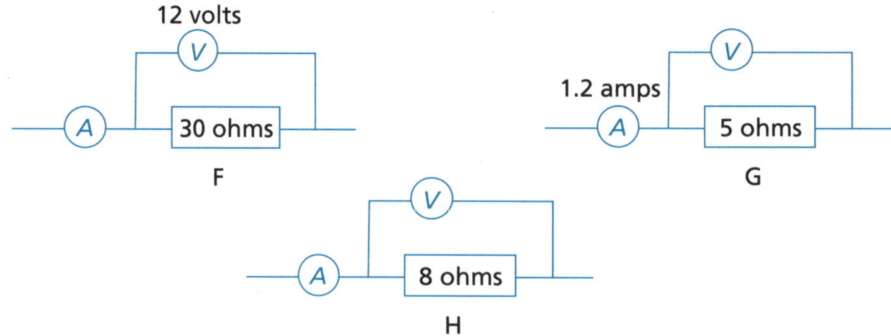

a In the diagrams what do the symbols –(A)– and –(V)– represent? (2 marks)
b What device is represented by a rectangle? (1 mark)
c In diagram F calculate the current reading on the ammeter. (2 marks)
d In diagram G calculate the potential drop (voltage) across the resistor. (2 marks)
e In diagram H give a set of possible readings for A and V. *Hint 2* (2 marks)

5 Consider part of an electrical circuit.

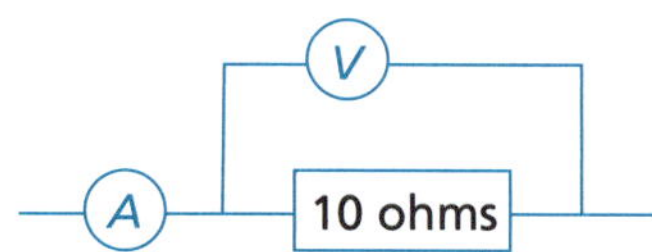

Use Ohm's Law ($V = IR$) to answer the following questions.

a The current passing through the resistor is at present 2 amps. What happens to the voltage if the resistance doubles, but the current remains the same? (3 marks)
b The voltage across the resistor is presently 20 volts. What happens to the current if the resistance doubles, but the voltage is kept the same? (3 marks)

6 Every material has an electrical resistance, some more than others.

a Explain electrical resistance. *Hint 3* (2 marks)
b What happens to the electrical energy when it encounters resistance? (1 mark)
c Comment on how each of the following might affect the resistance of a length of wire.
   i The material out of which the wire is made. (1 mark)
   ii The length of the conductor. (1 mark)
   iii The cross-sectional area of the conductor. *Hint 4* (1 mark)

7 Use these words to complete the statement: water pump; water pressure; top of the fountain; water. (4 marks)

If an electric circuit could be compared to a water fountain, then the:

a battery would be similar to the ____________
b positive terminal of the battery would be similar to ____________
c charge would be similar to ____________
d potential difference would be similar to ____________.

8 Consider the following diagram, and then complete the cloze passage. *Hint 5* (8 marks)

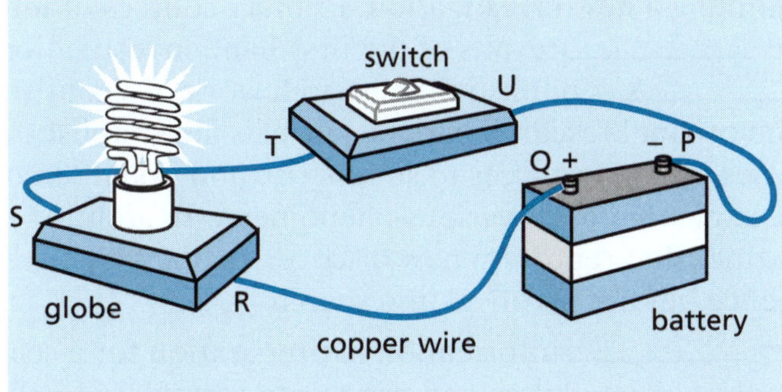

As a charge moves through the battery from P to Q, it ____________ (*gains, loses*) potential energy and ____________ (*gains, loses*) electrical potential. The highest energy point within a battery is the ____________ (*positive, negative*) terminal. As the conventional current moves through the external circuit from Q to P, it ____________ (*gains, loses*) potential energy and ____________ (*gains, loses*) electric potential. The point of highest energy in the external circuit is closest to the ____________ (*positive, negative*) terminal. A voltage drop ____________ (*should, should not*) occur between R and S, while a voltage drop ____________ (*should, should not*) occur between T and U.

Hint 1: The battery supplies a certain voltage. Three or four globes can 'feed' off it. What do you think this might do to the brightness of each globe?
Hint 2: There are many answers to this question; you only need to provide one.
Hint 3: As an analogy, imagine you are running down the school corridor but there are many students standing around in your path. What happens and why does this slow you down?
Hint 4: What happens at a bottleneck on the road where three lanes have to merge into one lane?
Hint 5: Use the answer to the previous question to help you with this one. Is a globe a type of bottleneck?

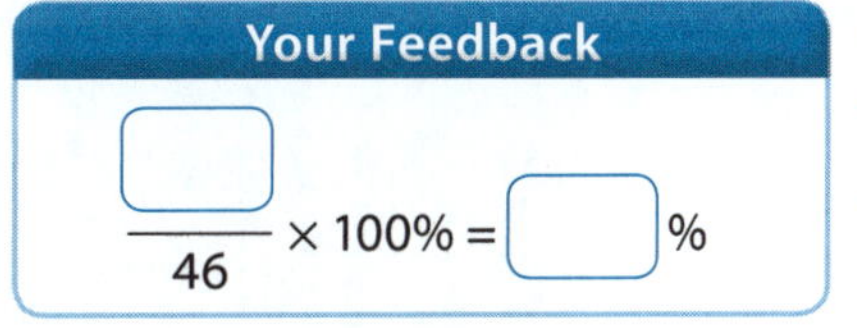

PAGES 241–242
PAGE 252

RESEARCH AND FIRST-HAND INVESTIGATIONS

Investigations and problem solving

QUICK REVISION

1 The __________ method is used by scientists when they undertake research and first-hand investigations. This procedure has been developed to ensure that the __________ and conclusions that are made are logical, __________ and as accurate as the equipment and procedures allow. Different scientists use different __________ to solve problems but their procedures and processes are always in accord with the scientific method.

A biologist is interested in how components of the __________ world behave and interact. A chemist researches the chemical properties of __________ and how atoms and molecules react. A geologist investigates the components of the __________ and the processes that occur in it. All these scientists collect and research published __________ about their area of interest before beginning a new investigation. Primary sources of information are very useful as they contain the __________ results of the first-hand investigations performed by a scientist. Various scientific __________ publish the research work of scientists and these are primary sources. Apart from original research papers, journals also publish __________ papers in which the author uses primary sources to summarise and update knowledge in this field or argue __________ explanations of various phenomena. In addition to research journals there are scientific magazines that report on new discoveries but which are written for those who are __________ in science but not involved in research.

When students are __________ information in preparation for a school research project, they need to summarise the collected data and use this to formulate the final problem that will be investigated experimentally.

2 The student needs to develop a __________ which can be tested experimentally. This hypothesis must be able to __________ the results of an investigation which is carried out to test the hypothesis. No experiments will be approved by the teacher which are __________ or which will harm animals. Here are some examples of hypotheses that can be tested in a first-hand investigation.

- Dark coloured materials will get hotter than light coloured materials when they __________ radiant energy.
- The solubility of ionic compounds increases as the temperature of the water increases.
- The weight of a body suspended in water decreases as the volume of the body increases.
- Watercress seedlings (see photo) grow faster when planted in loamy soil than in sandy soil.

RESEARCH AND FIRST-HAND INVESTIGATIONS

Investigations and problem solving

QUICK REVISION

Each of these hypotheses has been developed following the collection of data from primary and __________ sources. Each hypothesis is easily tested using simple first-hand investigations at school or at home. In each investigation many __________ need to be controlled and only one variable (the __________ variable) is tested. For example, some of the variables that would need to be controlled in the first hypothesis on coloured materials absorbing radiant energy are as follows.

- The materials should be made from the __________ fibre (e.g. cotton) and have the same thickness and weave.
- Each piece of material should have the same surface area.
- The same __________ of radiant energy (e.g. an infra-red lamp) should be used.
- The distance between the radiant energy source and the materials should be the same.
- The same type of __________ thermometers should be used.

The final experimental method or procedure is the result of some pre-trials to ensure the variables are controlled. The proposed method should be __________ by the teacher before proceeding. The method should be written as a series of __________ sentences.

3 It is important that your results are reliable. This means you will need to __________ each experiment a minimum of __________ times. If the results are clustered close together then the experiment is reliable. If the results are widely __________ then the reliability is low. In the case of experiments involving living things the number of __________ experiments needs to be much greater. Your results will also depend on how accurate any measuring instruments are. For example, a thermometer that has a scale marked in 0.2 degree divisions is more __________ than one marked in whole degrees. An electronic balance that has a scale marked in milligrams is more accurate than a whole gram scale. All measuring instruments need to be checked (__________) regularly to ensure they are measuring a quantity correctly.

Your experimental __________ must be fully analysed and discussed. This analysis often leads to new __________ to extend experiments to determine whether the hypothesis is supported or not. Some __________ variables may emerge that were not originally considered.

Answers 1 scientific; results; reliable; techniques; living; matter; Earth; information; original; journals; review; alternative; interested; collecting 2 hypothesis; predict; dangerous; absorb; secondary; variables; independent; same; source; calibrated; checked (approved); numbered 3 repeat; five; spread; repeat; accurate; calibrated; results; proposals; new

RESEARCH AND FIRST-HAND INVESTIGATIONS

Investigations and problem solving

REVISION SUMMARIES

1 Developing **research skills** is a very important part of becoming a good scientist. Scientists discover and test facts and principles by using a way of thinking called the **scientific method**. The scientific method is an orderly way of answering questions and solving problems. Scientists do not use the same techniques all the time, but they always use the scientific method.

The scientific method starts with **collecting information** from primary and secondary sources about an observation that you have made which has sparked your curiosity. You may have read, for example, a short article in a newspaper, science magazine or online about the effect of global warming on coral. You may have discussed this information with your family, friends or science teacher. Your friends may have holidayed near a coral reef and noticed that some sections of the reef were dead. You may have viewed a YouTube video about the effects of global warming on the oceans. You have decided to make this issue the problem that you will investigate for your science project. Your next step is to collect information from secondary sources about corals and global warming before you begin your own experimental research.

Primary and secondary sources can be used to obtain information. A primary source of information is often an original document created by the original researchers. In modern science this is usually a thesis or journal article written by the scientists conducting the research. When a science research article is published in a reputable journal (e.g. *Nature*) then it is usually considered highly reliable as it has been extensively peer reviewed. Review articles may also appear in science journals. These articles are often based on primary sources and therefore are secondary sources of information. Many articles written for science magazines such as *New Scientist* or *Scientific American* are secondary sources. Articles in science magazines also have usually been simplified for a wider group of non-specialist readers who are interested in science.

You will now need to **summarise** all the collected information and decide which aspect of the problem you will investigate experimentally.

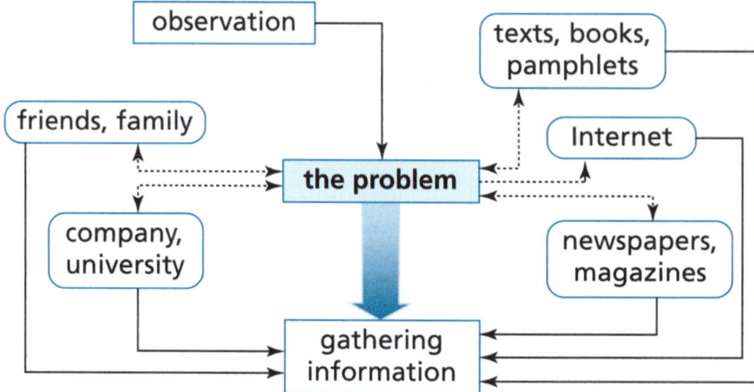

2 The next stage of the scientific method involves developing a hypothesis to investigate. Let us assume that the **hypothesis** to be investigated is:

The calcium carbonate skeletons of corals break down more readily when the salt water becomes acidic.

This hypothesis has been based on the secondary information that rising levels of carbon dioxide in the atmosphere has led to the acidification of the oceans. Carbon dioxide forms carbonic acid when it dissolves in water.

In the school laboratory, students have access to the required chemicals. These are salt, calcium carbonate (in the form of limestone or marble chips), water and carbonic acid (e.g. soda water). The plan is to determine how rapidly the limestone dissolves in acidified salt water

compared with normal salt water. Corals are living things and are not used. Marble pieces (also composed of calcium carbonate) are substituted. A number of **variables** need to be **controlled** in this experiment. These include the mass of limestone, the concentration of salt in the water, the water temperature and the volume of salt water, all of which should be the same in all repetitions of the experiment. The **independent variable** in this investigation is time and the **dependent variable** is the mass of limestone remaining.

The method for the experiment is finally determined and written prior to the practical work. The proposed method follows.

- Weigh out 5 g of salt (sodium chloride) using an electric balance and dissolve it in 150 mL of tap water in a 250 mL beaker.
- Weigh out 5 g of salt and dissolve it in 150 mL of fresh soda water in a 250 mL beaker.
- Weigh out two 2-g samples of limestone (marble chips).
- Use a thermometer to measure the temperature of each water sample and record this information.
- Add the 2-g samples of limestone to each beaker and record the time. Leave undisturbed for two days.
- At the end of the two days, filter off the limestone pieces and dry them thoroughly. Re-weigh each sample and record the masses.

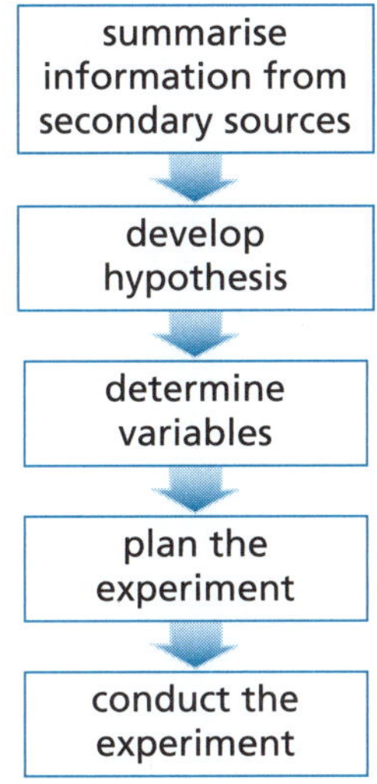

3 In order to achieve **reliability** involving experiments with non-living things the experiment should be repeated a minimum of five times and the results averaged and recorded. In experiments involving living things then very large sample sizes (> 1000) are required to determine whether conclusions are reliable. This is particularly true when trialling drug medications on people.

The next step in the marble/carbon dioxide experiment is to **analyse the results** of the experimental work and decide whether they support the original hypothesis or not. Whether or not the results support the hypothesis, the research should be continued to investigate the variables further.

The original experiments were conducted for only two days. This may be too short a time to obtain meaningful results. New experiments over longer time periods should be planned and conducted.

(cont.)

One problem with using soda water is that it loses carbon dioxide gas over time and so the carbonic acid becomes less acidic. Therefore, other acidic substances that retain their acidity could be investigated (e.g. vinegar contains acetic acid). The concentration of acid in the salt solution could also be investigated.

As corals normally grow in tropical waters, then the effect of temperature change could be investigated. The skeletons of corals are made from calcium carbonate but marble chips were substituted in the experiment. Both marble and limestone are made of calcium carbonate but marble is a metamorphic form and much harder. This difference could also influence the result and needs further investigation.

The following diagram shows examples of the extended experiments in which the temperature is varied and then the concentration of acid in the salt water is varied. In each case the control is a tube of salt water with the same mass of marble.

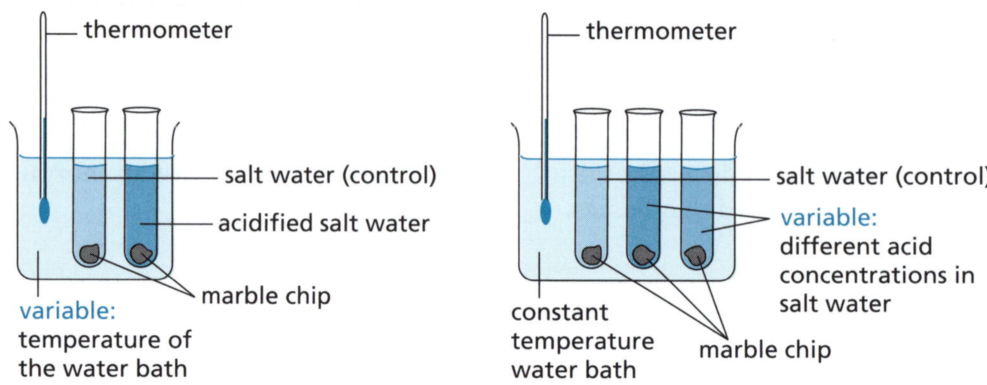

Checklist
Can you:

1. *Explain the importance of research prior to experimentation and the difference between primary and secondary sources of information?* ☐
2. *Explain the steps involved in developing a first-hand investigation plan that will test a hypothesis prior to performing the experiment?* ☐
3. *Explain why a critical evaluation of the results of a first-hand investigation will often lead to new investigations in which different variables are tested?* ☐

RESEARCH AND FIRST-HAND INVESTIGATIONS

Investigations and problem solving

REVISION TEST

1 During a recent holiday Liam noticed that different rivers had different coloured water. On closer inspection of some of these rivers he noticed that some had relatively clear water and some had quite brown water with lots of suspended solids. He collected samples of water from each river visited and made a note of the geographical location of each site and the typical surface rocks found there. Some of the rocks were igneous and quite hard. Other rocks were softer and crumbly when rubbed. At home he conducted a second-hand investigation of the geology of each area visited, using information from the internet.

- **a** Liam discovered that in each location there were different types of rocks through which the rivers were flowing. Explain the connection between this information and the different amounts of material suspended in each river. *Hint 1* (1 mark)
- **b** Liam developed a hypothesis after analysing his collected information. His hypothesis stated: Rivers have more suspended sediment if they flow through areas containing shales rather than granites.

  A good hypothesis must be able to be tested scientifically. Explain whether or not Liam's hypothesis can be tested. (2 marks)
- **c** Liam began his experiments to test his hypothesis. He selected one water sample from an area where shale rocks predominated and another sample from a granite area.
  - **i** Describe the method he will use to measure the amount of suspended solids. In your answer name the equipment he will use. (5 marks)
  - **ii** Explain whether or not his results will be reliable. (1 mark)

2 True or false?

- **a** A hypothesis is a scientific guess. (1 mark)
- **b** The accuracy of a first-hand investigation can be improved by repeating it several times. (1 mark)
- **c** *Scientific American* is an example of a secondary source of information. (1 mark)
- **d** Many variables need to be controlled in a first-hand investigation in order to ensure the validity of the conclusions made. (1 mark)
- **e** In an experiment where the rate of sugar dissolving in water as a function of stirring rate is investigated, the stirring rate is the independent variable. (1 mark)

3 Lawrence read the following information in a science magazine.

> Just as people can be right-hand or left-hand dominant, so they can be right-eye or left-eye dominant. In order to determine which eye is your dominant eye, follow these instructions: Hold a pen vertically in front of you with arm outstretched. With both eyes open, focus on the pen and line up the tip of the pen with some distant object. Close one eye at a time. The dominant eye is the eye which, when open, still sees the pen lined up with the distant object.

Lawrence used this method to test members of his family. His results showed that of the three right-hand dominant people two were also right-eye dominant and one was left-eye dominant. Both the left-hand dominant people were also left-eye dominant.

Discuss the results that Lawrence obtained and whether this experiment needs modification to make reliable conclusions. (2 marks)

(cont.)

4 Match the terms in column 1 to the terms in column 2 using the appropriate code letters. (5 marks)

Column 1		Column 2	
A	thesis	F	secondary source
B	controlled variables	G	allowed to change
C	independent variable	H	original source
D	*New Scientist* magazine	I	dependent variable
E	measured variable	J	remain constant

5 Betty conducted a first-hand investigation to determine the solubility of potassium nitrate crystals in water as a function of temperature. Her results are shown in the following graph.

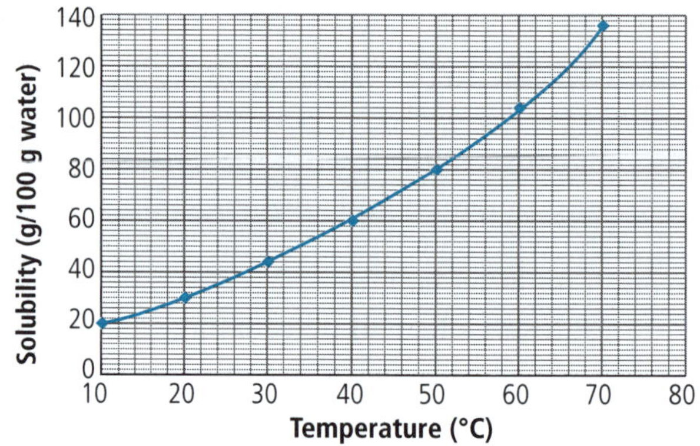

a Betty has not written a title for her graph. Write a suitable title. (1 mark)

b Identify the independent and dependent variables in this experiment. (2 marks)

c Identify a variable that would need to be controlled in this investigation. (1 mark)

d Write a suitable method for this investigation. *Hint 2* (4 marks)

e Write a suitable conclusion for this investigation. (1 mark)

f In her report, Betty wrote: 'The solubility of salts in water increases with temperature.' Is she justified in making this statement? Explain. (2 marks)

6 Felicity noticed that some of the nails in her tool shed showed evidence of rust. She wondered how she could protect her nails from rusting and so she collected information about rusting from the library and the internet. She summarised this information and decided to then investigate the effect of electric currents on the rusting of iron. She set up the apparatus as shown in the diagram. She made up a solution of salt water and placed two nails in the water so they were partly submerged. One nail was connected to the negative pole of a dry cell and a copper sheet was connected to the positive. This arrangement allowed an electric current to flow through the test nail.

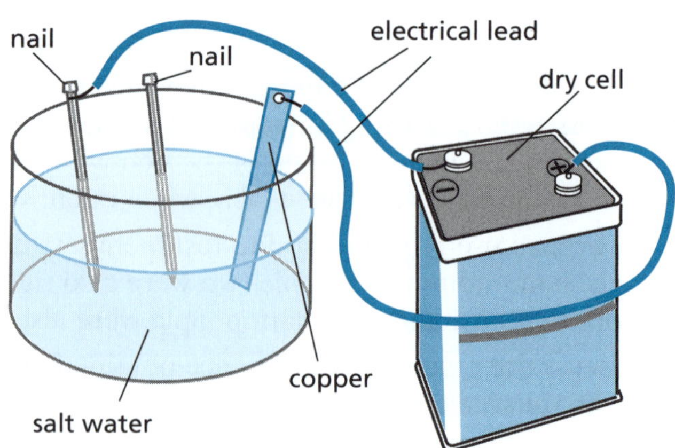

a What is the purpose of the second nail which is not connected to the dry cell? (1 mark)

b Identify three variables that need to be controlled. (3 marks)

c The experiment lasted several days and at the end of that time Felicity noticed that the nail connected to the dry cell had not rusted whereas some rust was found on the surface of the second nail. Felicity concluded that electrifying iron prevents resulting. Discuss the reliability of the experimental results. (1 mark)

7 In conducting second-hand research it is important that you understand the information you are reading. It is also important that you summarise the essential ideas that you read into a brief series of sentences or dot points. The text that follows concerns a useful polymer plastic called Teflon. Read this text and summarise the major ideas. *Hint 3* (5 marks)

> If one word can sum up Teflon, it is versatility. Since the 1960s, NASA has used Teflon in a number of roles for the space program. It is used to insulate wires and cables, to make space blankets, in seals and bearings; it is coated on landing gear and used for linings of the cargo door and hold. Teflon can withstand continuous flexing under high pressure. A braided hose made of Teflon can go through thousands of flex cycles without failure. Teflon is moisture resistant. Water doesn't wet Teflon. In fact, absorption is less than 0.01 per cent. Teflon is safe when in contact with food. Because no odour, colour or taste is imparted to food by Teflon it is safe to use in the kitchen. Almost nothing sticks to Teflon, thus making it very useful as coatings in pots and pans. Teflon is resistant to temperature changes. It can be used with very little change to its structure between temperatures of –70 °C to +260 °C. A large range of fluids can come into contact with Teflon at extremes of temperatures and pressures with no adverse effects.

Hint 1: Some types of rocks are usually more readily weathered and eroded than others.

Hint 2: Include the equipment used as well as amounts of different materials.

Hint 3: The summary of major ideas should be in your own words and the number of examples should be minimised.

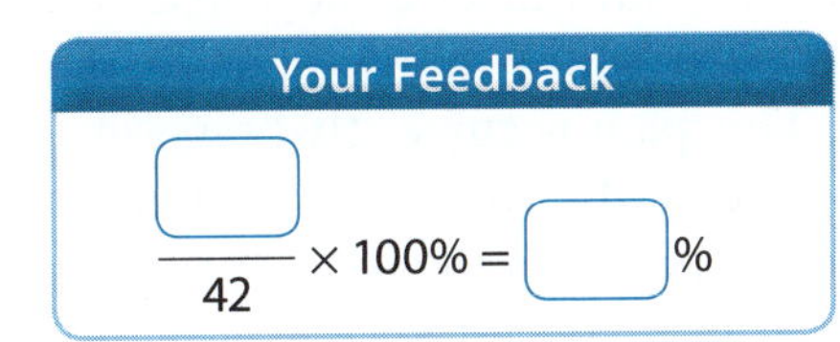

MODELS, SIMULATIONS AND THE MEDIA

Investigations and problem solving

QUICK REVISION

1 Some scientific ideas are so complex that __________ are constructed to help us to understand a structure or process. Biologists use light microscopes and electron microscopes to __________ cellular structures. These images are __________-dimensional. A better understanding of the cellular structure is obtained by constructing three-dimensional __________ models. Chemists investigate the structure and composition of matter. Writing a __________ formula for a compound (e.g. NaCl for sodium chloride) tells us that for every one sodium ion there is __________ chloride ion in the compound. By constructing a three-dimensional physical model of the NaCl crystal we can see how these __________ are arranged. The following diagram shows four different ways of modelling molecules of hydrogen peroxide (H_2O_2).

Drawings can also be used to model a scientific concept. For example, when Rutherford discovered the __________ nucleus, he developed the __________ model of the atom. In this model he placed the very small nucleus at the __________ of the atom and he placed the electrons in __________ around the nucleus, just like the __________ orbit the Sun.

2 Computers have greatly assisted scientists in their research. Computer __________ can visually display quite complex ideas in a more simplistic and visual form. Computer programs can be written to __________ a chemical or physical process. The more advanced and complex the program, the more __________ it is to accurately show the stages of the process being investigated. It is quite common for such programs to undergo various stages of __________ before being released. Simulations used by scientists include the progress of a chemical reaction, the trajectory of a rocket as it is __________ into space and the movement of electrons in a photoelectric cell. Simulations allow students to understand the individual __________ of a complex process. Simulations also remove the __________ to students when dangerous chemicals combine in a chemical reaction.

3 Scientific theories are not static. They can __________ as more data is collected. Since Darwin first proposed the theory of evolution in the mid-19th century, considerable experimental __________ has been collected which has led to important modifications of the __________ theory. This is the normal way __________ knowledge progresses.

Science is also driven by the needs of society. Increasingly these needs are __________ rather than local. For example, the loss of ozone from the ozone layer in the stratosphere was first __________ in the early 1980s. Scientists investigated this problem and showed that chemicals called chlorofluorocarbons (CFCs) were the __________. A scientific committee was set up by the United Nations and they produced an action plan (called the Montreal Protocol) for all countries to follow that would __________ the destruction of the ozone layer over many decades. This plan is currently being implemented. Sometimes groups in society do __________ agree with scientific theories. The current debate over climate change and global warming is such an example. When data is very complex and __________ to understand by non-specialists then __________ decisions and conclusions are very difficult to make.

MODELS, SIMULATIONS AND THE MEDIA

Investigations and problem solving

QUICK REVISION

4 Scientists working in universities, research institutes and in industry use the media to publish their __________ findings. Specialist journals publish scientific papers. These journal articles are __________ reviewed and so their __________ is high. These journals do not accept articles from the __________ public. Science writers are employed by media organisations to report on new __________ discoveries. Their interpretation of the __________ research can sometimes alter the public perception of the science involved as the writer will likely use less __________ language. Other non-specialist writers may comment on scientific __________ in blogs. Blogs often include the __________ views of the writer and so cannot be relied upon. Some blog sites exist to __________ the prevailing scientific viewpoints. Anti-vaccination campaigners provide __________ data which supports their views but they do not include data that refutes their views. Anti-evolution sites exist on the internet. These sites are __________ scientific as they do not use the scientific __________ to investigate an idea. School students should be careful to use reputable journals and internet sites when seeking information for an assignment.

Answers **1** models; investigate (study); two; physical; chemical; one; ions; atomic; nuclear; centre; orbit; planets **2** graphics; simulate; likely; development; launched; stages (steps); risk **3** change; evidence; original; scientific; global; observed (detected); cause; halt (stop); not; difficult; informed **4** research; peer; reliability; general; scientific; original; technical; discoveries; personal; counter (refute); selected; not; method

MODELS, SIMULATIONS AND THE MEDIA

REVISION SUMMARIES

1 Scientists use **models** to describe a structure or process. The model may be in the form of a diagram, a computer program or a three-dimensional physical construction. For example, scientists may build physical models of **crystalline structures** that help us to visualise the arrangement of atoms, molecules or ions in the crystal lattice (see crystal structures of diamond and sodium chloride in the diagram below), or **cells** or cellular structures (e.g. DNA) which assist in explaining how cells operate.

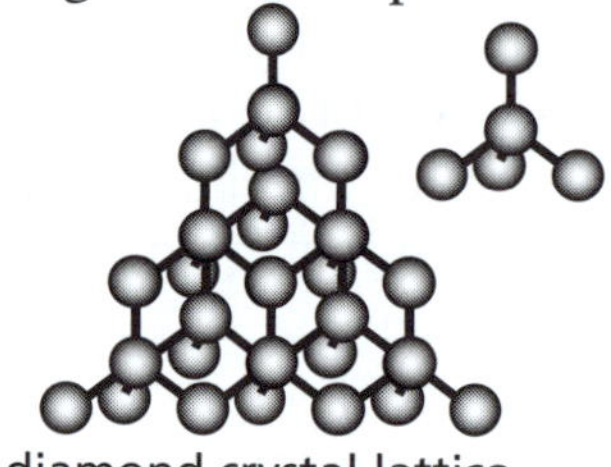

diamond crystal lattice

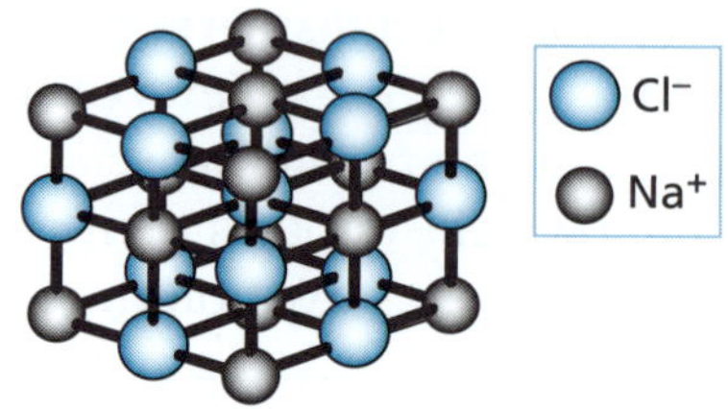

salt (NaCl) crystal lattice

The **Bohr model** of the atom is a useful model to explain the electron arrangements around the nucleus of an atom. This model can be described in a two-dimensional diagrammatic form showing the electron shells and the number of electrons in each shell (see the diagram below). Similarly the structure of the solar system can be shown diagrammatically with the Sun at the centre and planets orbiting at various distances from the Sun. The **solar system model** can also be physically constructed using plastic balls and wires to hold the planets in place.

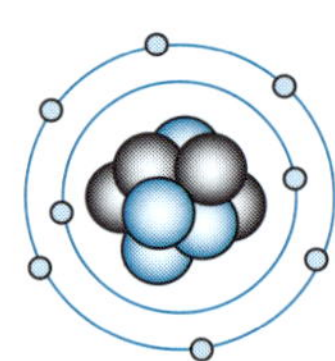

2 **Computer programs** are very useful in modelling or **simulating** a chemical or physical process. The more complex the computer program the more closely it simulates the actual event or process. The accuracy of the simulation depends on the information that is used to create the program. The computer program goes through various stages of **testing** and **modification** so as to improve the final simulation. The processes of physical weathering and erosion of rocks could be simulated using a computer program. Such a simulation has the advantage of speeding up a process that takes a very long time in nature. In a similar way, very fast chemical reactions can be simulated at a slower speed than they occur naturally, so the stages of the reaction can be examined.

Climate change is another example of the use of **mathematical models** to simulate past and future events. Scientists can develop computer models of the Earth's climate that can be tested to determine whether they correctly predicted past climate events. If the model is successful in predicting the past events then an assumption is made that these models will also be **predictive** for future climate change. Using models to predict the effect of increasing levels of carbon dioxide on global warming and sea-level change is quite difficult. For example, a model of sea-level rises produced by the Intergovernmental Panel on Climate Change in 1990 predicted that over the following two decades sea levels would rise by between 1 cm and 6 cm. Observations over that 20-year period showed that the model predictions were conservative and that the actual rise was slightly above the predictions.

At the school level, computer programs can provide examples of simulations of reactions and processes. For example, computer simulation programs are available to investigate the effect of

variables on the speed or rate of a chemical reaction. The rate of decay of a radioactive element can also be followed by computer simulation. This has the advantage of not exposing students to radiation while still conducting a simulated experiment. A simulation can also be performed in a much shorter time span.

3 **Scientific theories** develop from the work of many scientists, sometimes over a long time period, as they investigate a problem. Scientists continually debate and argue the merits and shortcomings of a theory. Theories often become **modified**, or discarded to be replaced by other theories, as new evidence is obtained. Over time a scientific theory may develop into a fundamental **law** of science. The law of energy conservation was eventually modified to become the law of mass–energy conservation when Einstein demonstrated the equivalence of mass and energy in his famous equation ($E = mc^2$).

Scientific arguments can be used to make decisions regarding personal and **community issues**. Science does not operate in isolation from the community but is often driven by community and global concerns. For example, the World Health Organization promotes **vaccination** programs against disease. These vaccination programs are based on scientific research and arguments which demonstrate that a large pool of vaccinated people in a population lowers the risk of disease outbreaks. Communities are protected when children are vaccinated at an early age. Unfortunately, in recent years, some families have refused vaccination of their children based on false logic or anecdotal evidence from anti-vaccination groups that vaccines can cause harm. Some internet sites use non-scientific information as evidence against vaccination. As a consequence, dangerous diseases are starting to reoccur in some populations.

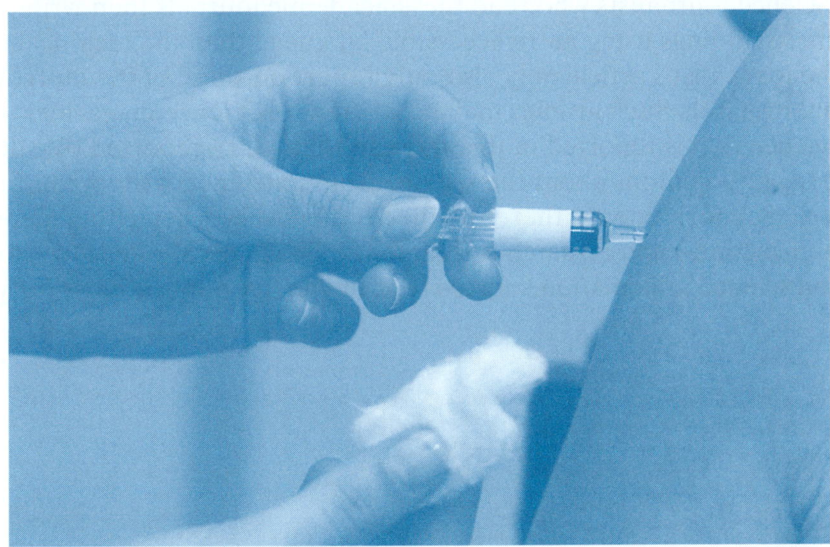

4 Studies have shown that the public generally hold the **science profession** in high regard. The major areas of science research cited by surveyed groups as having had a major impact on society are **advances in medicine**. The media reports about scientific discoveries, however, are often criticised as doing a poor job to educate the public about science. Science-based programs on television such as *Catalyst* (ABC) report on new discoveries. There are few science programs for children and they tend to focus on biology and nature rather than other sciences such as chemistry and physics. For example, there has been little mention in the media of the use of a form of carbon called graphene (see structural diagram on the next page) in computing.

(cont.)

The media news cycle can often **distort** the accurate reporting of new scientific discoveries. Scientists report their discoveries in **scientific journals** that are read by other members of their profession. Science journalists who read these academic journals then write short articles to summarise these findings using more accessible language than the technical terms used by the scientists. The journalist's article may also only report on some of the more interesting aspects of the original research. Such articles may be written for science magazines, newspapers and blogs. The science that is reported in this way can then be re-reported by other non-specialist media and personal opinions about the science from untrained writers can find their way into the mix as well. School students, therefore, need to be wary when citing media articles about science in their second-hand research assignments. The general rule is to only use reputable **peer-reviewed** science information sources (e.g. university publications) for your research.

Checklist

Can you:

	✓
1 *Explain how models can be used in science to aid understanding of scientific concepts?*	☐
2 *Explain how computer simulations are useful in understanding complex processes?*	☐
3 *Describe how scientific research is driven by global and community concerns?*	☐
4 *Describe how the media are involved in the dissemination of scientific discoveries?*	☐

MODELS, SIMULATIONS AND THE MEDIA

Investigations and problem solving

REVISION TEST

1 The following diagram shows a molecular model of butane. The larger atoms are carbon (C) and the smaller atoms are hydrogen (H).

a What is the molecular formula of butane? (1 mark)

b The model shows atoms joined by thin rods. What do these rods represent? (1 mark)

c What information does the model provide about butane which is not provided by its molecular formula? (1 mark)

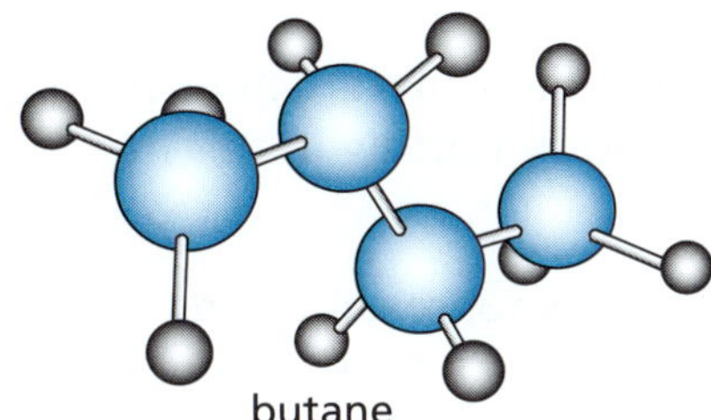

2 The following diagram shows models of the crystals of iron and copper.

a Identify two features that are common to each crystal lattice. (2 marks)

b Identify two features that are different about each crystal lattice model. (2 marks)

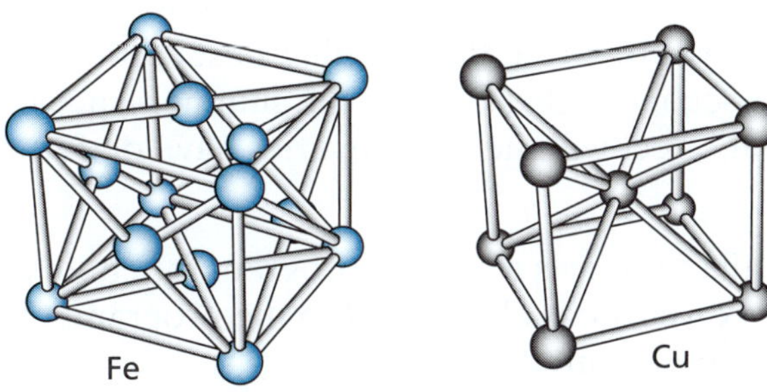

3 Read the following text and answer the questions below.

In 2010 a newspaper (*The Observer*) in the United Kingdom published an article under the title: 'Fish oil helps school children to concentrate'. The newspaper reported that 33 United States school boys aged from 8 to 11 had been given daily supplements of DHA (a type of omega-3 dietary fatty acid which can be found in fish oil capsules) had shown improvements in mathematical tasks requiring high concentration levels compared with students who had received placebos over an eight-week trial. The placebos contained no omega-3 acids. The article also stated that their behaviour had improved at school.

The original scientific paper on which the journalist had based his article showed that the scientists had not used fish oil supplements but omega-3 fatty acids extracted from marine algae. The original study had involved brain imaging technology in which the scientists were examining areas of the brain that became activated when various tasks were being performed. Some members of the group consumed different doses of omega-3 fatty acids prior to the brain imaging and others received none. Each test group had between 10 and 13 students. The results showed no differences in brain activity between groups.

a Has the journalist accurately reported the findings of the original scientific study? (1 mark)

b Did the scientists use fish oil supplements? Why did the journalist use this term? (2 marks)

(cont.)

c Were the scientists trying to show that academic performance and behaviour improved if omega-3 supplements were consumed by the students? (1 mark)

d What is a placebo? (1 mark)

e Were the sizes of each test group large enough to obtain reliable conclusions? Explain. *Hint 1* (2 marks)

f Later research using 450 students also showed that fish oil had no effect on learning. Suggest a reason why this research was never reported in a newspaper. (1 mark)

g The sales of fish oil capsules to the general public have increased to levels of over $2 billion per year. How might the type of reporting in this newspaper article affect the sales of fish oils? (1 mark)

4 Scientists develop climate models as representations of the interactions between the atmosphere, oceans, the land and solar energy. These models are mathematical models and are quite complex.

a What data do scientists use to test if their climate models produce reasonably accurate results? (1 mark)

b Climate models make predictions about future trends in sea levels. Explain how accurate these models have been in predicting rises in sea level. (1 mark)

c In September 2012 it was reported that satellite observations of Arctic sea ice showed that it was melting at the second highest rate in the summer since measurements were first started in 1979. What effect will these discoveries have on future developments in climate modelling? (1 mark)

5 Network Ten in Australia produces a science program for children called *Scope* which is screened once a week. A website is also linked to the program. On this website past episodes, videos and experiments are featured.

a How do producers of a children's science TV program ensure that the content is educationally appropriate as well as scientifically accurate? *Hint 2* (2 marks)

b Identify the main branch of science that is related to the following program content.

i Zinc has many important uses in society; it is used in sunscreens, galvanising of steel and as electrodes in some batteries. (1 mark)

ii Different technologies for communication include mobile phones, the internet and communications satellites. (1 mark)

iii Plants have many uses including agriculture, oxygen production and providing biomass energy. (1 mark)

iv Our solar system is located in the Milky Way galaxy which is one of many galaxies in the universe. (1 mark)

6 In 2008, following a long investigation, the New South Wales Health Care Complaints Commission (HCCC) issued a health warning about the Anti-Vaccination Network (AVN) and its website which provides inaccurate and misleading advice on vaccination. Following a court case in 2012 the NSW Supreme Court stated that the HCCC did not have the right to investigate the practices of the AVN as the original complaints had not shown that anyone had been adversely influenced by their practices. Other anti-vaccination groups exist in Australia and overseas. They generally believe that vaccines are poisons and that governments, doctors and pharmaceutical companies have colluded over this issue. In 2012 the Bill and Melinda

Gates Foundation funded an internet-based initiative to counter the misinformation of anti-vaccination groups so that the foundation's efforts to vaccinate people in the Third World is not compromised.

Discuss how the media can influence the work of science to improve the health of the population. *Hint 3* (3 marks)

7 The following photograph shows a flight simulator for a commercial jet.

- **a** Before trainee pilots take off in a real aeroplane they spend many hours in a flight simulator. How is such a simulation created? (1 mark)
- **b** Why is it important to learn on a flight simulator before flying a real aeroplane? (1 mark)
- **c** In 1909 flight training usually involved the trainee sitting in a barrel-shaped object (representing the cockpit) mounted on a steel frame which could be made to pitch and roll using various levers and wheels. How is such a model different to a modern flight simulator? (1 mark)

Hint 1: In experiments involving living things the number of repeats needs to be much higher than investigations involving non-living things.

Hint 2: Identify the people or groups who are experts in this area.

Hint 3: You should consider both positive and negative aspects.

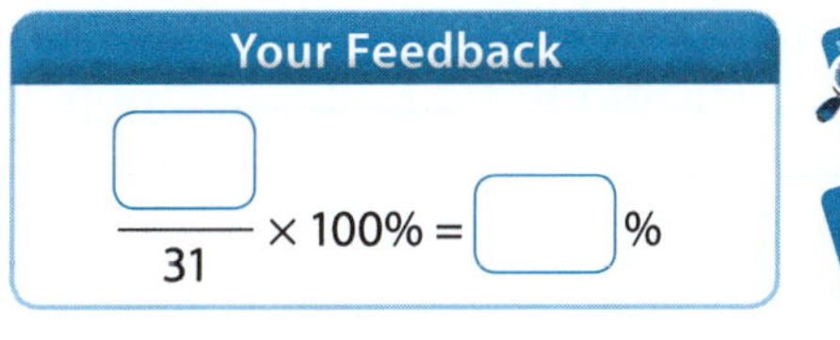

QUANTITATIVE ANALYSIS OF RESULTS

Investigations and problem solving

QUICK REVISION

1 *Qualitative* refers to descriptions or distinctions based on some __________ or characteristic rather than on some quantity or measured value. Qualitative data is data that is not quantified and not measured __________. This deals with descriptions and includes colours, textures, smells, tastes and appearance.

Quantitative refers to information based on quantities or data that deals with __________ and can be measured. Examples include length, height, area, volume, speed, time, temperature, humidity, weight and sound levels.

2 Many experiments yield results that can be measured. For example, the time taken to complete a certain __________ in a race, or the __________ rise in a solution as it is warmed. An experiment should be repeated __________ times and scientists repeat experiments to show that their results can be replicated. Repetition makes the results more __________ and reduces the possibility of __________ or irregular results.

Suppose a scientist repeated an experiment __________ times and obtained these values: 60.5, 54.7, 59.8, 60.0, 60.3.

All these results, except 54.7, are very __________ to 60. The slightly different value can be referred to as an __________, which is an observation that lies an abnormal distance from other values in a sample. The best way to deal with an outlier is to ignore it and make the __________ again. When this is done the results could be: 60.5, ~~54.7~~, 59.8, 60.0, 60.3 and 59.7.

The outlier has been removed, or __________ out, in the data.

The __________ of a set of data is: highest value – lowest value. For this example the range = 60.5 – 59.7 = 0.8.

The mean (or __________) of a set of data is the sum of all the scores divided by the __________ of scores. For this example: (60.5 + 59.8 + 60.0 + 60.3 + 59.7) ÷ 5 = 60.06.

Given that these measurements were made to an accuracy of only one __________ place, the average cannot be any more accurate than that. So, writing the mean correct to one decimal place gives __________.

3 Results are often represented in table format or as __________. Graphs or charts are often used to simplify understanding of large quantities of data and the relationships between parts of the __________. They can be created by hand (often on __________ paper) or by computer using a charting application, such as __________. Certain types of charts are more useful for presenting a given data set than others.

- A column or bar graph has rectangular __________ with lengths proportional to the values that they represent. The bars can be plotted __________ (column graphs) or horizontally (bar graphs).
- A pie chart (__________ graph) and a divided bar graph show values as a __________ or percentage of the whole.
- A line graph is a __________-dimensional scatter plot of ordered observations where the observations are connected and continuous.

Typically a graph contains very few words, and the meaning is inferred faster than from reading the text. The graph's __________ appears above the main graphic and provides a concise __________ of what the information in the graph refers to.

QUANTITATIVE ANALYSIS OF RESULTS

Investigations and problem solving

QUICK REVISION

The following diagram shows some examples of graph types. These graphs are not yet complete as the titles, axes and scales have not been included.

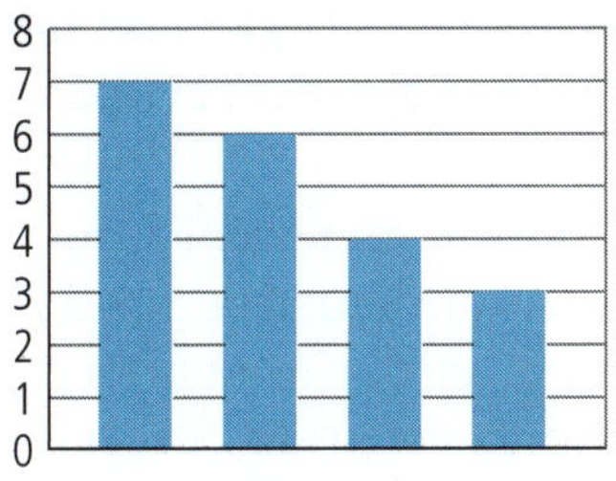

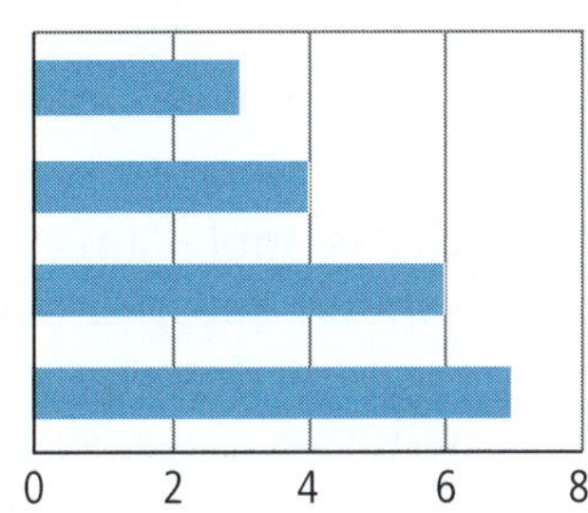

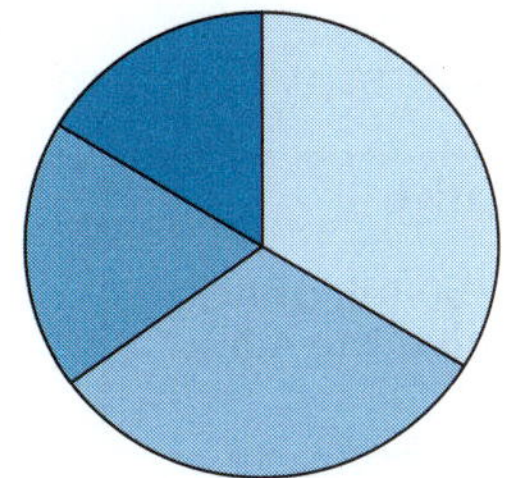

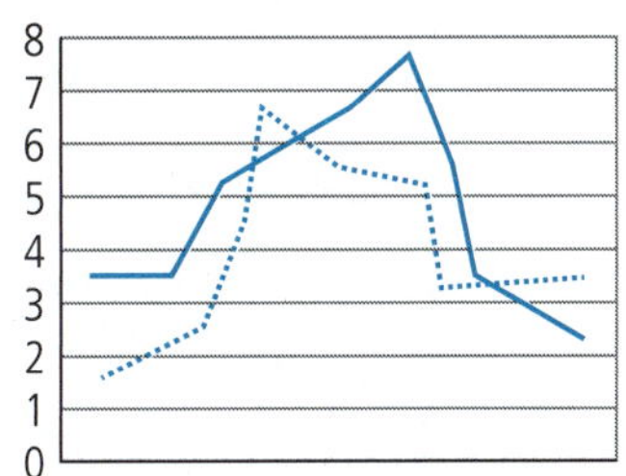

4 Interpolation means to estimate a value ____________ the values already known or determined. Extrapolation means to extend the relationship ____________ the given values to an unknown situation by assuming that the existing pattern holds true. It is always risky to extrapolate from a sample very far ahead. While you are assuming the same ____________ still applies, it may not.

For example, in this table:

Variable A	2	4	6	8	10
Variable B	6	12	18	24	30

- if variable A has a value of 7 then, by interpolation, variable B should be ____________.
- if variable A has a value of 20 then, by extrapolation, variable B should be ____________.

Answers **1** quality; numerically; numbers **2** distance; temperature; several; reliable; errors (mistakes); five; close (near); outlier; measurement; crossed; range; average; number; decimal; 60.1 **3** graphs; data; graph (grid); Excel; bars; vertically; sector; fraction (proportion); two; title; description (statement) **4** between; beyond; relationship (pattern, link); 21; 60

QUANTITATIVE ANALYSIS OF RESULTS

Investigations and problem solving

REVISION SUMMARIES

1 Research methods in science fall into two fundamental categories: **quantitative** or **qualitative**.

- Quantitative research gathers data in **numerical** form which can then be categorised, ranked or measured in some way. This can be used to draw **graphs** and **tables** of raw data. Experiments normally produce quantitative data, as they are concerned with measuring things.
- Qualitative research assembles information that is not in numerical form, for example, descriptions, open-ended questionnaires, observations and unstructured responses. Qualitative data is usually **descriptive** and therefore is harder to analyse than quantitative data. Analysing qualitative data is difficult and requires accurate descriptions. Sometimes the objectivity of the researcher can play a part since scientists necessarily have attitudes and values which they bring to their research.

Many experiments can produce both quantitative and qualitative information.

2 In the following example quantitative data is generated.

Students add powdered or finely divided zinc to solutions of hydrochloric acid in a foam cup and measure the temperature changes. The volume and concentration of the acid remains **constant**, only the mass of the zinc added changes. The reaction is:

$$\text{zinc} + \text{hydrochloric acid} \rightarrow \text{zinc chloride} + \text{hydrogen} + \text{heat}$$

(Appropriate eye protection must be worn throughout the experiment and during any clearing up session. Flammable hydrogen gas is released during each reaction, so care should be taken to make sure there are no naked flames or other ignition sources in the laboratory.)

The results of this experiment are recorded in the following table.

Mass of zinc added (g)	1.4	2.7	3.3	4.5	5.1	6.5	7.3	7.9	8.5
Temperature increase (°C)	2.1	4.4	6.0	8.5	9.2	18.3	15.0	15.5	16.6

Looking through these results it can be seen that increasing the mass of zinc increases the temperature of the final solution. All other factors are **controlled** (such as the size of the cup, volume and concentration of acid, and point in the reaction when the temperature is taken). However, the temperature results given for 6.5 g zinc appear in error. A temperature of 18.3 °C

was recorded while common sense suggests it should lie between 9.2 °C and 15.0 °C. This value is an **outlier**, and should be checked. An outlier is an observation that is **numerically distant** from the rest of the data. It lies outside the main body or group that it should be a part of. Outliers can occur by chance in any distribution but they often indicate measurement error.

On repeating the experiment for 6.5 g zinc, a more appropriate temperature of 13.3 °C was recorded. The table can now be corrected.

Mass of zinc added (g)	1.4	2.7	3.3	4.5	5.1	6.5	7.3	7.9	8.5
Temperature increase (°C)	2.1	4.4	6.0	8.5	9.2	13.3	15.0	15.5	16.6

3 These results can now be graphed. A **line graph** would be appropriate. You can do this manually using a sheet of graph paper, or use a computer spreadsheet program such as Excel.

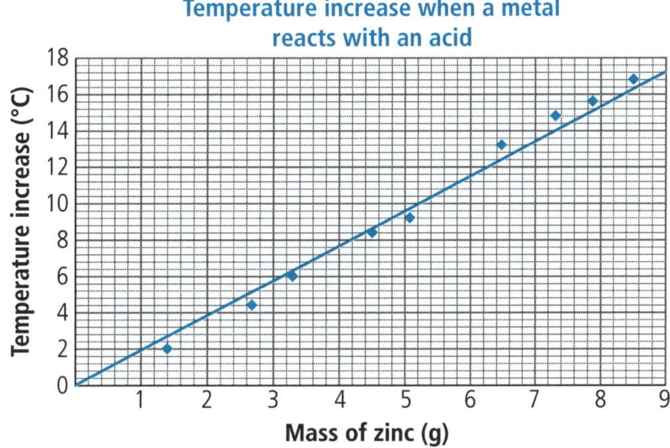

Notice that this graph:

- has a **title** describing what it shows
- has appropriately **labelled axes**, with units indicated
- occupies most of the area on the grid provided (your graphs should be at least ⅓ to ½ page in size)
- has a **line of best fit** drawn through the data points.

A line of best fit is a common way for showing how two variables may be related. It is drawn through the middle of a group of data points plotted on a scatter plot, dividing the points plotted roughly equally on either side of the line if they don't fit on the line.

4 The graph is a summary of the experimental results. It can also allow predictions to be made.

- **Interpolation** allows you to construct new data points **within the range** of a discrete set of known data points. For example, there was no zinc sample with a mass of 6.0 g used. But from the graph you can read 11.5 °C as being a close temperature increase that would have been obtained.
- **Extrapolation** is the process of estimating **beyond** the original observation interval on the basis of the connection found so far. For example, if 10.0 g zinc was used then the temperature increase should be around 19.1 °C. This was found by extending the graph beyond the plotted points.

(cont.)

Interpolation produces estimates between known observations that you can be reasonably certain of. Extrapolation is subject to **greater uncertainty** because you are assuming the same relationship between the two variables holds as you go beyond the plotted points. Extrapolating data is often less valid and has a higher risk of producing meaningless results than interpolating data.

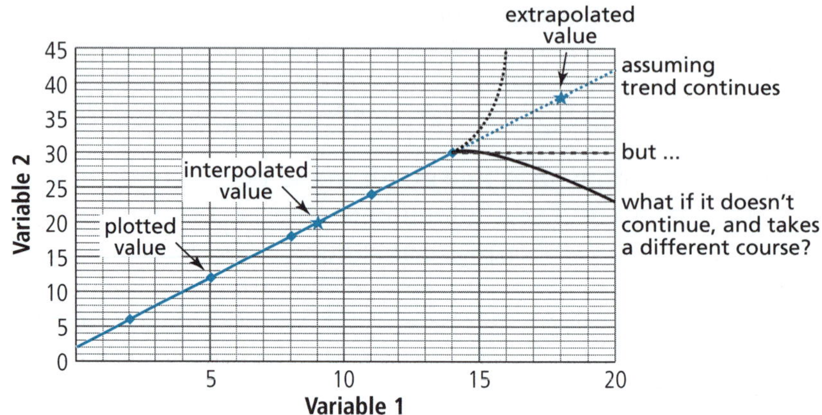

Checklist

Can you:

1. *Distinguish between the terms 'quantitative' and 'qualitative', giving examples?* ☐
2. *Tabulate experimental data (place data in a table), calculate means and ranges, and recognise and deal with outliers?* ☐
3. *Use and model spreadsheets to present data in tables and graphical forms and to carry out any other mathematical analyses?* ☐
4. *Extract data from graphs by extrapolation and interpolation?* ☐

QUANTITATIVE ANALYSIS OF RESULTS

Investigations and problem solving

REVISION TEST

1 Liquid water and its vapour can coexist in equilibrium. The vapour pressure is a measure of the tendency of water molecules to escape as a gas from the liquid. The vapour pressure/temperature curve for water is shown.

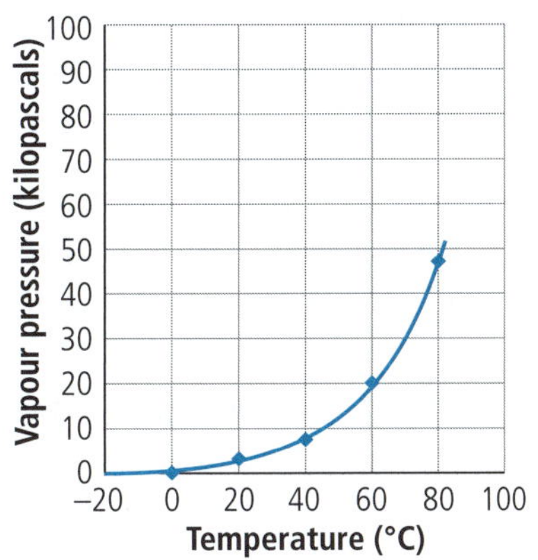

a What is the vapour pressure of water at 60 °C? (1 mark)

b What is the vapour pressure of water at 70 °C? What process are you using to determine this value? (2 marks)

c Estimate the vapour pressure of water at 90 °C? What process are you using to determine this value? (2 marks)

d At what temperature will the vapour pressure reach 100 kPa (kilopascals)? What are you assuming when you do this? *Hint 1* (2 marks)

2 In order to measure the temperature of 250 mL distilled water as it is being heated, a student made the following measurements.

Time (min)	0	1	2	3	4	5	6	7	8
Temperature (°C)	20	24	33	45	56	67	78	89	

a **i** Which value is placed along the horizontal axis? Why? (2 marks)

ii Which value is placed along the vertical axis? Why? (2 marks)

b Use a grid, or a computer graphing program, to draw a graph of this data. Label the axes appropriately, and give the graph a title. (5 marks)

c What was the initial water temperature? (1 mark)

d Estimate the water temperature after 4.5 minutes. (1 mark)

e The water temperature at 8 minutes was not measured. What would this temperature be? What assumption are you making? (2 marks)

f By extrapolating the graph to 10 minutes, the student obtained a temperature of 122 °C. Explain whether or not the student made a valid estimate of the temperature of the water. *Hint 2* (2 marks)

3 Jenna made the following measurements of the radioactivity in a sample of material. She used a Geiger counter to count the number of audible clicks over a 1-minute period when the counter was placed 1 m away from the sample. She repeated the measurements five times, and obtained these results: 32; 27; 69; 33; 29.

a She realised one of these values is in error. Which one and what could she do about it? *Hint 3* (2 marks)

(cont.)

b She made another measurement and obtained 31 clicks per minute. She now has these values: 32, 27, 33, 29, 31.
   i What is the range of these values? (1 mark)
   ii What is the mean of these values? (1 mark)

c Why is it important that a number of measurements be taken and a mean (average) calculated? (2 marks)

4 Wally collects a series of six groundwater samples from a well. He measures the dissolved oxygen concentration in these. His observations in mg/L are: 8.8; 3.1; 4.2; 6.2; 7.6; 3.6.

a He calculates the sample average for these six values and obtains 5.583333. How should Wally report his result? (1 mark)

Wally understands that he can reduce the uncertainty in the estimate of the mean results by having a larger number of observations, so he takes another six samples. His results are: 8.8; 3.1; 4.2; 6.2; 7.6; 3.6; 5.2; 8.6; 6.3; 1.8; 6.8; 3.9.

b i What is the range of these values? (1 mark)
   ii What is the mean of these values? (1 mark)

c Is there any significant difference between the mean for the six values and the mean for the 12 values? (1 mark)

5 The graph shows the population of Australian states and territories in 2012.

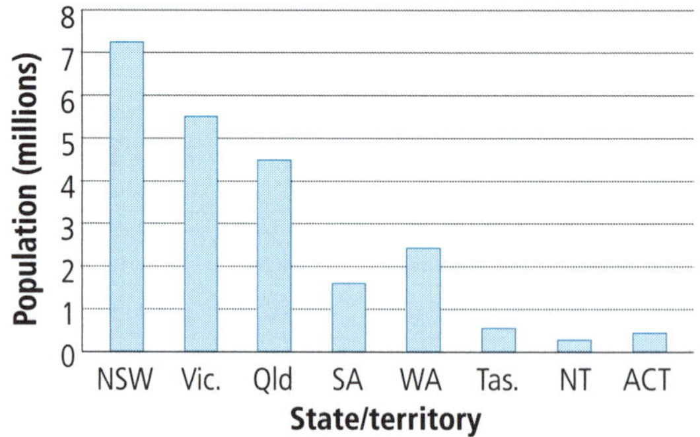

a What type of graph is shown? (1 mark)

b Explain why a line graph is not suitable for this data. *Hint 4* (1 mark)

c Another way to show this information is with the graph below. What type of graph is this? (1 mark)

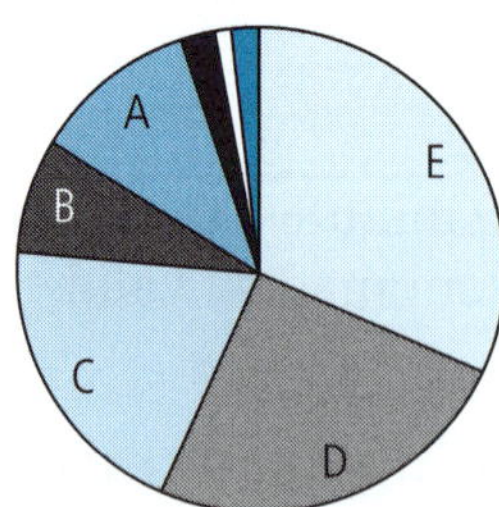

d Why is it suitable to represent this data using this sector graph? (1 mark)

e The state which grew the fastest during the year was Western Australia. Which letter in the graph above shows Western Australia? (1 mark)

f New South Wales is the most populated state. Given the total population of Australia in 2012 was 22.6 million, what percentage of residents live in New South Wales? (2 marks)

6 The following photo shows crystals of the element bismuth.

Bismuth has the unusual property of expanding while cooling. It is, therefore, introduced in many alloys to lessen or check shrinkage in the mould. One typical alloy, called cliche metal, is composed of 48% tin, 32.5% lead, 9% bismuth and 10.5% antimony. It possesses a considerable degree of hardness and it wears well.

a Use these values to draw up a divided bar graph of the metals in this alloy. (4 marks)

b What is the chemical symbol for antimony? (Look at the periodic table on the inside cover of this book.) (1 mark)

c If 60 kg of this alloy is made, what mass of antimony is used? (1 mark)

Hint 1: You will need to continue the curve, carefully following the same smooth curve.

Hint 2: Be careful! What happens at the boiling point of water?

Hint 3: This value is an outlier.

Hint 4: What properties of the variables are needed so that the line graph would be suitable?

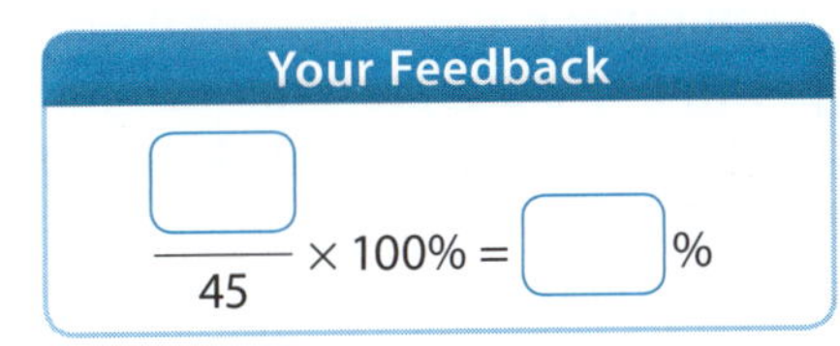

PAGES 245–246
PAGE 252

PROBLEM SOLVING

Investigations and problem solving

QUICK REVISION

1 Much of science is concerned with ____________, so standards of measurement are needed. In Australia we use a metric ____________ of measurement. SI units are now the sole legal units of measurement in Australia, although in a few other countries the old imperial measure is still used. The metre, kilogram and second were defined as some of the ____________ units in this system and units formed from combinations of these base units are known as ____________ units. So for example, 1 ____________ (cm) equals one hundredth of a metre (m), or 1 m = ____________ cm.

The main prefixes are shown below.

Prefix	Symbol	Multiplier
kilo-	k	× 1000
mega-	M	× 1 000 000
giga-	G	× 1 000 000 000

Prefix	Symbol	Multiplier
centi-	c	× 0.01
milli-	m	× 0.001
micro-	μ	× 0.000 001

The Greek letter mu (μ) is the symbol used for micro/one millionth.

For example, 1 μm is one millionth of a metre, so 1 000 000 μm = 1 m, and 1000 mm = 1 m. From this you can determine that ____________ μm = 1 mm.

Appropriate units should be used when measuring. For example, the length of a piece of A4 paper is best measured in ____________, the length of your school is best measured in ____________, while the distance from Adelaide to Perth is best measured in ____________. Distances in space are so large that even the kilometre is too small a unit. Astronomers often use a distance called the ____________-year, which is the distance that light travels in one ____________.

2 No measurement is exact. How ____________ you can take a measurement will depend on the precision of the measuring ____________. The number of digits that can be assumed for any measurement is also limited. When reading a measuring instrument, you should do so to its smallest scale division. You may ____________ within a part of a division. The figures you record for the measurement are known as significant figures.

For instance, recording 45.0 mL means that you were able to estimate the ____________ decimal place of the volume and you are implying an error of 0.05 units. If you just write 45 mL, you are stating that you were not capable of determining the first decimal ____________ and are therefore implying an error of 0.5 units.

This implies the true measurement lies somewhere between 44.95 mL and 45.05 mL.

This implies the true measurement lies somewhere between 44.5 mL and 45.5 mL.

3 Scientific notation was developed to deal with very large and very ____________ numbers. When written in scientific notation, a number like 6.89×10^7 has only ____________ digit before the decimal point. For example, children have roughly 4.5 billion (thousand ____________) red blood cells for each millilitre of blood. This number, 4 500 000 000, can be written as ____________ $\times 10^9$. On the other hand, the typical thickness of a chicken's egg shell is 0.000 34 m. In scientific notation this number is ____________ $\times 10^{-4}$. Sometimes it is easier to use a smaller unit and write it as an ordinary number. So 0.000 34 m = 0.34 ____________.

PROBLEM SOLVING

Investigations and problem solving

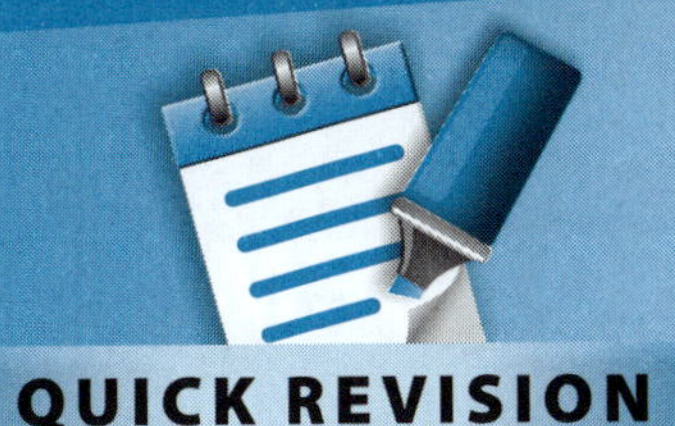

QUICK REVISION

4 Simple calculations are often required in science. For example, suppose the final velocity, v, of an object t seconds after it was thrown upwards with an initial velocity, u, is given by the ________:

$$v = u - 10t$$

If the initial velocity $u = 120$ m/s, then after $t = 3$ seconds its final velocity is $v = 120 - 10 \times 3$ $=$ ________ m/s.

You need to ________ the appropriate numbers into the equation to work out the value you need.

As another example, suppose a particle is moving at 200 cm/s. How fast is it moving in km/h? First realise that the particle covers 200 cm each ________. This is 2 m each second, since ________ cm = 1 m. Now in 1 min the particle has covered 2 × ________ = 120 m, and in 1 h 120 × 60 = ________ m. That is, it is travelling at 7200 m/h = 7.2 km/h, since 1000 m = 1 km. Hence the speed 200 cm/s equals 7.2 ________.

Answers **1** measuring; system; base (fundamental); derived; centimetre; 100; 1000; millimetres (centimetres); metres; kilometres; light; year **2** exactly (precisely); instrument (device); estimate; first; place **3** small; one; million; 4.5; 3.4; mm (millimetre) **4** equation (formula); 90; substitute (place, insert); second; 100; 60; 7200; km/h

1 The **SI system** (from the French *Système International d'Unités*) developed during the French Revolution is the system of measurement we use in Australia today. This **metric system** has units of like kind defined by powers of ten. Prefixes are used to separate units. For instance, 1 kilogram = 1000 grams, since *kilo* is the prefix for 1000.

The main **fundamental units** you are concerned with at the moment are metre (m), kilogram (kg) and second (s). Other units can be derived from these.

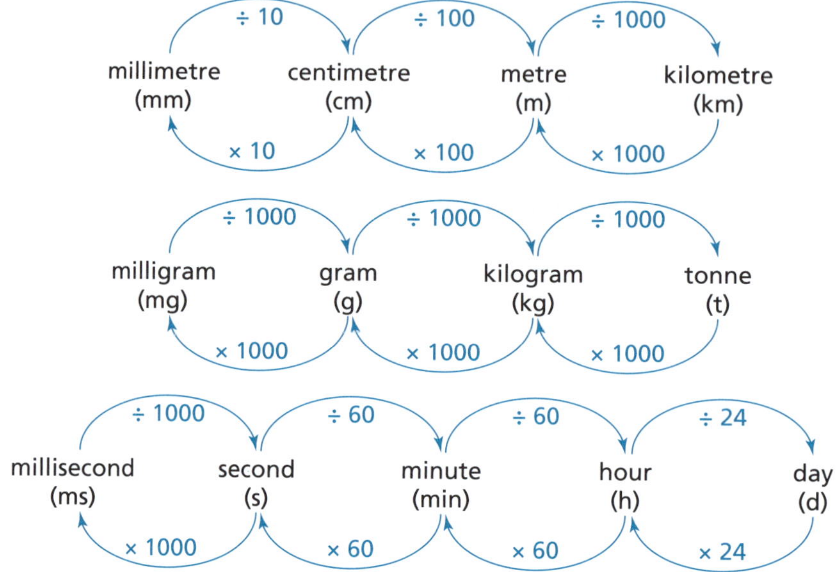

However, just because 1 cm = 10 mm, an area of 1 $cm^2 \neq 10$ mm^2. You need to square both sides to give 1 cm^2 = 100 mm^2.

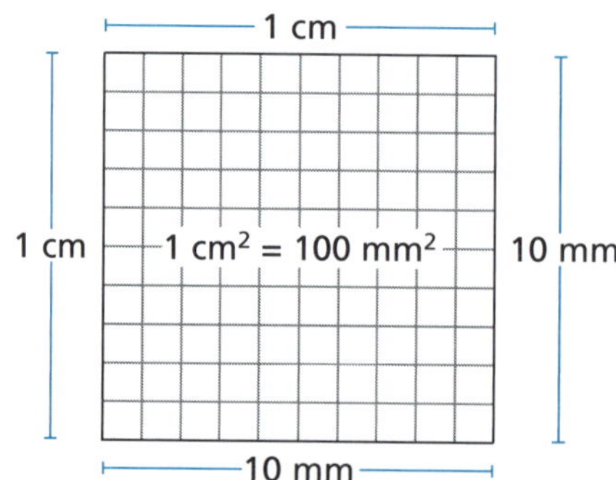

2 Counting can be exact. For example, there are 26 students in the classroom; a matchbox contains 50 matches; a cube has six faces. Measurement is not exact, even though steps are taken to be as accurate as possible. For example, the mass of a lump of lead could be 25.7 g. This was measured to an **accuracy** of one decimal place; that is, there is one digit after the decimal point. If more accurate weighing scales were used then a value of 25.73 g or maybe 25.69 g might have been obtained. But this is the best our scales can do. All we can say is that the true value lies somewhere between 25.65 and 25.75.

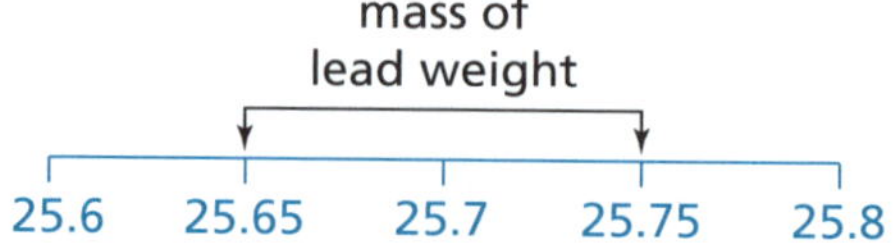

The **limit of accuracy** is half the smallest unit. In order to indicate that the true mass lies between these two extremes, it is sometimes recorded as 25.7 ± 0.05 g, since 25.65 = 25.7 – 0.05 and 25.75 = 25.7 + 0.05. (The symbol ± is read as 'plus or minus'.)

3 **Scientific notation** is used for very large numbers or for very small numbers. An example is 3.46×10^6. Only one digit is written before the decimal point, with all others after it. Your calculator has a $\times 10^x$ key that allows you to write numbers in scientific notation. For example, $345\,000\,000 = 3.45 \times 10^8$. The power that the ten is raised to is found by counting the number of places from between the 3 and 4 to after the last zero (where the decimal point is found in a normal number). Numbers smaller than 1.0 have a **negative index** (power). For example, $0.000027 = 2.7 \times 10^{-5}$. The power that the ten is raised to is found by counting the number of places from between the 2 and 7 to where the decimal point is.

One number used in science is called Avogadro's constant and has the value of 6.022×10^{23}. This is a very large number and represents the number of water molecules in 18 g water. On some calculators this may be shown as 6.022E23. At another extreme, the mass of an electron is very small at 9.1×10^{-31} kg. On some calculators this may be shown as: 9.1E – 31.

4 Consider a question like: 'Calculate the average speed of a car that covers 500 km in 7 h 15 min'. (Note that 7 h 15 min = 7.25 h, or 7¼ h.)

The formula for average speed is:

$$\text{speed} = \frac{\text{distance covered}}{\text{time taken}}$$

This can be written in symbols as:

$$v = \frac{d}{t}$$

Scientists often use appropriate letters in equations and **formulae** to save writing out the words in full. In this equation v stands for velocity which can be taken here to be a measure of the speed in a straight line. So to calculate the average speed (or average velocity) in this question:

$$v = \frac{500}{7.25} = 68.9655...$$

Now keep in mind that not all the digits after the decimal point are **significant**, even though a calculator fills its screen with them. After all, the distance was **rounded** to 500 km, and the time to the nearest quarter hour. So the answer should be rounded appropriately. In this question, giving speed = 69 km/h will be accurate enough.

A rough rule is to not use more decimal places in your answer than there are in the numbers in the question. In this example, 500 km was given to the nearest whole number, so speed correct to the nearest integer will be suitable.

Another point to remember is to use **appropriate units**. Distance was measured in kilometres and time in hours, and so speed should be given as km/h.

Consider another question: 'Between noon and 2:30 pm a car averaged 60 km/h, and then until 6:00 pm it averaged 80 km/h. What is the total distance covered in this time?'

This time you need to calculate distance, so rearrange the formula above to make distance the subject:

$$d = v \times t$$

(cont.)

Now you need to recognise there are two parts to this question.

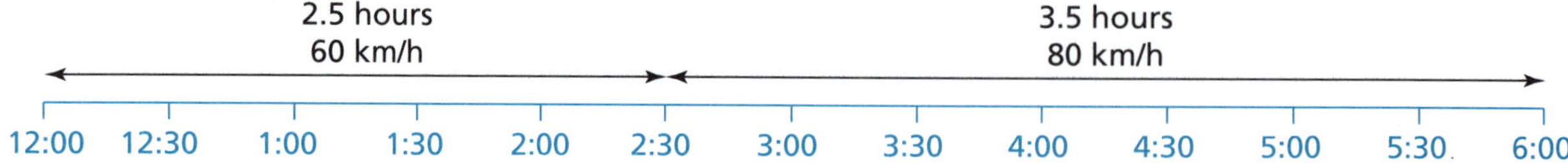

Call the distance travelled in the first part of the journey distance 1 (d_1): $d_1 = 60 \times 2.5 = 150$ km.
The distance travelled in the second part of the journey is distance 2 (d_2): $d_2 = 80 \times 3.5 = 280$ km.
The total distance travelled $= d_1 + d_2 = 150 + 280 = 430$ km.

Sometimes questions are asked from graphical information. The following graph shows the motion of a particle over 14 seconds.

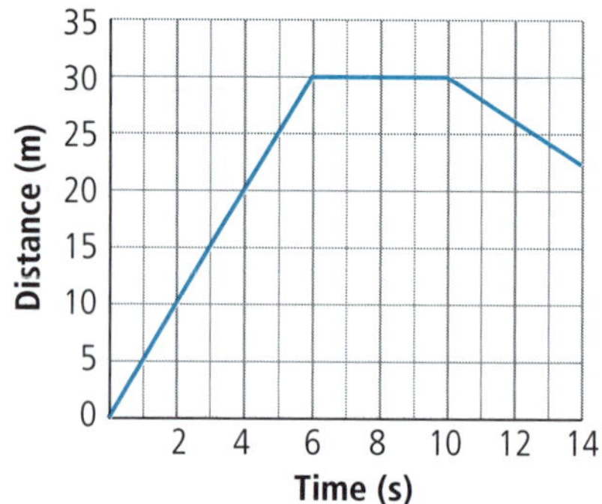

- For how long was the particle stationary?
  Stationary indicates that the particle was not moving, or at rest. This occurs when the curve is horizontal (flat) as distance is neither increasing nor decreasing. It was stationary for 4 seconds (between the 6th and the 10th second).
- Calculate the speed of the particle during the first 6 seconds.
  The particle covered 30 m in 6 s, so speed = 30 ÷ 6 = 5 m/s.
- When did the particle begin travelling in the opposite direction?
  The particle was moving outwards for the first 6 seconds, and then remained stationary for the next 4 seconds. It first started travelling in the opposite direction at 10 seconds.

Checklist

Can you:

1. *Identify the key features of the metric or SI system of measurement, and convert between simple measures of length, mass and time?* ☐
2. *Explain why all measurements have inherent errors and make measurements to an appropriate degree of accuracy?* ☐
3. *Deal with large and small numbers written in scientific notation?* ☐
4. *Perform simple calculations using a formula or from information supplied from a graph?* ☐

PROBLEM SOLVING

Investigations and problem solving

REVISION TEST

1 The dimensions of an A4 sheet of paper can be written as 297 ± 0.5 mm by 210 ± 0.5 mm.

- **a** What is meant by ± 0.5 mm? (2 marks)
- **b** The length of paper was written as 297 mm, but what is the largest length it could be? (1 mark)
- **c** What is the smallest width the paper could be? (1 mark)
- **d** Explain why the largest possible area of the paper is $297.5 \times 210.5 = 62\,623.75$ mm^2. (2 marks)
- **e** What is the smallest possible area of the A4 sheet of paper? *Hint 1* (1 mark)

2 Nicholas was answering the question: 'A force of 400 N is applied to a mass of 17 kg. Calculate the acceleration produced'. He knows the formula:

$$\text{force} = \text{mass} \times \text{acceleration}$$
$$\text{or} \quad F = ma$$

He rearranges this to get $a = F \div m = 400 \div 17 = 23.52941$ m/s.
Comment on Nicholas' answer. *Hint 2* (2 marks)

3 Write the numbers in bold print as ordinary numerals.

- **a** African elephants are the largest of land animals with a typical adult male weighing between $\mathbf{1.8 \times 10^3}$ kg and $\mathbf{6.3 \times 10^3}$ kg. (2 marks)
- **b** One light year = $\mathbf{9.46 \times 10^{12}}$ km. (1 mark)
- **c** The speed of light in a vacuum is $\mathbf{3.0 \times 10^8}$ m/s. (1 mark)
- **d** Most bacteria range from $\mathbf{2.0 \times 10^{-5}}$ m to $\mathbf{2.0 \times 10^{-6}}$ m in diameter. (2 marks)
- **e** An A4 sheet of 80 gsm paper has a thickness $\mathbf{1.03 \times 10^{-2}}$ cm. (1 mark)
- **f** The radius of a hydrogen atom is $\mathbf{2.5 \times 10^{-11}}$ m. (1 mark)

4 Write the numbers in bold print in scientific notation.

- **a** The typical speed of an alpha particle is **20 100 000** m/s. (1 mark)
- **b** The average distance from the Earth to the Sun is **149 600 000** km. (1 mark)
- **c** The Earth moves at a speed of **107 300** km/h around the Sun but still take one whole year to make one revolution. (1 mark)
- **d** Tectonic plates move at speeds of **0.05** m per year to **0.1** m per year. (2 marks)
- **e** The time taken by light to travel 1 m is roughly **0.000 000 003** seconds. (1 mark)
- **f** The wavelength of green light is **0.000 000 55** m. (1 mark)

5 The diagram shows two organic molecules.

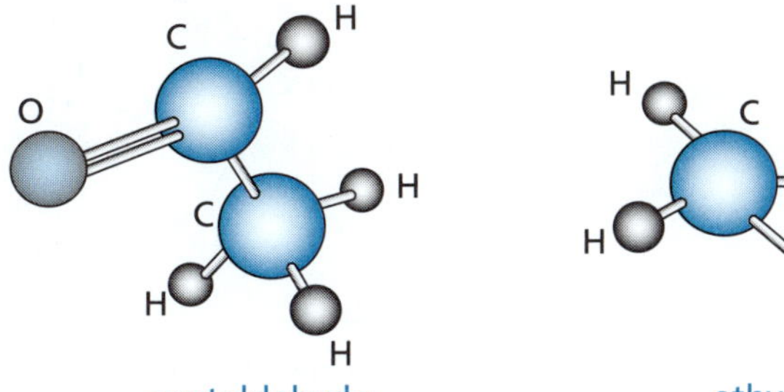

acetaldehyde

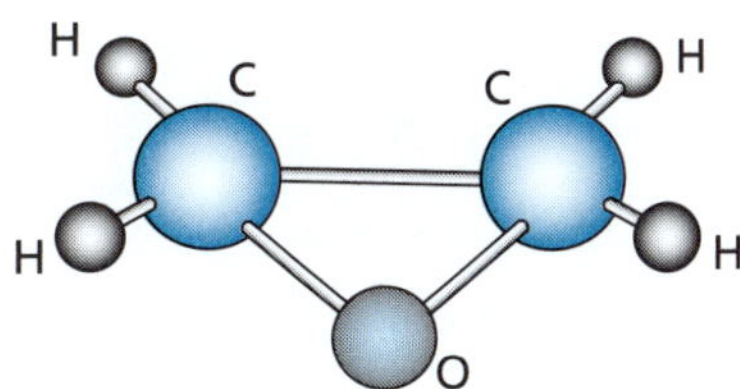

ethylene oxide

- **a** Use the following atomic weights to calculate the molecular weights of each of the molecules in the diagram: C = 12.01 u; H = 1.008 u; O = 16.00 u. (2 marks)
- **b** What do you notice about the values you obtained? (1 mark)

(cont.)

PROBLEM SOLVING *(continued)*

Investigations and problem solving

REVISION TEST

6 The average human heart beats 72 times each minute. Given a lifespan of 80 years, calculate the number of beats in this time. Write your answer in scientific notation. *Hint 3* (3 marks)

7 The average distance from the Earth to the Sun is 150 000 000 km. If light travels at 3.0×10^8 m/s calculate how long, in minutes and seconds, it takes a light beam to travel to Earth from the Sun. (4 marks)

8 The following diagram shows a glucose molecule.

glucose

a The formula for glucose is $C_xH_yO_z$. What are the values of x, y and z? (3 marks)

b Use the atomic weights from Question 5, part a to calculate the molecular weight of this molecule. (2 marks)

c What is the percentage mass of carbon in glucose? *Hint 4* (2 marks)

9 The wave equation is velocity = frequency × wavelength. In symbols, $v = f\lambda$.
A certain orange light has a wavelength of 6.15×10^{-7} m. Given the speed of light is 3.0×10^8 m/s, calculate its frequency. (2 marks)

10 GPS satellites orbit at a height of 20 200 km above the Earth's surface and complete one orbit of the Earth every 12 hours. The following photo shows a GPS satellite.

a Given the radius of the Earth is 6400 km, how far above the centre of the Earth is a GPS satellite located? (1 mark)

b Use the formula for the circumference of a circle, $C = 2\pi r$, to calculate the distance travelled in one orbit. (1 mark)

c Calculate the average speed of a GPS satellite. (1 mark)

11 The following graph shows the motion of a particle.

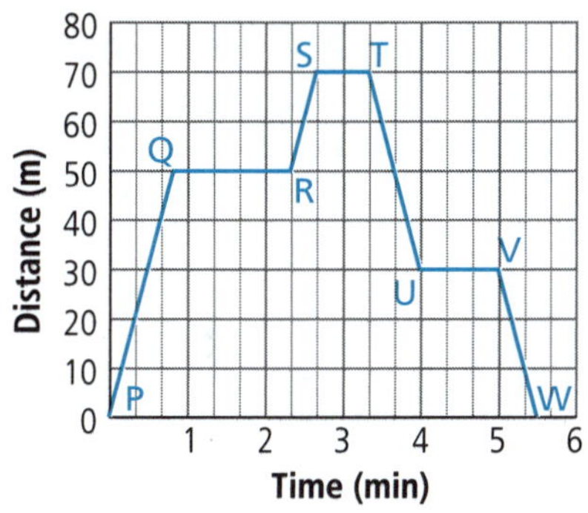

a What does each division on the horizontal axis represent? (1 mark)
b For how long had the particle been travelling when it reaches point Q? (1 mark)
c Calculate the average speed between P and Q, in m/s. (2 marks)
d Between which points on the graph was the particle stationary? (2 marks)
e Approximately how long was the particle stationary altogether? (2 marks)
f Calculate the total distance travelled by the particle in the first 5½ minutes. *Hint 5* (2 marks)

12 The guillotine on a paper production line has a frequency of 25.0 Hz. That is, it can make 25 cuts each second. A4 sheets of paper are to be cut from a long roll. Each A4 sheet has a length of 297 mm.

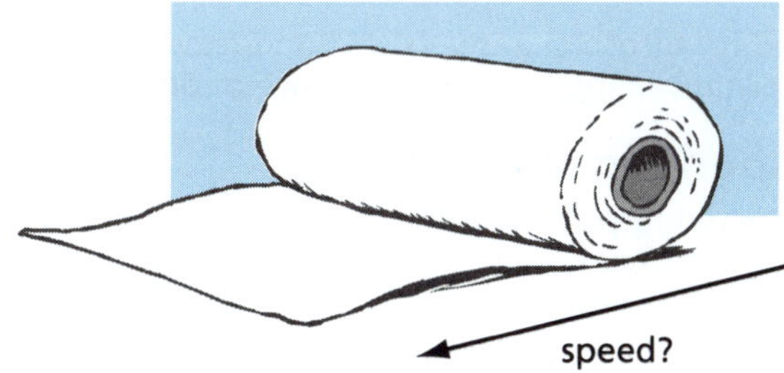

How fast must the paper on the roll be fed through so that appropriate lengths are cut? Give the speed in m/s, correct to two decimal places. (4 marks)

Hint 1: Use the two smallest extremes for the length and width.
Hint 2: The units are correct, and so is his calculation.
Hint 3: You need to change minutes to hours to days to years. Whether you use 365 days for a year or 365¼ days won't make much difference to the answer. The solution uses 365¼ days.
Hint 4: Remember from maths: percentage mass = mass ÷ total mass × 100 ÷ 1.
Hint 5: The particle travels outwards from its starting point and then back to it.

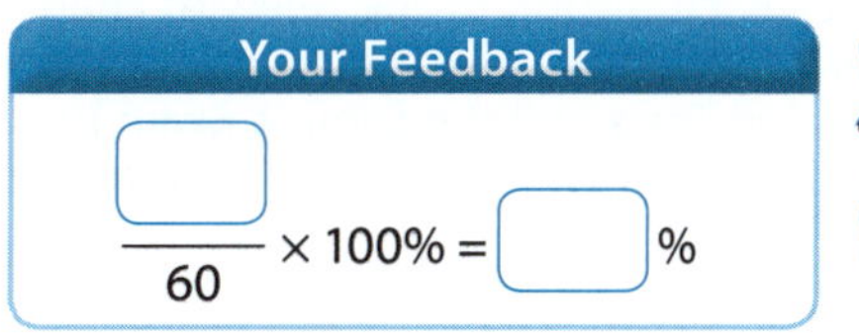

TIPS FOR THE SAMPLE EXAM PAPERS

STUDY TIPS

Scope of work covered in the Sample Exam Papers

- There are two Sample Exam Papers in the next section.
- The first one covers content from chapters in the first half of the book.
- The second one covers the whole book content but with greater emphasis on the chapters in the second half of the book.

Your school's science program

- Your school may teach the content in a different order to that presented in this book.
- You will need to identify those questions that are relevant to the content you have studied in class as you prepare for major examinations.

Suggested times

- The suggested time for each Sample Exam Paper is 50 minutes.
- Keep track of time when you do each exam.
- If you have not completed the exam in that time, continue on and record the time you have taken.
- While finishing within the time given is important, it is equally important to answer the questions correctly.

Part A: Multiple-choice questions

- Part A of each Sample Exam Paper has 15 multiple-choice questions. Read the stem of the question carefully to check what is being asked.
- If a diagram is supplied, check the labels carefully as they may give important clues.
- Choose the best response. If you are unsure of the correct answer, make the best logical choice that you can.
- No marks are deducted for incorrect answers, so don't leave any unanswered. Sometimes you can eliminate a wrong alternative, improving your chance of deciding on the correct answer.

Part B: Restricted-response questions

- Part B contains 10 restricted-response questions where one-word answers are required to fill the missing space in a sentence. These questions generally test basic recall of knowledge.
- Often the answer is a key word that forms part of the definition or point being examined.

Part C: Written knowledge and skill questions

- Part C contains written knowledge and skill questions that test your deeper understanding of the topics and your ability to process new data. These exams often test your mathematical and graphing skills, as well as your ability to interpret or draw scientific diagrams and analyse information.
  - Use the mark value of each question as a guide to the length of the response required.
  - Look for key words in each question.
  - Do not restate the question in your answer.
  - You should always present your answer in a logical order.
  - If you draw a labelled diagram as part of your answer, make sure you refer to the diagram in any explanation. Always use a pencil and ruler for diagrams.
- Sometimes you will be asked a question that involves using common knowledge and applying thinking ability to a novel scientific situation.

Preparation for the Sample Exam Papers

- Study each section of the relevant chapters again as you prepare to do the Sample Exam Papers.
- It is important to revise throughout the year. As the school year progresses, go back and redo the Revision Tests and the Sample Exam Papers. This will help you revise for major school exams. Your results will show where more revision is needed.

SAMPLE EXAM PAPER

PAPER 1

PART A *Multiple-choice questions* 15 marks

1 Which system are the lungs and diaphragm parts of?
A digestive system
B circulatory system
C excretory system
D respiratory system (1 mark)

2 What is the role of motor neurons?
A transmit nervous impulses to muscles or glands
B transmit impulses via the spinal cord to the brain
C relay messages from sense organs to the spinal cord
D transmit impulses to the central nervous system (1 mark)

3 Which of the following is an example of a non-infectious disease?
A measles
B heart disease
C cholera
D tinea (1 mark)

4 Select the correct statement about the use of electromagnetic radiation in medicine.
A Gamma rays can be used to detect cancerous breast tumours.
B X-rays can be used to treat sore muscles and stiff joints.
C UV radiation can be used to sterilise equipment in operating rooms.
D IR rays can be used to kill and treat certain types of cancers. (1 mark)

5 Which of the following is a biotic component of an ecosystem?
A wind speed
B relative humidity
C soil pH
D competitors (1 mark)

6 Which of the following best describes scavengers?
A Animals that feed off the remains of animals left by carnivores.
B The carnivores that eat the herbivores.
C Microbes that break down dead remains.
D Plants, algae or phytoplankton that absorb sunlight. (1 mark)

7 Select the true statement about the seasonal effects on ecosystems.
A Amphibians migrate as the winter approaches.
B Fish hibernate to conserve food reserves in their tissues.
C Some animals use food stored in their burrows during the wintry months.
D Evergreen trees lose leaves in the spring and summer. (1 mark)

(cont.)

8 Identify the correct statement about John Dalton.

A He proposed that atoms consisted of electrons embedded in a positive sphere.

B He proposed that atoms were like a solar system in which the electrons orbited a central nucleus.

C He proposed that electrons were located in shells surrounding the nucleus.

D He proposed that atoms were indivisible, and they could not be destroyed or created.

(1 mark)

9 Which of the following is the electron configuration of aluminium (Z = 13)?

A 2,11

B 2,8,3

C 2,8,8,3

D 3 (1 mark)

10 An element X has the nuclear symbol $^{249}_{97}X$. Which of the following statements about the atoms of this element is correct?

A There are 97 neutrons present in the nucleus.

B There are 249 neutrons present in the nucleus.

C There are 97 protons located in the shells surrounding the nucleus.

D There are 152 neutrons present in the nucleus. (1 mark)

11 Select the correct statement about the discovery of radioactivity.

A Marie and Pierre Curie discovered the radioactive elements polonium and radium.

B Becquerel discovered and named alpha and beta radiation.

C Rutherford first showed that uranium was radioactive by using a photographic plate.

D Villard discovered X-rays. (1 mark)

12 Select the true statement about C-14 dating.

A Dinosaur bones can be dated using carbon-14 isotopes.

B Carbon dating can be used to date material up to about 50 000 years.

C C-14 isotopes make up 98.9 % of natural carbon.

D C-14 is used to date the rock minerals that surround a fossil. (1 mark)

13 Which of the following reactions could be classified as a synthesis reaction?

A Magnesium reacts with oxygen when heated to form white magnesium oxide.

B Gold oxide produces gold and oxygen when heated to a high temperature.

C Methane burns in air to form carbon dioxide and water.

D During cellular respiration, glucose is converted to carbon dioxide and water. (1 mark)

14 What salt forms when zinc dissolves in sulfuric acid?

A zinc sulfide

B zinc oxide

C zinc sulfite

D zinc sulfate (1 mark)

15 The following diagram is a model of a chemical reaction.

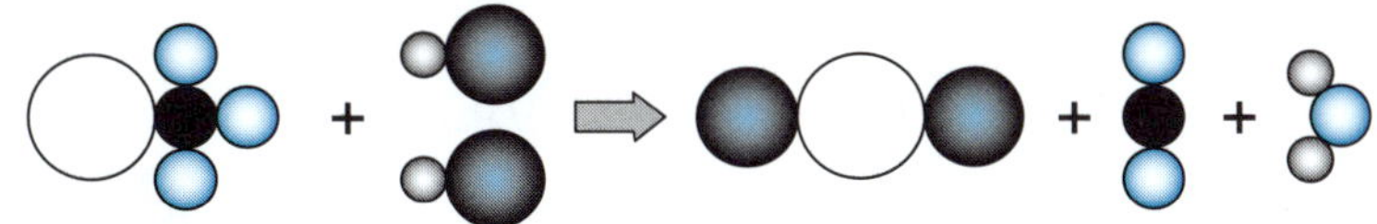

Which of the following reactions is this model demonstrating?

A complete combustion of a hydrocarbon
B neutralisation of a carbonate
C neutralisation of an oxide
D synthesis of silver oxide from its elements (1 mark)

Your Score

15

PART B *Restricted-response questions* 10 marks

Insert the missing word to complete each sentence.

16 In a food chain, on average only about __________% of the total energy at one trophic or feeding level is passed on to the next level. (1 mark)

17 There are strict requirements in Australia that all vaccines are rigorously tested for effectiveness and monitored for __________ before being released for general use. (1 mark)

18 In vertebrates such as humans, the nervous system consists of the brain, spinal cord and all the __________ that run to various parts of the body. (1 mark)

19 Nuclear medicine uses __________ to supply diagnostic information about the functioning of specific organs in a person, or to treat them. (1 mark)

20 Inside a Geiger-Müller tube an inert gas briefly conducts an electrical charge when a nuclear particle or ray __________ the gas. (1 mark)

21 Photosynthesis is the process plants use to change light energy into __________ energy. (1 mark)

22 One indicator of a chemical change is when an insoluble solid (or __________) forms when two solutions are mixed. (1 mark)

23 The test for __________ gas that is produced in a chemical reaction is the rekindling of a glowing wooden splint. (1 mark)

(cont.)

24 Pathogens of plants can destroy vegetation communities and several plant species can be at risk of ____________. (1 mark)

25 The kidneys ____________ the blood and remove nitrogenous waste such as urea. (1 mark)

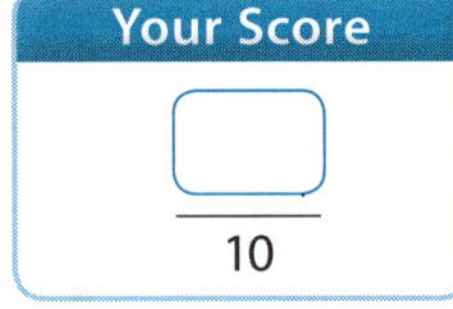

PART C *Written knowledge and skill questions* 25 marks

26 The following diagram shows an experiment in which a student tried to measure the energy content of an almond. He placed the almond on a needle stuck in a cork. 80 mL of water was placed in a beaker and the temperature of the water recorded. The almond was set alight and placed under the beaker of water.

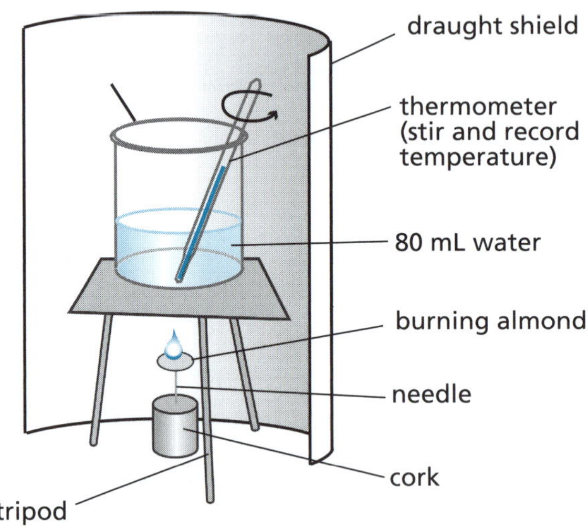

a What type of chemical reaction is occurring in the experiment? (1 mark)

b Is this reaction endothermic or exothermic ? (1 mark)

c Explain whether or not all the heat produced during the burning of the almond is absorbed by the water in the beaker. (2 marks)

d The heat energy (E) absorbed by the water can be calculated using the formula:

$$E = 4.2 \times \text{mass of water} \times \text{change in temperature}$$

Calculate the heat energy absorbed if the temperature rise of the water was 34.0 °C and the density of the water is 1 g/mL. (2 marks)

e The published chemical energy content for almonds is 24 000 J/g. If the almond had a mass of 1.2 g, calculate the energy content of the almond. (1 mark)

f Has the chemical energy content of the almond been fully converted into the heat energy absorbed by the water? Explain. (2 marks)

27 The following diagram shows a food web in an estuary.

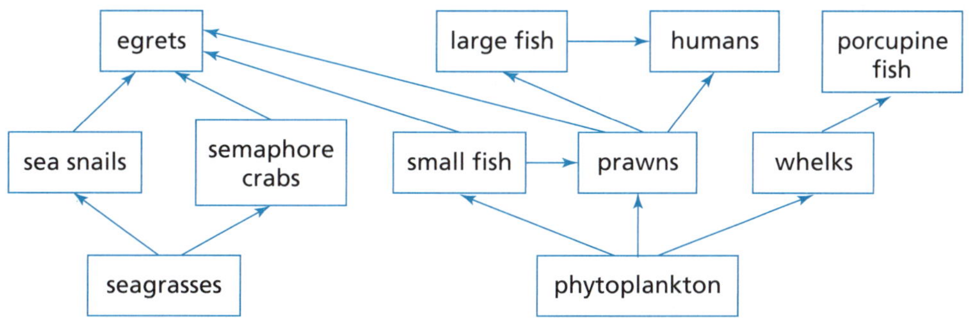

a How many first order consumers are present in this food web? (1 mark)

b Name a first order consumer that is also a second order consumer. (1 mark)

c Explain why humans are second and third order consumers in this web. (2 marks)

d If a disease kills all the semaphore crabs, what will happen to the population of sea snails and small fish? Explain your answer. (2 marks)

28 The following diagram shows an experiment in which a student added drops of white vinegar into a beaker containing a mixture of ammonia solution and phenolphthalein.

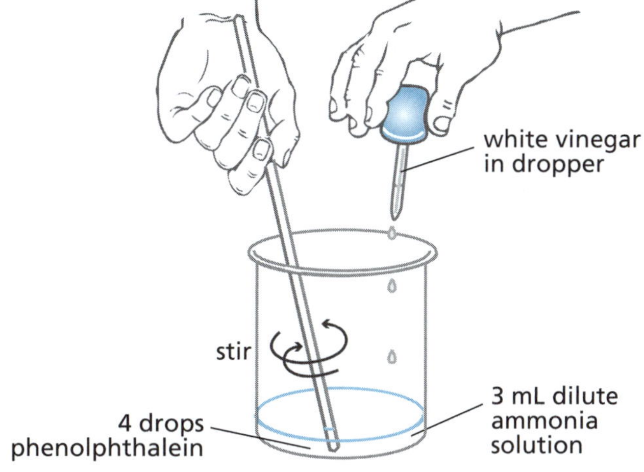

a Name the acid present in vinegar. (1 mark)

b What is the colour of the phenolphthalein indicator in the ammonia solution? (1 mark)

c Describe what will be seen as an excess of vinegar is slowly added to the ammonia solution. (1 mark)

d How will the pH of the mixture change during the reaction? (1 mark)

e The beaker felt warmer at the end of the experiment than it was at the start. Explain. (1 mark)

(cont.)

29 A sandy beach is a hostile habitat for living things. Suggest three factors which make it such an inhospitable environment. (3 marks)

30 State the functions of the following hormones.

a insulin (1 mark)

b adrenaline (1 mark)

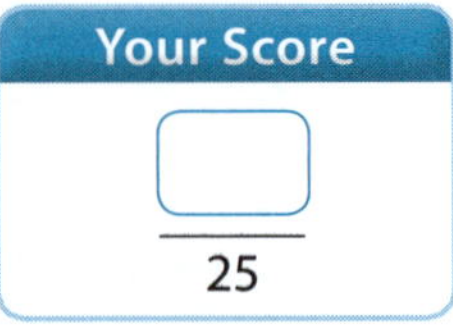

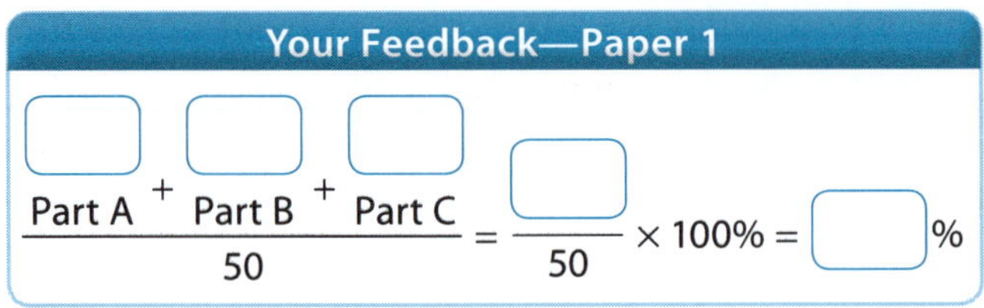

SAMPLE EXAM PAPER

PAPER 2

PART A Multiple-choice questions

15 marks

1 What must be inserted between P and Q to maintain a steady current flow?

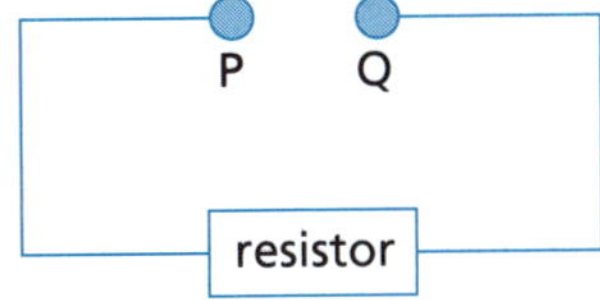

A switch
B source of potential difference
C voltmeter
D ammeter

(1 mark)

The next two questions refer to the following diagram which shows a light ray entering a piece of transparent plastic.

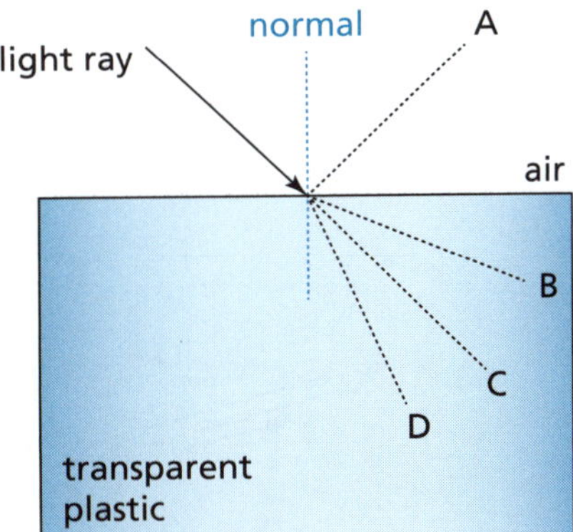

2 The path of the refracted ray is best shown by
A line A.
B line B.
C line C.
D line D.

(1 mark)

3 When the light ray emerges from the plastic it will be
A parallel to the incident ray.
B perpendicular to the incident ray.
C parallel to the normal.
D perpendicular to the normal.

(1 mark)

4 A light ray is reflected from a plane mirror as shown in the diagram. What is the angle of reflection of this ray?

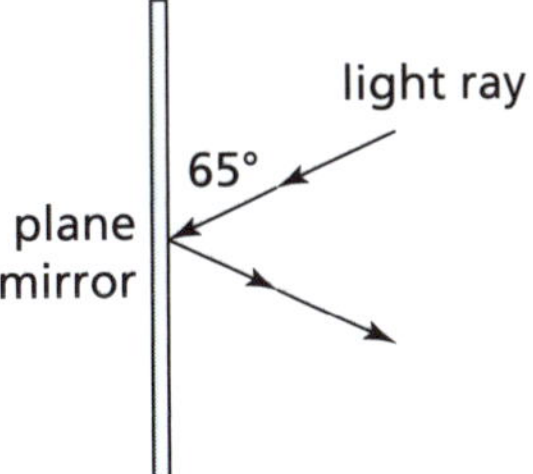

A 25°
B 35°
C 50°
D 65°

(1 mark)

(cont.)

5 Which of the following shows the best example of erosion?
- A shale breaking apart as water freezes in a crack
- B bedrock crumbling in the process of soil formation
- C a pebble rolling along the bottom of a stream
- D rock particles on a marble statue being dissolved by acid rain (1 mark)

6 A student turns a hand crank on an electrical generator and makes a light globe glow. The generator converts
- A electrical energy to mechanical energy.
- B mechanical energy to electrical energy.
- C electrical energy to chemical energy.
- D chemical energy to electrical energy. (1 mark)

7 The following diagram shows three ways heat can be transmitted.

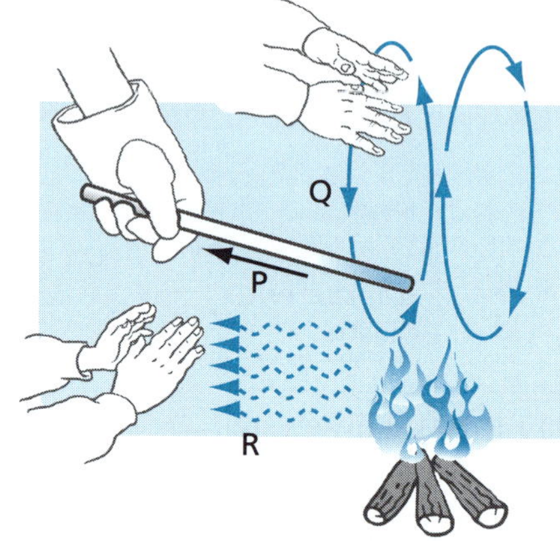

The letters P, Q and R represent, in order
- A conduction, convection, radiation.
- B convection, conduction, radiation.
- C radiation, convection, conduction.
- D conduction, radiation, convection. (1 mark)

8 The following diagram shows a periodic wave moving through a uniform medium.

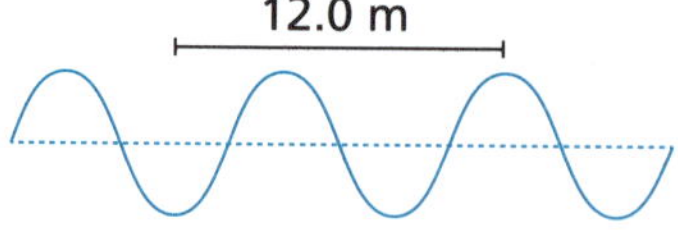

Given the frequency of the wave is 4.0 Hz, its speed is
- A 16.0 m/s.
- B 24.0 m/s.
- C 32.0 m/s.
- D 48.0 m/s. (1 mark)

9 In the following seismogram, the time difference between the arrival of the first P waves and the arrival of the first S waves is closest to

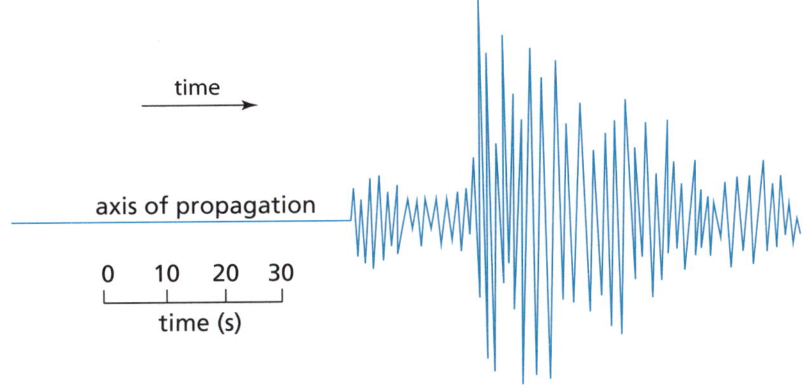

A 10 seconds.
B 15 seconds.
C 20 seconds.
D 25 seconds.
(1 mark)

10 In the electromagnetic spectrum, the correct order from low frequency (long wavelength) to high frequency (short wavelength) is
A X-rays, ultraviolet, visible light, microwaves.
B microwaves, visible light, infra-red, gamma rays.
C visible light, gamma rays, ultraviolet, infra-red.
D radio waves, infra-red, ultraviolet, X-rays.
(1 mark)

11 The following graph shows the relationship between the current in a metallic conductor and the potential difference across the conductor (at constant temperature).

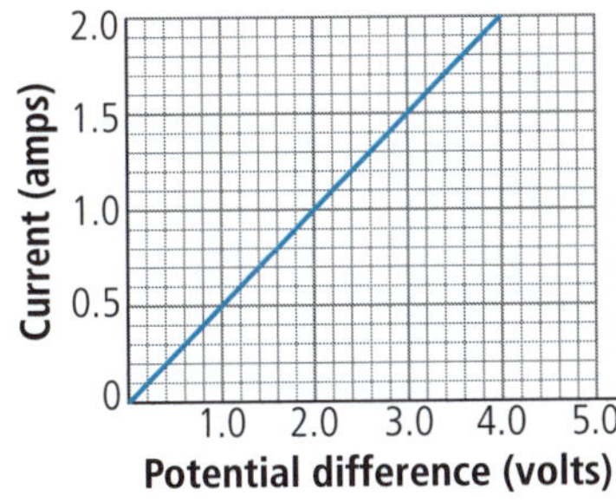

The resistance in the conductor is
A 0.5 Ω.
B 1.0 Ω.
C 2.0 Ω.
D 4.0 Ω.
(1 mark)

(cont.)

12 Which of the following layers of the Earth is where most convection currents that cause seafloor spreading are thought to be located?

A crust
B asthenosphere
C inner core
D outer core

(1 mark)

13 What are the parts of the eye labelled X and Y on the following cross-section of the eye?

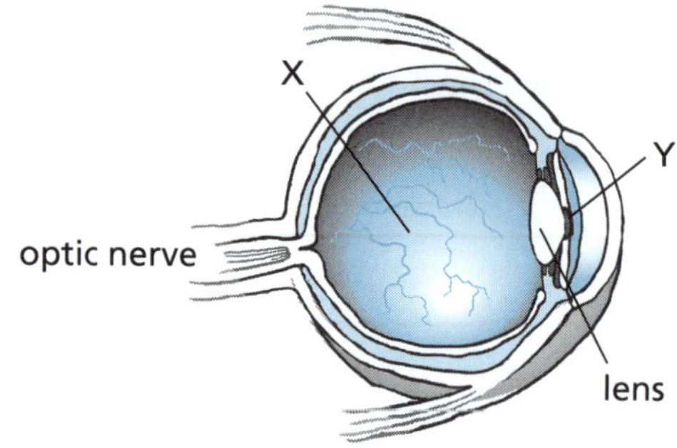

A aqueous humor, pupil
B aqueous humor, cornea
C vitreous humor, pupil
D vitreous humor, iris

(1 mark)

14 Radio waves, light and gamma rays travelling through space have the same

A speed.
B frequency.
C wavelength.
D period.

(1 mark)

15 The diagram shows six light globes connected in a circuit.

Suppose the globe marked X burns out. How many globes will continue to glow?

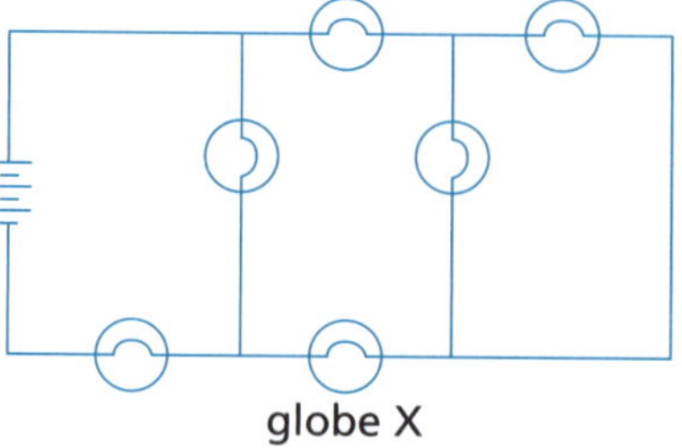

A 2
B 3
C 4
D 5

(1 mark)

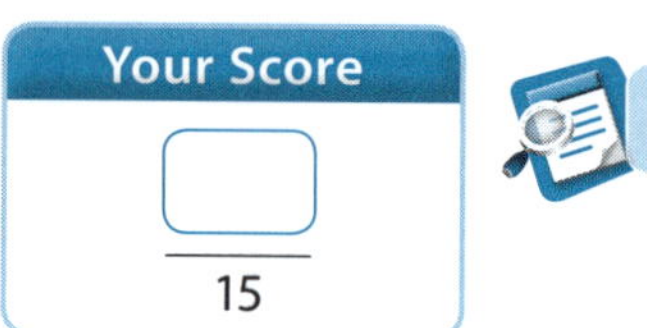

PAGES 249–250

SAMPLE EXAM PAPER

PAPER 2

PART B *Restricted-response questions* 10 marks

Insert the missing word to complete each sentence.

16 The earliest evidence of an animal having teeth comes from a fish fossil known as a placoderm that is 3.8×10^8 years old. In ordinary numerals, this number is ____________. (1 mark)

17 In the following diagram movement occurs along line XY which is known as a ____________. (1 mark)

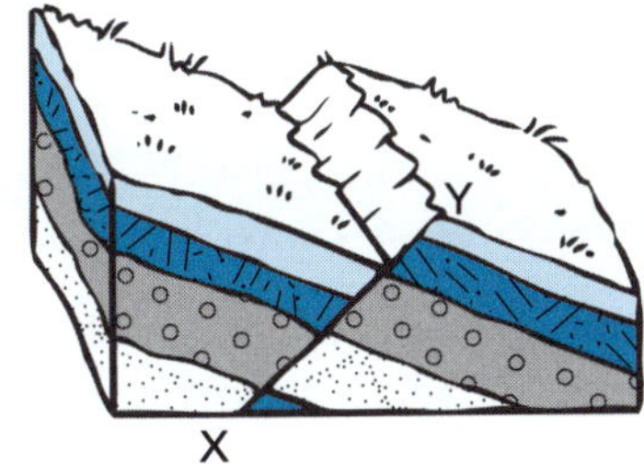

18 Scientists have inferred the structure of Earth's interior mainly by analysing data obtained from ____________. (1 mark)

19 An earthquake's magnitude can be determined by analysing the seismic waves recorded by an instrument called a ____________. (1 mark)

20 When sound waves travel they push air particles together in compressions and the particles are pulled apart at ____________. (1 mark)

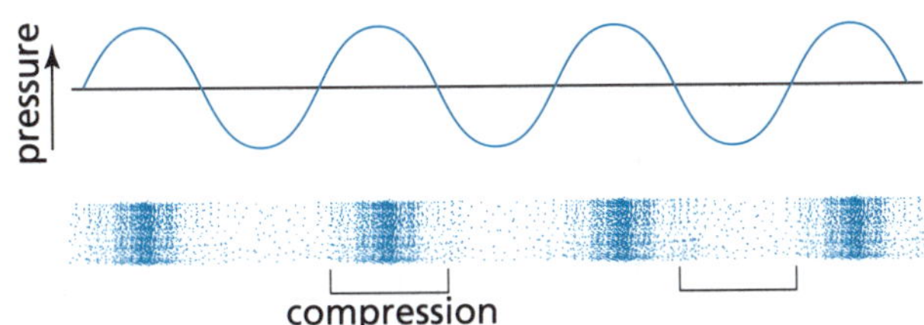

21 A student used a process called ____________ analysis to determine that a particular ore contained 22.68% platinum by mass. (1 mark)

22 Rod and cone cells in the ____________ of the eye allow conscious light perception and vision including being able to distinguish colours. (1 mark)

23 Light travels through optical fibres, which are used to transmit photons as signals or to carry energy for long distance communication, using total internal ____________. (1 mark)

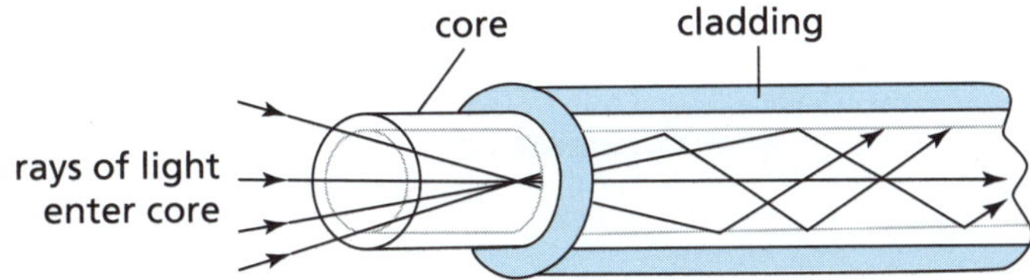

(cont.)

24 A substance that does not readily allow the passage of heat or electricity is termed an ____________. (1 mark)

25 The ____________ scale is a numerical scale for expressing the magnitude (the strength or total energy) of an earthquake on the basis of seismograph oscillations. (1 mark)

PART C *Written knowledge and skill questions* 25 marks

26 In 2012 a parachutist jumped from a height of 39 km above the Earth, reaching a speed of 1342 km/h. If the speed of sound is 1225 km/h, how many times the speed of sound is this (to two decimal places)? (1 mark)

27 The diagram below represents a model of a radioactive sample with a half-life of 15 000 years. The grey circles represent undecayed radioactive material and the blue circles represent the decayed material after the first half-life.

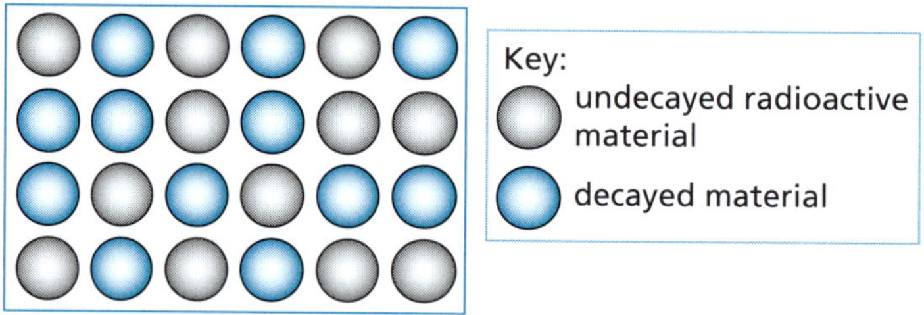

a How many more boxes should be blue to represent the additional decayed material formed during the second half-life? (1 mark)

b Originally all circles were grey. How much time will have elapsed at the point when only three circles are grey? (2 marks)

28 The cross-section below shows the direction of movement of an oceanic plate over a mantle hot spot. This produces a chain of volcanoes labelled P, Q, R and S.

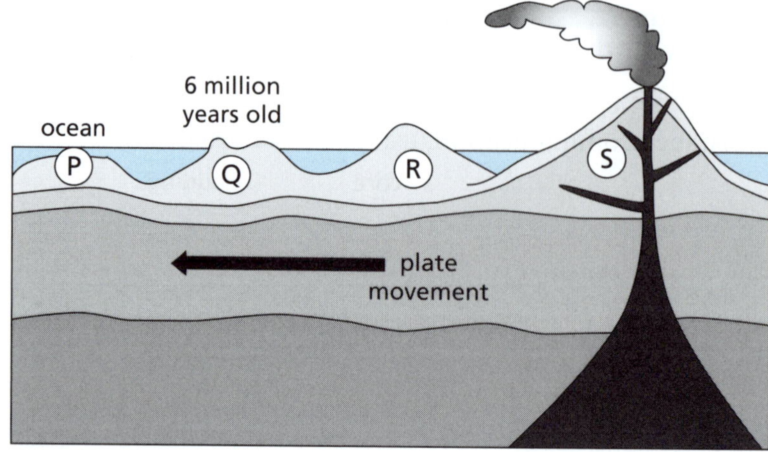

The geologic age of volcano Q is given.

a How old are the volcanoes P, R and S? (3 marks)

b What assumption are you making when giving these answers? (1 mark)

29 Fossils of trilobites, graptolites and eurypterids are found in the same bedrock layer. What can be inferred about the ages of these fossils? (1 mark)

30 The graph shows the rate of decay of the radioactive isotope K-40 into the decay products Ar-40 and Ca-40.

a What is meant by K-40? (1 mark)

b Name the decay products. (1 mark)

c A sample of basalt rock indicated that 75% of its K-40 has decayed. How old is the basalt? (1 mark)

d A sample of rock is around 4.0 billion years old. Approximately what percentage of K-40 should remain undecayed? (1 mark)

31 Two seismograms are shown, both taken at the same seismic station.

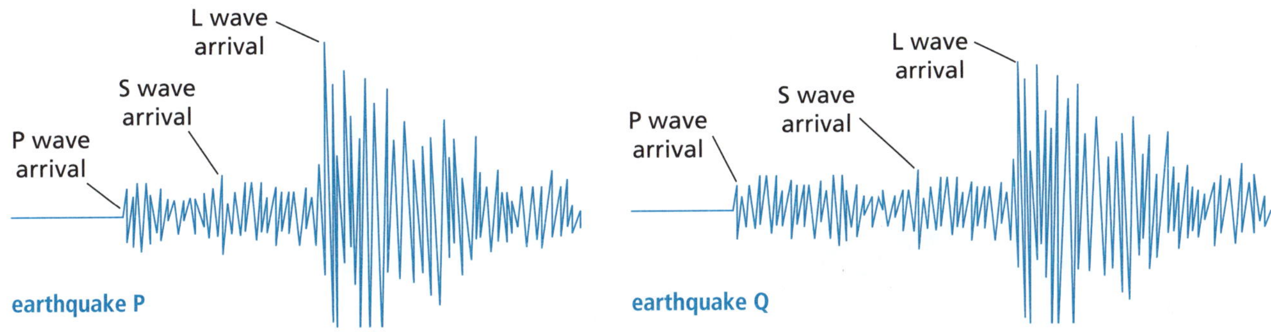

The time scale on both graphs is the same. Which earthquake is closer to the seismic station? Explain your answer. (2 marks)

32 The following diagram shows an Earth feature.

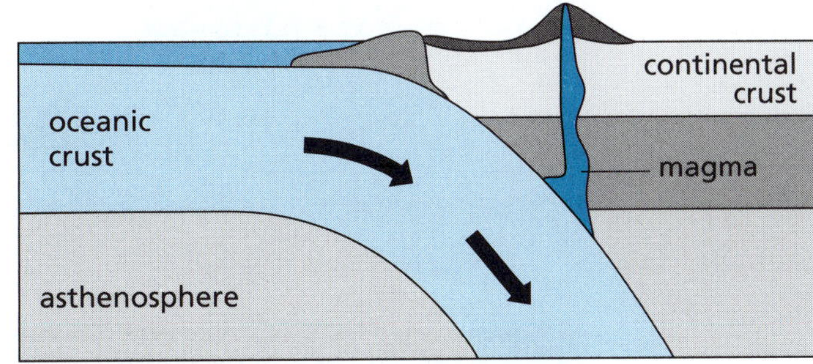

a What is shown by this diagram? (1 mark)

b Identify the features that occur in the region and explain how they form. (3 marks)

(cont.)

33 In an experiment Phil applied various downward forces to a vertical spring by attaching weights at the end of the spring. The applied forces and the corresponding elongations (stretching) of the spring from its equilibrium (unstretched) position are shown in the table. He then produced the following graph, drew a line of best fit and extended his line to the end of the graph paper.

Force (N)	Extension (cm)
0.0	0.0
0.5	1.2
1.0	2.5
1.5	3.5
2.0	4.9
2.5	6.2
3.0	7.4

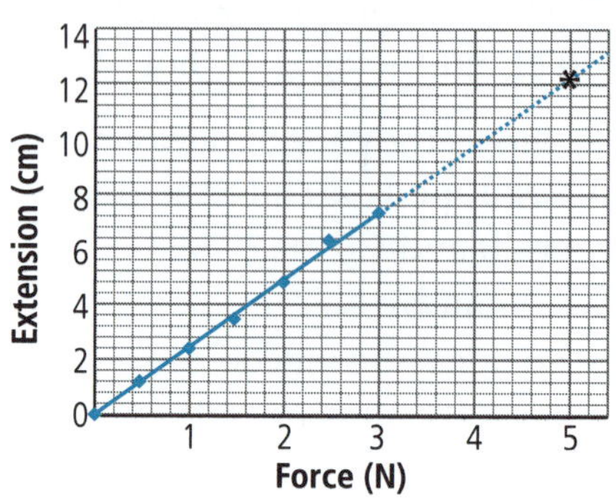

a Explain why the extension is placed on the vertical axis. (1 mark)

b What is extending a line beyond the plotted points called? (1 mark)

c Phil used his graph to predict that a force of 5.0 N should produce an extension of just over 12.0 cm. He showed it with an asterisk (*) on his graph. Comment on the validity of Phil's prediction. (1 mark)

34 Write a brief geological history to explain this cross-section. (3 marks)

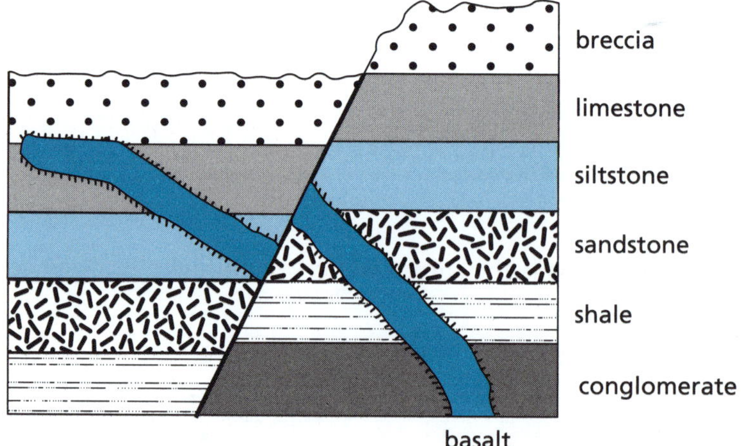

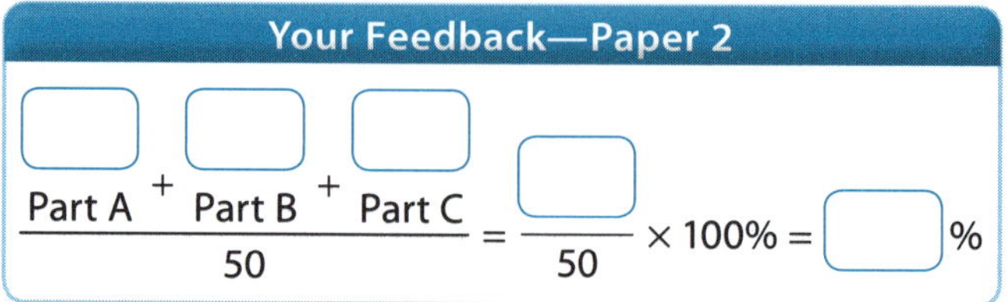

PAGES 250–251

PAGE 252

REVISION TESTS
Answers

CHECK YOUR ANSWERS

Strand: Biological sciences

REVIEW OF MAJOR BODY SYSTEMS
The body systems Pages 5–6

1 **a** liver ✓; **b** mouth ✓; **c** circulatory ✓; **d** excretory ✓; **e** oxygenated ✓

2 a = F ✓; b = B ✓; c = A ✓; d = D ✓; e = E ✓; f = C ✓

3 A = F ✓; B = D ✓; C = E ✓

4 **a** A = vena cava ✓; B = right atrium ✓; C = left ventricle ✓; D = aorta ✓

b **i** transports oxygenated blood back to the heart from the lungs ✓

ii pumps deoxygenated blood into the pulmonary artery ✓

c The oxygenated blood mixes with deoxygenated blood and this mixture is pumped around the body. The bluish colour of the deoxygenated blood makes the baby look blue. ✓

d The diagram is drawn from the perspective of the person looking at the body. ✓

5 **a** false ✓ The diaphragm is part of the respiratory system.

b true ✓

c false ✓ The tubes that transport urine from the kidneys to the bladder are called ureters.

d true ✓

e true ✓

6 **a** The alveoli increase the internal surface area of the lung which in turn increases the rate at which gaseous exchange occurs with the blood. ✓

b The fluid in the lungs decreases the alveoli surface area and so less oxygen can be absorbed into the blood. ✓

7 See the diagram below.

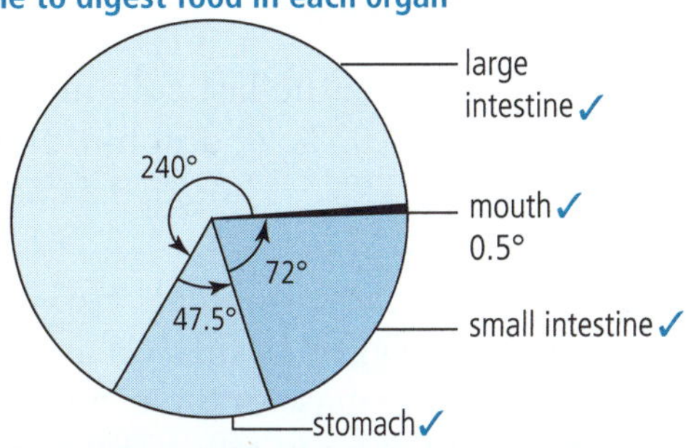

NERVOUS AND ENDOCRINE SYSTEMS
The body systems Pages 12–14

1 **a** reflex arc ✓

b A = cerebrum ✓; B = cerebellum ✓

c X = motor neuron ✓; Y = connector neuron ✓; Z = sensory neuron ✓

d The action is involuntary as it is a reflex. The brain does not coordinate this action but the pain from the burn will be registered by the brain. ✓

2 **a** synapse ✓

b A chemical is released from the end of one neuron which quickly diffuses across the synapse and stimulates the next neuron to fire an electrical signal. ✓

3 **a** **i** false ✓; **ii** true ✓; **iii** true ✓

b While the size of the impulse in each case is the same, it is the frequency of the pulses that is different. ✓ There are more pulses coming from the right hand than from the left hand. ✓

4 **a** See the table below.

	Hormone	Function
A	secretin ✓	Causes pancreas to release hydrogen carbonate (HCO_3^-) ions. Also causes gall bladder to release bile. ✓
B	cholecystokinin ✓	Acts on the pancreas to release trypsinogen. ✓
C	gastrin ✓	Causes HCl (hydrochloric acid) and pepsinogen to be released from the stomach wall and into the stomach. ✓
D	enterogastrone ✓	Turns off stomach acids and enzymes. ✓

b The dot indicates where the hormone is produced ✓; the arrow indicates to where it goes to function or act ✓; the +/– indicates whether it promotes the release ✓ of other chemicals or inhibits ✓ them.

5 **a** The values in the last column are: elephant, 0.02% ✓; dolphin, 1.18% ✓; human, 2.33% ✓; horse, 0.24% ✓; budgerigar, 5% ✓.

b Although larger animals are expected to have 'bigger' brains, on average, the ratio of the mass of the brain to the mass of the body tells a different story. Students may be surprised to learn that the human brain does not occupy the greatest percentage. It's not size that counts. ✓

c false ✓ evidence supplied in the above calculations in part a ✓

d Although it occupies over 2% of the mass of the body, the budgerigar brain occupies at least twice that. Size alone is not the only determinant of intelligence. In fact, intelligence may not even be related to size at all! ✓

e budgerigar ✓

MICROBES AND DISEASE
The body systems Pages 19–21

1 **a** true ✓ Most visits to GPs in the western world are for colds.

b false ✓ The common cold and flu are viral infections.

c false ✓ This is a common misconception as common colds and flu usually occur in the autumn and winter months. Cold weather itself does not cause colds. However, in colder months people spend more time indoors and in close proximity to each other, making it easier for cold and flu viruses to spread.

d false ✓ Antibiotics treat bacterial infections, not viral infections. Currently there is no cure for the common cold.

e true ✓ Rhinovirus C, for example, was only discovered in 2007 and is now found globally.

f true ✓ This is one of the most effective ways to stop the spread of contagious diseases, not just colds and flu.

g true ✓ For antibiotics to be successful, the correct dose must be taken for the length of time recommended.

h true ✓ Some of these diseases are malaria, yellow fever, dengue fever, Ross River virus, Murray Valley encephalitis and Japanese encephalitis. Nevertheless, there is a low risk of catching a serious mosquito-borne disease in Australia.

i false ✓

j true ✓

2 **a** virus ✓

b They had no natural resistance to the disease. ✓ There were no medicines at the time that were effective. ✓

c The smallpox vaccine helps the body develop immunity to smallpox. (Vaccination within three days of exposure will usually prevent or lessen the severity of smallpox symptoms.) ✓

d People in cities are in closer proximity to each other than people in the country, so it was easier for the disease to spread in the cities. ✓ Smallpox can also be spread through direct contact with infected body fluids (blood, saliva, vomit) or personal items such as bedding or clothing that have been used by an infected person—with more people in cities there was more likelihood of coming into contact with body fluids and personal items. ✓

3 **a** the surgical removal of some of a patient's blood for therapeutic purposes ✓

b Bloodletting remained popular as many thought it was better to give any treatment rather than no treatment at all. ✓

c Leeches were attached to the patient's body where they gorged themselves with blood. ✓ (In some reports up to 50 leeches were used at a time.)

d Removing a lot of blood from a person would weaken them. ✓ It is also possible that other infections could also take hold in this weakened state. ✓

e Total volume withdrawn
$= 13 + 20 + 20 + 40 + 32 = 125$ ounces. ✓
This is $125 \times 30 = 3750$ mL ✓ $= 3.75$ L.
For comparison, blood donors give around 470 mL at each donation. Some researchers argue there is very little chance that he would have survived even if he hadn't been bled since there were no antibiotics in the 18th century with which to treat infections.

f $90 \times 75 = 6750$ mL $= 6.75$ L of blood ✓

g $\frac{3750}{6750} \times 100 = 56\%$ ✓ (Removing more than half of Washington's blood volume within a few hours contributed his death.)

4 Viruses are not made of cells—they have no cytoplasm, cell membrane or anything else apart from just a core of nucleic acid (DNA or RNA) surrounded by a covering layer of protein. ✓ Viruses are very simple and do not carry out most of the processes which other living organisms perform. ✓ Viruses are inactive when outside of a living cell, but once their nucleic acid is inside the host cell they take over the cell's activities. ✓ Usually they make many copies of themselves, then break out of the cell to infect others.

5 *(any four of the following)* The skin stops microbes from getting into the body. Scabs form on the skin after a cut or scrape, stopping microbes from getting into your body. Acid in the stomach kills many microbes. Sticky mucus in the lungs traps microbes and then cilia sweep it out of the lungs. Tears contain substances that kill bacteria. Earwax (cerumen) provides some protection from bacteria, fungi, insects and water. ✓✓✓✓

6 **a** See the diagram below.

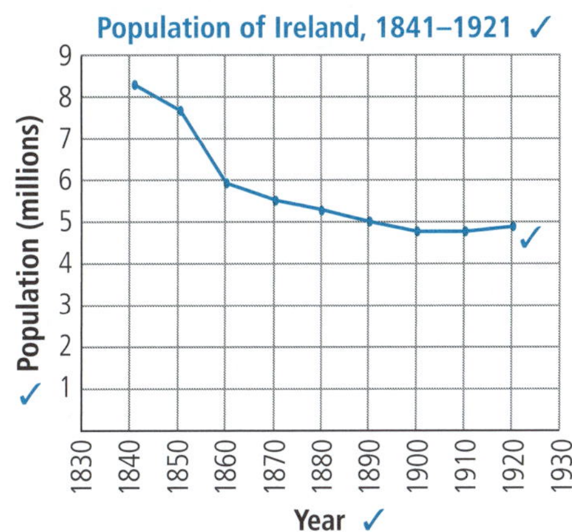

b The graph shows a decrease in population.
With no food and/or livelihood people were forced to leave Ireland or slowly die of starvation, causing the population to decrease. ✓ (In the decades following the Irish Potato Famine there were over a million deaths and substantial emigration. Some two million people migrated to Britain, the United States and Canada, among other places, just in the period 1845 to 1855.)

7 **a** v; ii; vii; viii; i, vi; ix; iii; iv. ✓✓✓✓ *(Take one mark off for each out of sequence.)*

b methods *(any two)*: removing bodies of stagnant water in which mosquitoes breed should lessen the number of mosquitoes; using insecticidal sprays in and around human populations; sleeping under an insecticidal bed net (since the infected mosquitoes that transmit malaria bite only at night) ✓✓

REVISION TESTS

Answers

CHECK YOUR ANSWERS

HUMAN HEALTH
The body systems Pages 27–29

1 **a** It is the safest and most successful way of providing protection against disease. ✓ After being immunised, the child has a greatly reduced chance of catching the infection if there are cases in the population. ✓ If enough people are immunised, the infection can no longer be spread between people and the infection dies out altogether. ✓ (This is how smallpox was eradicated from the world and polio has vanished from many countries.)

b A number of vaccinations are necessary in the first few years of life to immunise the child against the most serious childhood infections. ✓ The immune system in young children is immature and does not work as well as the immune system in older children; hence more vaccine doses are required. ✓ (In the first months, a baby is protected from most infectious diseases by the mother's antibodies, which are transferred to the baby during pregnancy. As these antibodies wear off, the baby is at danger of severe infections and so the first immunisations are given before these antibodies have vanished.) New vaccines against serious infections keep on being developed. ✓ (The number of injections can be reduced by the use of combination vaccines, where some vaccines are pooled into one injection.)

2 *(any five of these methods)* person-to-person contact (through blood (e.g. from sharing needles), saliva (kissing) or through contact with clothes or bedding); through the air (coughing or sneezing); contaminated food (bacteria may survive in food that is not cooked properly or that is reheated to give food poisoning); by water (contaminated water may spread diseases like typhoid or cholera); by insects (the bubonic plague was carried by fleas living on rats); animals (Microbes can get into a person who is scratched or bitten by an animal carrying harmful microbes. Rabies is a virus spread by infected saliva that enters the body through a bite by an infected dog or fox. New Zealand and Australia have never had rabies.) ✓✓✓✓✓

3 **a** Some children experience minor side-effects following immunisation, but these last a short time and the child gets better with no problems. ✓ (Common side-effects include redness, soreness and swelling at the site of an injection, mild fever and being grizzly or ill at ease.) Serious reactions to immunisation are very rare, so the advantage of protection against the disease far outweighs the slight risks of immunisation. ✓

b Natural immunity and vaccine-induced immunity are both normal responses of the immune system, and the body's response in both circumstances is the same. ✓ Antibodies produced through immunisation are ready to act far more quickly than waiting for the actual disease to cause the body to produce antibodies. ✓ (Vaccines are far safer than the diseases they prevent, which have a high risk of severe illness and sometimes death.)

c All vaccines available in Australia must pass rigorous safety tests before being approved for use. ✓ (This is a legal requirement and is usually done over many years as the vaccine is developed. As well, once they are in use their safety is scrutinised.) Before vaccines are made available they are thoroughly tested in large clinical trials and monitored for safety. ✓ (The approval process can take up to a decade. As a result, a number of vaccines failing these early tests are not released.)

4 **a** The immune system is designed to defend you against millions of pathogens and toxins that would otherwise invade your body. It kills microbes if they get past the body's natural barriers. ✓

b There are different types of white blood cells in the immune system, each with different functions. ✓ Some white blood cells engulf germs and kill them. ✓ Some white blood cells produce substances called antibodies that attach to microbes. ✓ Microbes produce chemicals called antigens. ✓ Different microbes produce different antigens. ✓ Antibodies attach to microbes if they match the antigen on the microbe. ✓ When this happens the microbes can be killed, or clumped together to make it easier for other white blood cells to kill them. ✓

5 When infected by a pathogen, it takes time for your body to respond and begin fighting the infection (0 to 2 days). ✓ It does this by creating white blood cells with the proper antibody (2 to 10 days). ✓ During this time, you keep feeling ill. You start to get better when enough antibodies have been produced (10 to 16 days). ✓ After the invading microbes have been destroyed, the number of antibodies decreases (16 to 22 days). ✓ But some of the white blood cells that produce that particular antibody remain in your blood (22 days and onwards). ✓ This allows your body to produce the right antibodies quickly should you be infected again by the same pathogen, so the microbe doesn't get an opportunity to make you sick this time. ✓ You are now immune to the microbe and the disease it causes. (The numbers of days given here are only approximate.)

6 **a** The influenza virus mutates (alters) constantly ✓ so scientists need to obtain a sample of the new, mutated, strain before they can develop a vaccine to fight it ✓.

b Yes, once the vaccine was distributed, the number of cases of people reporting flu symptoms to the doctor decreased significantly. ✓ Of course, not everyone took the vaccine, so there were still people being affected by the flu even after many doses were distributed. ✓

c A vaccine can lessen the severity of the flu, even if someone has been immunised. ✓ After vaccinations there are fewer germs to spread around in the community. ✓ The flu can be especially dangerous and lethal to the very young, very old and to those who may have compromised immune systems. So protecting them from the illness makes sense. ✓

7 Artificially acquired immunity from vaccines is simpler and less hazardous that waiting for natural immunity to develop. ✓ Vaccines can avoid a disease from occurring altogether, rather than attempting to cure it afterwards. ✓ It is also more economical to prevent a disease than to treat it. ✓ Vaccines protect not only the individual but also others around them. ✓ With a vaccine-primed immune system an illness is prevented from starting and the individual will be contagious for a shorter time period, or perhaps not at all. ✓

8 **a** around 400 000 to 500 000 ✓

b 1967 ✓

c early 1960s ✓ (actually 1963)

d the sudden decrease in the number of reported measles cases ✓

9 **a** to initiate the chemical process and allow the photopolymer to harden ✓

b The dentist did not have control over how quickly the material cured. ✓ This meant the dentist had to act quickly and appropriately place the material into the tooth. ✓ That is, he had a short time to fill the cavity. If the material was not properly positioned, then it had to be excavated and the process started over again.

c The new malleable resin materials do not need to be mixed, can be dispensed directly into the site and can only be fully cured/hardened with a dental curing light. So now the time constraint is lifted ✓ and the dentist can guarantee that the material is properly placed ✓ as there is no need to rush to get the tooth filled.

10 **a** Optical fibres are narrow tubes, so they can be inserted down the throat and into the stomach to light up the inside of the stomach for the doctor to view. ✓

b Only a small entryway incision needs to be made to insert the arthroscope and the optical fibre can illuminate the joint to be examined. ✓

c Surgery would have been required. Optical fibres have made methods easier for doctors and procedures less invasive for patients. ✓ They let doctors view and work inside the body, requiring no surgery and no extensive recuperation (recovery) period or a lengthy hospital stay. ✓ Optical fibres have paved the way for an entire new field of surgical procedure, called laparoscopic surgery (or, more commonly, keyhole surgery).

STUDYING AN ECOSYSTEM
Ecology Pages 34–36

1 **a** abiotic ✓; **b** 77% ✓; **c** 16 °C ✓

d The air in the leaf litter is quite moist and prevents the invertebrates from drying out. The leaves help to retain moisture. ✓

2 **a** abiotic ✓

b The more turbid the water (i.e. the more suspended solids it contains) the less light penetrates, so the number of aquatic producers is decreased. ✓ Lower numbers of producers leads to fewer herbivores. ✓ Lower numbers of herbivores (first-order consumers) means that the population of higher order consumers is lower. ✓

c Method

1 Collect a known volume of water. ✓
2 Weigh a filter paper (mass 1). ✓
3 Use the filter paper to filter the water. ✓
4 Dry the paper containing the solid residue. ✓
5 Re-weigh paper and residue (mass 2). ✓
6 Calculate mass of the solid: mass 2 – mass 1 = mass of solid. ✓

3 **a** These numbers are magnification factors—the number of times the drawing has been magnified. The drawing of the *Paramecium* is 100 times larger than the organism's true size and the *Chlamydomonas* drawing is 1000 times larger. In order to find the real size of each invertebrate, divide the measured length in the drawing by the magnification factor. ✓

b The *Paramecium* is larger as it is magnified to a smaller extent. ✓

4 **a** false ✓; **b** true ✓; **c** false ✓; **d** true ✓; **e** true ✓; **f** true ✓

5 Environment: the abiotic and biotic parts of the total surroundings of an organism. ✓
Habitat: the actual place where the organism lives. ✓

6 A Chironomidae (gnat or midge larvae) ✓;
B Dixidae (midge larvae) ✓;
C Oligochaeta (freshwater worm) ✓;
D Simuliidae (sand or black fly larvae) ✓

REVISION TESTS
Answers

FOOD CHAINS AND FOOD WEBS
Ecology Pages 40–41

1 **a** crabs ✓
b The population of all the consumers would decrease. ✓ Sea snails that feed on the seagrass would decrease in numbers as there is less food to eat. In turn, the crabs have fewer sea snails to feed on and their numbers would decrease. In turn, the egret numbers would decrease if crab numbers decrease. ✓
c bacteria (or moulds) ✓

2 **a** producer ✓; **b** herbivores ✓;
c carnivores ✓; **d** true ✓; **e** false ✓;
f young fish ✓

3 **a** See the diagram below.

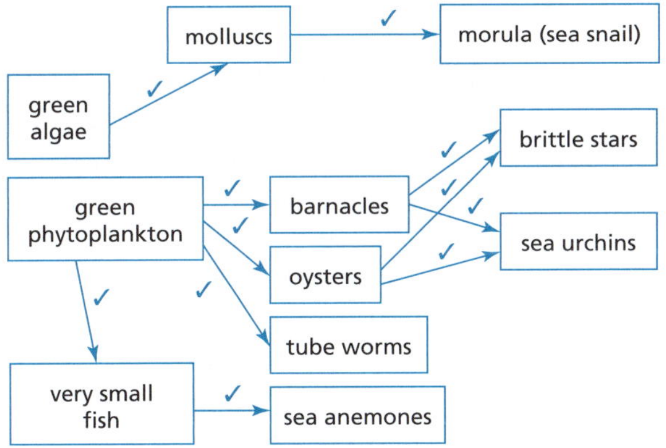

b There would be an increase in the population of herbivorous molluscs. ✓

4 **a** 7 ✓; **b** spiders ✓, carnivorous snails ✓;
c second-order consumer ✓; **d** false ✓;
e true ✓; **f** herbivore ✓

ECOSYSTEMS AND NATURAL CHANGE
Ecology Pages 47–49

1 **a** true ✓
b true ✓
c false ✓ Carnivores are part of the biotic (living) component of an ecosystem.
d true ✓ Plants store this energy for later use. If the plant is eaten, the energy is used by the animal that ate it.
e true ✓
f false ✓ Some ecosystems are large, others small.
g false ✓ Herbivores (plant-eaters), carnivores (meat-eaters), insectivores (insect-eaters) and omnivores (both plant- and animal-eaters) are all examples of consumers.
h true ✓
i true ✓ This is one of several functions of a decomposer.
j false ✓ Matter is recycled, not energy. Energy is eventually dissipated (lost to the system).
k true ✓ An energy pyramid is a graphic representation of the total energy in each trophic level of an ecosystem.
l false ✓ Energy flows from bottom to top.

2 **a** the direction of flow of energy (and nutrients) ✓
b the Sun ✓
c foxes, owls and snakes ✓
d foxes, owls and hawks ✓
e increase ✓, as they are only eaten by snakes
f a bird that eats (derives its energy from) insects ✓
g *(any one of the following)* rabbits, squirrels, foxes, owls, hawks ✓

3 **a** Changing the numbers of organisms at the top of the food chain (wolves) has had a flow-on effect on all other organisms at lower (trophic) levels in the food chain. ✓
b Once their predator was removed, the number of elk multiplied greatly as there were no wolves to keep them in check. ✓
c More elk meant less plants as they fed off them. Elk also trampled seedlings

in their zeal to feed. These meant less food for birds, insects and rodents. ✓ (This decline in elk after the wolf was re-introduced has resulted in changes in flora, most specifically willows, cottonwoods and aspens along the periphery of heavily timbered areas.)

d Regardless of whether a wolf-specific disease wiped wolves out or they were killed off by hunters and trappers, the outcome would be the same. In either case wolves would have been exterminated from the area allowing elk to multiply unchecked. ✓

4 **a** R ✓; **b** P ✓; **c** U ✓; **d** Q ✓; **e** S ✓; **f** T ✓

5 The timing of these events may change in some parts of the country. ✓ For example, warmer springs may lead to earlier nesting for migratory bird species ✓; and birds that winter in the north are returning south in the spring days earlier than they did before ✓. The same is true for migratory insect species. Changes like these can lead to mismatches in the timing of migration, breeding and food availability. ✓ Growth and survival are reduced when migrants arrive at a location before or after food sources are present. ✓

6 **a** algae → zooplankton → arctic cod → seals → polar bears ✓✓

b Only about 10% of the energy at one trophic level is available for the next level.

i $\frac{10}{100} \times 6\,000\,000 = 600\,000$ kJ ✓

ii $\frac{10}{100} \times \frac{10}{100} \times \frac{10}{100} \times \frac{10}{100} \times 6\,000\,000$
$= 600$ kJ ✓

c Fewer algae mean less food (and energy) for each of the levels higher up the food chain. This leads to a decline in the number of polar bears, as fewer of them can be sustained. ✓

d Consumers at each level commonly change only about 10% of the chemical energy in their food to their own organic tissue ✓ so very little energy remains after about five or six levels ✓. Food chains begin at trophic level 1 with producers such as green plants and phytoplankton (e.g. marine algae), move to herbivores at level 2, predators at level 3 and usually finish with carnivores (also predators) at level 4 or 5.

IMPACTS ON ECOSYSTEMS
Ecology

Pages 55–57

1 **a** trapping or shooting or otherwise removing feral cats and foxes ✓; restricting grazing land near populations of wallabies ✓

b The fence is designed to keep out feral animals which attack and kill rock wallabies. ✓

2 Harmful *(any two)*: kill animals and plants; destroy houses and other structures; soil erosion due to loss of vegetation; ash and sediment can foul waterways, killing organisms that live there. ✓✓
Useful *(any two)*: clear undergrowth and litter; return nutrients to soil; allow new shoots and seedlings to grow; some plants need fire to set seed; new succulent growth encourages animals into an area. ✓✓

3 **a** The number of species (birds and mammals) adversely affected decreases rapidly at first ✓ as the distance from the road increases, then flattens out ✓.

b Reasons *(any two)*: Building the road can interfere with the animals' habitat; trees and other plants would have to be cut down for the purpose; water courses could be interrupted; greater access for humans and predators to previously inaccessible land; noise on the road can scare away animals; road kill is now more likely as animals move about. ✓✓

c 5½ km ✓

4 **a** An invasive species is a species that does not naturally occur in a particular area ✓and whose introduction does or is likely to cause financial or environmental damage. The fox has been able to survive and spread rapidly. ✓

b Opportunistic feeders are species adapted for exploiting changeable, unpredictable or passing environments to obtain food. That is, they can tolerate variable environmental conditions and food sources. Some opportunistic species can thrive on almost any available nutrient source and can quickly take advantage of favourable conditions when they occur. ✓

c By preying on many species of Australian native wildlife. Foxes include birds, small mammals and reptiles in their diets. ✓ They have brought a number of small native mammals close to extinction.

d Farmers have had many of their animals, including lambs, calves, poultry, water fowl and goats killed by foxes. ✓ This is costly to the farmer, not only from the lost stock, but also because the farmer needs to take steps to prevent foxes from doing this again. ✓

e $400\text{ g} \times 365\text{ days} = 146\ 000\text{ g}$ ✓
$= 146\text{ kg/year}$

f Predation on other pest vertebrates does not provide effective control for these pests, because, unlike native animals, rabbits and mice can breed faster than the foxes can eat them. ✓

g *(any three)* fumigation of dens; baiting; fencing (effective for small areas only as it is an expensive option, and requires regular maintenance); shooting ✓✓✓

5 features *(any three)*: saline surroundings; fierce heat during summer and cold winter nights; long dry spells; short periods of water availability ✓✓✓ (Other organisms are also well adapted to floods. The life cycle of plants and animals within these areas is geared towards high productivity in a short time period. There is a large population increase after a flood and a considerable decrease in numbers when the area is dry for an extended time.)

6 **a** The release of liquid petroleum into the environment by the action of human activity, either deliberately or accidentally. ✓ Spills may occur on land or at sea.

b On the surface of the sea they can spread for hundreds of kilometres in a thin oil slick ✓ eventually covering beaches a long distance away. These spills can kill bird and mammal life as well as crustaceans and any other organisms they coat. Land spills are often more easily enclosed and controlled before any oil escapes, and animals can avoid them more easily. ✓

c effects *(any three)*: oil blinds animals so they are defenceless; animals that rely on scent can become confused; oil penetrates the birds' feathers and impairs their ability to fly; oil in the fur of animals and feathers of birds reduces their insulating ability and they can die from cold (hypothermia); oil prevents animals from sourcing food, and makes them more easily preyed on; ingested oil causes poisoning and internal problems with metabolism and can kill the animal; some sea animals and birds can become heavier with the oil and sink and drown. ✓✓✓

d These floating masses are insoluble in water making their removal from the surface easier. ✓ They hold the oil in their pores until it is disposed of properly. ✓

e $40\ 900\ 000\text{ L} \times \frac{1\text{ ha}}{2.9\text{ L}} \times \frac{0.01\text{ km}^2}{1\text{ ha}}$ ✓
$= 141\ 000\text{ km}^2$ ✓

7 **a** They had never been threatened on their native land, so had few defences, and had adapted to co-exist with other organisms on the island. ✓ Also, being isolated, there were no significant predators. ✓

b Humans killed the birds for food ✓; humans destroyed the forest habitats of the bird ✓; introduced species either competed with them for food, or ate them and/or their eggs ✓, which could be easily taken from ground nests.

c Living happily in isolation for a long time, the dodo was completely wiped out within 60 to 70 years. ✓ This was all due to human intervention. (The last dodo bird was killed in 1681.)

8 **a** The damming of the Murray River means that periodic flooding no longer occurs. ✓ The gum has developed an association with the water where it needs periods of partial flooding ✓ and its trunk is inundated. Seeds are washed to high ground during a flood and germinate to take root and grow before the next flood submerges the new tree.

b This will bring back the periodic flooding that the tree has come to depend on. ✓

Strand: Chemical sciences

HISTORICAL DEVELOPMENT OF ATOMIC THEORY
Atomic theory Pages 62–63

1 **a** The nucleus has 11 protons (11 positive charges) so 11 negative electrons are required to keep the atom neutral. ✓

b true ✓

c No, it is not to scale. The nucleus shown is extremely large relative to the whole atom, when in fact it is proportionally very tiny. ✓

d neutron ✓

2 **a** John Dalton ✓; **b** Ernest Rutherford ✓; **c** Niels Bohr ✓; **d** Ernest Rutherford ✓; **e** JJ Thomson ✓

3 The drawing below shows two electrons in K shell and two electrons in L shell ✓; nucleus contains four protons ✓ and five neutrons ✓.

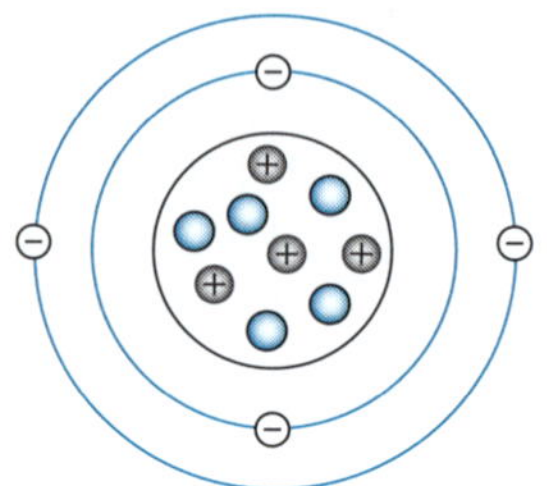

Beryllium atom model

Key: neutron ⊕ proton ⊖ electron

4 **a** $2(3)^2 = 18$ ✓; **b** $2(4)^2 = 32$ ✓; **c** $2(5)^2 = 50$ ✓

5 C ✓ (Protons are positive and neutrons are neutral.)

6 **a** B ✓; **b** D ✓

c The atom contains a very tiny and dense nucleus at its centre. ✓
The rest of the atom in which the electrons are located is largely empty. ✓

7 **a** $\frac{0.0016}{240} \times \frac{100}{1}$ ✓
$= 6.67 \times 10^{-4}\%$ ✓

b $6.67 \times 10^{-4} \div 100 \times 200 = 1.33 \times 10^{-3}$ m ✓
$= 1.33 \times 10^{-3} \times 1000 = 1.33$ mm ✓

ATOMIC STRUCTURE
Atomic theory Pages 69–70

1 **a** 2,7 ✓; **b** 2,8,3 ✓; **c** 2,8,6 ✓

2 **a** 23 protons ✓, 51 – 23 = 28 neutrons ✓, 23 electrons ✓

b 92 protons ✓, 238 – 92 = 146 neutrons ✓, 92 electrons ✓

3 **a** nucleus A has seven protons so it is nitrogen (Z = 7) ✓
nucleus B has 12 protons so it is magnesium (Z = 12) ✓

b The nucleus of nitrogen has 7 protons and 7 neutrons, This means Z = 7 and A = 14, so the nuclear symbol is ${}^{14}_{7}N$. ✓ The nucleus of magnesium has 12 protons and 13 neutrons. This means Z = 12 and A = 25, so the nuclear symbol is ${}^{25}_{12}Mg$. ✓

4 **a** In-115 has two more neutrons than In-113. ✓

b 114.8 u is much closer to 115 than 113 ✓ so the isotope in the greater proportion in natural indium is In-115 ✓

5 **a** 2,8,18,18,8 ✓

b Xenon is a noble gas with a stable octet of electrons in its outer shell. ✓

6 **a** 3 protons ✓; **b** 2,1 ✓; **c** ${}^{6}_{3}Li$ ✓, ${}^{7}_{3}Li$ ✓

d average atomic weight

$$= \frac{7.4}{100} \times 6 + \frac{92.6}{100} \times 7 \checkmark$$

$$= 0.444 + 6.482 = 6.926 \text{ u} \checkmark$$

e Li^{+} ✓ electron configuration = 2 ✓

7 **a** true ✓; **b** false ✓; **c** true ✓; **d** false ✓; **e** false ✓

RADIOACTIVITY
Atomic theory Pages 76–78

1 **a** After a certain time period a radioactive nucleus has either decayed or not decayed. Coins can be used to model this as there are only two ways the coin can fall ✓—with the Queen facing upwards (representing an undecayed nucleus) or facing downwards (representing a decayed nucleus) ✓.

b No. The outcome of tossing a coin is a matter of chance—there is a 50% chance it will land face up and a 50% chance it will land face down. For any particular coin Louisa has no way of knowing which way it will land. However, if she tosses a number of coins she can be more certain of the outcome—the more coins she tosses, the closer the results will be to half the coins facing up and half facing down. ✓ The same is true for radioactive particles—when there is only a small number, or a single particle, the inherent randomness of the decay process is more obvious and unpredictable. When there are a large number of particles, they follow a predicted behaviour very closely. ✓

c Yes. After one half-life, half of all the radioactive particles have decayed. ✓ When tossing coins there is a 50% chance that half of them will end up face down (representing the decayed state). ✓

d No. Eventually the last remaining radioactive particle will decay away ✓, although this may take a relatively long time.

2 P = alpha particle ✓; Q = beta particle ✓; R = gamma ray ✓

3 **a** They do not carry a charge. ✓ Only charged particles are deflected by electric or magnetic fields.

b Alpha and beta particles carry opposite charges. ✓ Alpha particles carry a positive charge, while beta particles carry a negative charge. ✓

c Beta particles are much lighter than alpha particles and can be turned more easily. ✓

4 C ✓ The half life of radioactive substances is only dependent on the type of matter.

5 If a compound is radioactive, at least one of the elements in the compound must be radioactive. Since QS is not radioactive, Q and S are not radioactive elements. If PR and P_2S are radioactive then P must be the radioactive element. As for R, not enough information is given to say anything about it. P_3 and P_2R are radioactive substances because of radioactive element P. While S and Q are not radioactive, we don't know

whether R_2S and QR_3 are radioactive or not as this depends on R.
a radioactive ✓; **b** radioactive ✓; **c** can't be determined ✓; **d** can't be determined ✓

6 The difference between 15 and 20 years is $\frac{m}{8} - \frac{m}{16} = 7$. So $\frac{m}{16} = 7$ kg ✓ (as decreasing by half means there is half left).
Hence $m = 7 \times 16 = 112$ kg ✓

7 **a** 5 counts per minute ✓
b It takes 10 days ✓ for the count to drop from 80 to 40. This is the half-life. ✓
c It takes 20 days, which is two half-lives. ✓

8 **a** true ✓ When the mass decreases from M to $\frac{M}{4}$, it does so in two half-lives: $M \rightarrow \frac{M}{2}$ (one half-life); $\frac{M}{2} \rightarrow \frac{M}{4}$ (one half-life). These two half-lives total 80 years, so each half-life is 40 years.
b true ✓ The rate of decay is proportional to mass; the more mass, the greater is the decay rate. As the mass after 40 years is less than the initial mass, the rate of decay is less. Also, the slope of the curve indicates the decay rate. The slope is steeper when time equals 0 than when time equals 40.
c false ✓ During the first half-life the amount of matter that decays is $M - \frac{M}{2} = \frac{M}{2}$. During the second half-life the amount of matter that decays is $\frac{M}{2} - \frac{M}{4} = \frac{M}{4}$.

9 **a** See the graph top of next column. (✓ for values on axes; ✓ for correctly labelled axes; ✓ for correctly drawn graph; ✓ for title to graph)
b around six hours ✓
c between 15% and 20% ✓
d so it does not remain in the body for long ✓

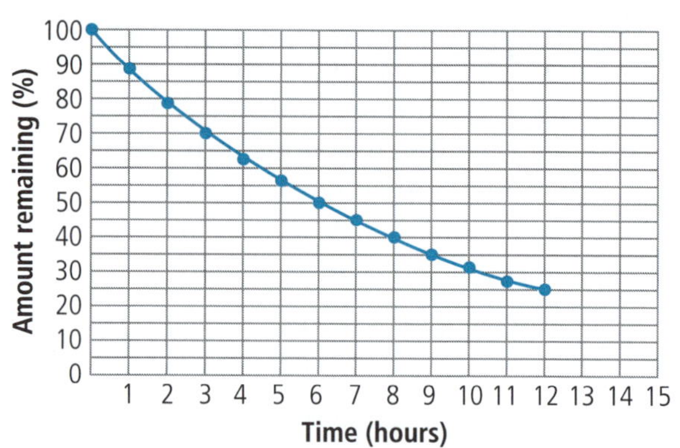

APPLICATIONS OF RADIOACTIVITY
Atomic theory
Pages 82–84

1 **a** It is an occupational health and safety issue. People working with radioactive isotopes have the potential to be exposed to the radiation. ✓ The dosimeter monitors potential exposure. The film is developed regularly to determine whether or not the person has been exposed ✓ and, if so, to what extent and to what radiation.
b Different types of radiation penetrate to different extents. ✓ This is one way to determine the kind of radiation, if any, the person wearing the badge has been exposed to. ✓ Film badges do not provide information in real time, but they can detect previous radiation exposure if the gadget was worn at the time.
c Not likely. This is because alpha particles can only travel a few centimetres through air ✓ and may be also be stopped by the light shielding ✓ around the photographic film. Dosimeters are typically used to detect X-rays, gamma rays, beta particles and neutrons.

2 **a** If they decay slowly they have longer half lives. ✓ Therefore, appreciable amounts will be around for a far longer time to date rocks. ✓

b The top fossils are between 395 and 410 million years old ✓; the bottom fossils are between 480 and 500 million years old ✓.

3 **a** Sterilisation refers to any process that eliminates (removes) or kills all forms of microorganisms or pathogens.

b Many materials can be damaged by heat. ✓ Imagine heating a pair of surgical gloves to sterilise them. Dry heat can be used to sterilise items, but the heat takes much longer to be transferred to the organism, so both the time and the temperature frequently must be increased.

c Gamma rays are very penetrating ✓ and can reach microorganisms deep within the material. They are commonly used for sterilising disposable medical equipment.

d Pathogens can enter the food preparation process at any step along the way until it is packaged. ✓ Once the food is packaged, no microbes can enter. Irradiating earlier will kill microbes then, but does not prevent others from entering further along the process. ✓ Food irradiation won't prevent illnesses connected with poor handling after the irradiation has taken place. What happens to the food after irradiation still puts people at risk, which is why food must still be stored and handled properly.

4 Levels of carbon-14 become difficult to measure after about 50 000 years. This is around eight to nine half-lives, and where only 1% of the original carbon-14 remains undecayed. ✓

5 **a** Alpha particles don't have a high penetrating power ✓ and would be stopped by just the thickness of paper.

b The isotope used must have a long half-life (at least few years). ✓ This is so the source does not need to be replaced regularly ✓ and also to maintain a constant supply of beta particles.

c to maintain a constant thickness of the paper ✓

d Decreasing radiation indicates that the paper is too thick ✓ (thicker paper reduces the number of beta particles getting through). The detector sends a signal to the rollers to apply more pressure ✓ and thin out the paper ✓.

6 **a** A radioactive tracer is added to the fluid (liquid or gas) and allowed to flow through the pipe. ✓ A detector is then moved along the ground above where the pipe is buried. ✓ Where the pipe is leaking a larger amount of liquid containing the radioactive source will pool there. ✓ The count rate will increase where the pipe is leaking.

b The radioactive source would be a gamma emitter. ✓ Gamma radiation can penetrate through the ground (and metal pipe) and be detected at the surface. ✓ Alpha and beta radiation would not penetrate through to the surface. ✓

c In order to minimise any possible danger to living things near to the pipe. ✓ A short half-life means that the quantity of radioactivity emitted from the pipe would decrease rapidly. ✓ There would be very little radioactivity in the material exiting the pipe at the other end.

7 **a** seven years ✓

b Approximately 18 years are needed to remove 50% of Cs-137. This is 11 years longer. ✓

c Around 11 years. ✓ At 11 years the strontium activity is around 35% while the caesium activity is around 70%.

d The graph shows the concentrations of the Sr-90 and Cs-137 dropped quite rapidly, so the plants have been effective in removing the contaminants from the soil. ✓ For example, 50% of the Sr-90 was removed in the first seven years. Radioactive decay alone would mean that it would take almost 29 years for half to decay away. ✓ So, the majority removal would be due to the action of plants. ✓ A similar story can be given for Cs-137. Even though the half-lives for the two radioactive contaminants are similar, the graph shows that Sr-90 is removed faster than Cs-137. ✓

OBSERVING AND MODELLING CHEMICAL REACTIONS
Chemical reactions Pages 88–89

1 a 1 = elements ✓; 2 = compound ✓
b *(1 mark for showing reactants correctly, 1 mark for showing product correctly)* ✓✓

2 a synthesis ✓
b A = crucible ✓; B = tripod ✓; C = Bunsen burner ✓; D = heat mat ✓
c a new substance has formed ✓
d

reactants

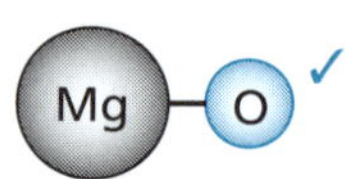

product

3 a physical ✓
b chemical ✓
c chemical ✓
d chemical ✓
e chemical ✓
f chemical ✓
g chemical ✓
h physical ✓
i physical ✓
j physical ✓

4 a C_2H_2 ✓
b A = oxygen ✓; B = carbon dioxide ✓; C = water ✓
c i The atoms of each element are not equal for the reactants and products. ✓
ii See the diagram below.

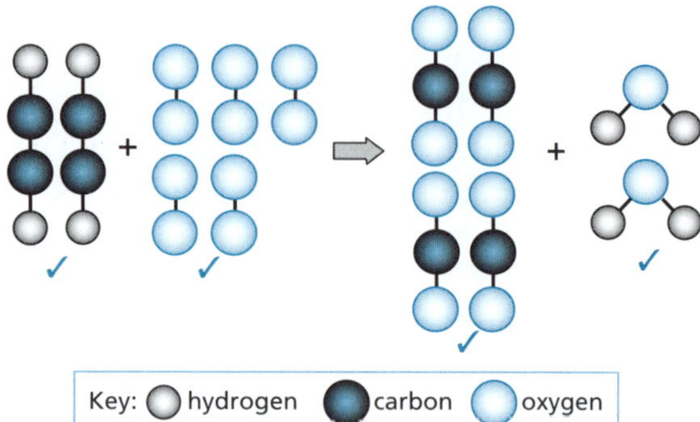

5 The final solution is a darker blue than solution 2. A colour change indicates a new substance has formed. ✓
Heat is produced as shown by the rise in temperature. ✓

CHEMICAL EQUATIONS
Chemical reactions Pages 94–95

1 a iron + oxygen → iron oxide ✓
b $6Fe + 4O_2 \rightarrow 2Fe_3O_4$ ✓✓✓

2 a true ✓; **b** false ✓; **c** true ✓; **d** true ✓;
e false ✓; **f** false ✓; **g** true ✓; **h** false ✓;
i false ✓; **j** false ✓

3 a carbon dioxide ✓; hydrogen gas ✓
carbon monoxide ✓; water ✓
b CO_2 ✓; H_2 ✓; CO ✓; H_2O ✓
c $CO_2 + H_2 \rightarrow CO + H_2O$ ✓

4 a $C_2H_4O_2$ ✓
b It more readily shows the arrangement of atoms within the molecule. ✓
c i hydrogen ✓ H_2 ✓
ii $2Na + 2CH_3COOH \rightarrow 2NaCH_3COO + H_2$ ✓✓✓✓

5 a calcium nitrate ✓
b cobalt sulfate ✓
c aluminium chloride ✓

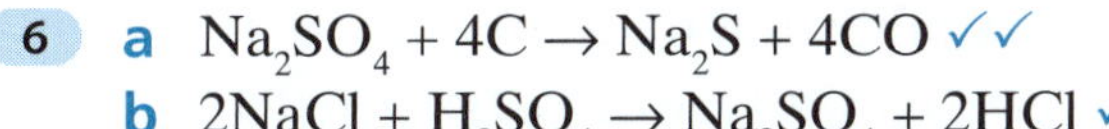

6 **a** $Na_2SO_4 + 4C \rightarrow Na_2S + 4CO$ ✓✓
b $2NaCl + H_2SO_4 \rightarrow Na_2SO_4 + 2HCl$ ✓✓

NEUTRALISATION
Chemical reactions Pages 101–103

1 neutralisation ✓

2 Jim is correct. ✓ In chemistry the term *salt* is used for an ionic compound (a substance consisting of ions) produced by reacting an acid with a base. ✓ Geoff is referring to one type of salt, table salt, which is sodium chloride. ✓

3 **a** **i** barium chloride ✓
ii aluminium nitrate ✓
iii calcium sulfate ✓
iv sodium sulfate ✓
b **i** $2HCl + Ba(OH)_2 \rightarrow BaCl_2 + 2H_2O$ ✓
ii $3HNO_3 + Al(OH)_3 \rightarrow Al(NO_3)_3 + 3H_2O$ ✓
iii $H_2SO_4 + Ca(OH)_2 \rightarrow CaSO_4 + 2H_2O$ ✓
iv $H_2SO_4 + 2NaOH \rightarrow Na_2SO_4 + 2H_2O$ ✓

4 **a** hydrochloric acid + zinc → zinc chloride + hydrogen ✓✓
b sulfuric acid + iron → iron sulfate + hydrogen ✓✓

5 **a** calcium hydroxide ✓
b basic ✓
c This is due to the insoluble suspension of calcium carbonate. ✓
d acids acting on carbonates ✓; acid + carbonate → salt + water + carbon dioxide

6 **a** Some dilute sulfuric acid and copper oxide is placed in a beaker and heated. ✓ When the reaction is complete, the solution is filtered and a blue filtrate is obtained. ✓
b Many reactions proceed faster when heated. ✓ Some reactions don't occur at all unless heated.
c The copper oxide will stop dissolving. ✓ As long as there is a reaction, some copper oxide will be observed to dissolve.
d Copper oxide is insoluble and this is the sample remaining after the reaction is complete. ✓
e copper sulfate ✓
f sulfuric acid + copper oxide → copper sulfate + water ✓; $H_2SO_4 + CuO \rightarrow CuSO_4 + H_2O$ ✓

7 **a** carbonic acid ✓ which makes seawater more acidic
b As the water becomes more acidic, the calcium carbonate mineral from which corals are made (aragonite) becomes more soluble in water. ✓
c When carbon dioxide from the atmosphere is absorbed by the ocean, it reacts with the water to produce carbonic acid. The more carbon dioxide in the atmosphere the more that can be dissolved in the oceans ✓ (Some studies suggest that the oceans have taken up around half of all carbon dioxide emitted from fossil fuel burning.)

8 **a** false ✓ Lemon juice is acidic.
b true ✓ By diluting the tomato juice you are making it less acidic.
c false ✓ They react together in a process called neutralisation.
d true ✓ Bases feel slippery, and change litmus to blue.
e true ✓
f true ✓ Acids have a pH < 7, bases have a pH > 7.
g true ✓ Not all bases are soluble, however.
h false ✓ The more acidic a substance, the lower is the pH.

9 **a** either eggs or baking soda ✓
b decrease ✓ It moves closer to pH = 7, which is neutral.
c ammonia ✓

10 **a** Both change from red to yellow as the pH increases, but methyl red does it around 1.5 units later than methyl orange. ✓
b blue ✓
c any value from around 10 to 14 ✓
d colourless ✓
e methyl orange ✓

COMBUSTION
Chemical reactions Pages 107–109

1 **a** false ✓ Fossil fuels formed from the remains of living things many millions of years ago, and include petroleum, coal and natural gas.
b true ✓
c true ✓
d true ✓
e true ✓
f false ✓ In order to melt ice, energy is absorbed.
g false ✓ Hydrogen and oxygen are produced.
h true ✓ Since its boiling point is below room temperature, LPG (liquefied petroleum gas) will evaporate rapidly at room temperatures and pressures. It is usually supplied in pressurised steel vessels.
i true ✓ But the chemistry is complicated.
j false ✓ Many fuels are hydrocarbons.
k false ✓ An endothermic reaction takes in energy.
l true ✓ Plants use the energy from the Sun to convert water and carbon dioxide to glucose and oxygen.

2 **a** Each molecule of methane consists only of carbon and hydrogen atoms (one carbon atom and four hydrogen atoms chemically bonded together). ✓
b With the hole open, air (particularly oxygen) can enter and mix with the fuel gas. ✓ This will allow complete combustion to occur as there is a plentiful air mixture with the methane. ✓
c Complete combustion occurs when there is a plentiful air supply and produces carbon dioxide and water. ✓ During incomplete combustion carbon monoxide and even carbon particles (soot) form, indicating that there was not an ample air supply to burn the gas completely. ✓ (With the air hole closed, air can only mix with the methane once it comes out the top of the nozzle. This doesn't allow for enough opportunity for thorough mixing.)
d carbon monoxide and carbon dioxide ✓
e With the hole open, complete combustion produces more energy (the flame is hotter) ✓, and the carbon dioxide produced is clean, unlike the sooty (due to carbon particles) yellow flame that can coat glassware black ✓.
f The flame is hot (heat is produced) ✓ and you can see it (although more visible when in the 'standby' mode during incomplete combustion) ✓.

3 **a** During complete combustion carbon dioxide forms, which is a carbon atom attached to the greatest possible number of oxygen atoms. ✓ During incomplete combustion carbon monoxide and soot form. ✓ The greatest amount of energy is released when the greatest number of oxygen atoms react. ✓
b Ventilation will ensure an abundant air supply which is necessary for complete combustion, and so ensure that only harmless water and carbon dioxide are produced. ✓ A poor air supply will result in incomplete combustion and the production of carbon monoxide, a harmful gas. ✓
c Faulty appliances can result in incomplete combustion. ✓ This results

in the production of carbon monoxide which is colourless and odourless, and even at low concentrations in the air can be fatal. ✓ It is important gas appliances are checked regularly.

4 *(any three)* the low pH causes plant damage, particularly to trees; acid rain harms wildlife and kills certain life forms; acid rain damages energy and material cycles, as well as food chains; acid rain increases the corrosion rates of building stone (particularly limestone and marble) ✓✓✓

5 **a** crude oil ✓

b mostly converted to petroleum products, in the form of petrol and diesel ✓

c true ✓

d *(any one)* hydro-electricity; bagasse (sugar cane waste); wood; solar; wind energy ✓

6 **a** Fuels produced from renewable biological sources ✓, such as plant biomass, vegetable oils and treated municipal and industrial wastes.

b Current fossil fuels have a finite supply so it is only a matter of time before they run out. ✓ Alternative fuels need to be found. Also traditional fuel sources are becoming more expensive ✓ as demand is outstripping supply.

c These are energy-rich sources which would otherwise go to waste ✓, and they are also renewable ✓.

d Laboratory experiments are done on a small scale in order to develop and streamline the process ✓ with very little thought to costs or efficiency. However, for this to succeed in any large-scale industrial process the procedure needs to be made more efficient and cost-effective. ✓ It is the job of the chemical engineer to overcome these hurdles.

e A enzymes; B sugars; C yeast; D lipids (fats); E hydrogen ✓✓

7 Biofuels are considered neutral with respect to the emission of carbon dioxide because the carbon dioxide given off by burning them is balanced by the carbon dioxide absorbed by the plants that are grown to produce them. ✓ This reduces the carbon footprint of the airline as it is not adding more carbon to the atmosphere. Also, biofuels are renewable ✓ unlike fossil fuels which were laid down millions of years ago.

8 Cleaner burning refers to carbon dioxide and water being the main products. ✓ With fossil fuels there is some incomplete combustion resulting in emission of carbon monoxide and particulates. ✓

9 **a** $Fe_2O_3 + 2Al \rightarrow Al_2O_3 + 2Fe$ ✓

b The reaction produces extreme heat ✓ and bright light ✓.

c This reaction supplies its own oxygen, so doesn't require air for the reaction to occur. ✓ Water will boil before reaching the reaction ✓ because of the tremendous heat generated. (Enough water will remove heat and may stop the reaction. Even so, thermite is used for welding underwater.)

10 **a** from its formula C_3H_8—it consists of only hydrogen and carbon ✓

b carbon ✓ and carbon monoxide ✓

c **i** one propane molecule reacts with five oxygen molecules ✓ (from the balanced equation)

ii two propane molecules react with seven oxygen molecules, so one propane molecule reacts with 3.5 oxygen molecules ✓

d More oxygen molecules are used in the complete combustion of propane ✓ than in its incomplete combustion.

REVISION TESTS
Answers

CHECK YOUR ANSWERS

Strand: Earth and space sciences

TECTONIC PLATES
Plate tectonics Pages 115–116

1 a transform fault zone ✓
b North American Plate ✓ and Pacific Plate ✓

2 a false ✓; b false ✓; c true ✓; d true ✓; e false ✓

3 This similarity in fossils is consistent with the theory of continental drift. Originally these continents were part of Gondwana and when it broke up the west coast of Africa separated from the east coast of South America. Plants and animals living near this boundary would have been similar. ✓

4 This region is a subduction zone where an oceanic plate dives under a continental plate, resulting in mountain building. The sedimentary layers (together with their fossils) that were once at sea level are pushed higher during this mountain building. ✓

5 a constructive plate boundary ✓
b seafloor spreading forming a mid-ocean ridge ✓
c A is younger than B ✓ as it is closer to the ridge where lava emerges to form new oceanic crust ✓.

6 a A = oceanic crust ✓; B = subduction zone ✓; C = volcano ✓; D = continental plate ✓; E = frictional heating ✓
b east of the New Zealand's North Island ✓; west coast of South America ✓
c convergent ✓

7 lithosphere = stone ball (the outer rocky layer of the Earth) ✓
asthenosphere = weak ball (the soft plastic layer of the mantle which is 'weaker' in structure than the rest of the mantle) ✓

VOLCANOES
Plate tectonics Pages 122–123

1 a X = dyke ✓; Y = sill ✓
b Igneous rocks are composed of interlocking crystals that have formed as the melted rock solidifies. ✓ These rocks are usually quite hard and resistant to erosion compared with sedimentary rock. ✓

2 a i W ✓; ii Y ✓
b The tuff is the oldest of the sedimentary rocks shown. The fact that the magma has intruded across all the horizontal layers of existing sedimentary rocks, including the most recent, suggests that the intrusion happened after all these layers were laid down. ✓

3 a true ✓; b true ✓; c false ✓; d false ✓; e false ✓

4 A/H ✓; B/F ✓; C/J ✓; D/G ✓; E/I ✓

5 An active hot spot exists under the oceanic crust and magma rises to the surface through fissures in the crust. ✓ The lava that emerges forms a volcano on the sea floor, the top of which eventually rises above sea level. ✓ As the tectonic plates move, new volcanoes form. Older volcanoes become extinct and erode to form the sea mounts. ✓

6 a plug ✓
b When volcanoes become extinct there is solidified lava still present in the vent. ✓ As the sides of the volcano erode away the plug (which is usually harder) is exposed and forms a tall column. ✓

7 The water is acidic as many volcanic gases (e.g. sulfur dioxide) can dissolve in the lake water to form acidic solutions. Alternatively the acidic gases can dissolve in rain to produce acid rain that can collect in the lake. ✓ The death of the fish and invertebrates can be caused by thermal

pollution. The hot ashes will warm the lake water and reduce the concentration of dissolved oxygen that these animals require for respiration. ✓
The aquatic plants can die for several reasons. The ashes can severely reduce sunlight and the thermal pollution of the lake can reduce carbon dioxide concentrations. Plants need both sunlight and carbon dioxide to photosynthesise. ✓

EARTHQUAKES
Plate tectonics Pages 129–131

1 **a** false ✓ Tension pulls rocks apart, compression causes rocks to fold.
b false ✓ Faulting is when forces in the Earth's crust cause large bodies of rock to break.
c true ✓
d false ✓ Compression causes reverse faults; tension causes normal faults.
e true ✓
f false ✓ This is the focus.
g true ✓
h false ✓ They travel the slowest.
i true ✓
j true ✓

2 A reverse fault ✓; B normal fault ✓; C strike-slip or transform fault ✓

3 Earthquakes can occur in all parts of the world, but they are not randomly distributed around the Earth. ✓ Most earthquakes occur in well-defined belts that match up with active plate tectonic zones. ✓ Some 80% of all seismic energy is generated from a belt bordering the Pacific Ocean. A large number of volcanoes are found in this Pacific belt (also called the Ring of Fire), and volcanoes set off many earthquakes.

4 **a** They occur in remote areas ✓ where there are few people, and/or have very small magnitudes ✓ which many people don't notice.
b Most earthquakes are small tremors. As the Richter magnitude increases, the number of earthquakes occurring each year decreases. ✓ Also it appears that for each increase in magnitude category there is a ten-fold decrease in the number of earthquakes. ✓
c The strong earthquakes, say magnitude ≥ 6.0, are already being detected, so more seismographs would have least effect on these. ✓ However, they would pick up the slight tremors that occur in remote locations that often go undetected. (Improved detection can give the wrong impression that earthquakes are on the increase.)

5 Between 5.8 and 6.1 there is a difference of 0.3 units. Ground motion increases 2.0 times. ✓ Energy released increases 3.0 times. ✓

6 A seismograph, or seismometer, detects and records earthquakes. Generally, it consists of a free-swinging mass attached to a fixed base. In an earthquake, the base, being firmly bolted into the ground, moves. ✓ The heavy mass remains stationary because of its inertia. ✓ The mass has a writing implement attached that marks the shaking paper under it, which is moving through the seismograph at constant speed. ✓ (Sometimes the motion of the base with regard to the mass is transformed into an electrical voltage. The electrical voltage is recorded on paper, magnetic tape or another recording medium.) This record can be used to determine how far away and how intense the earthquake is. ✓

7 **a** A = P waves ✓; B = S waves ✓; C = L waves ✓
b 5 min ✓
c around 4100 km ✓ (This distance is obtained from the graph by finding the distance where the time between the two curves is 5 minutes.)

d i The distance the earthquake is located from that city. ✓

ii From the seismograms they've recorded, they measure the time difference between the arrival of the first P waves and the arrival of the first S waves. ✓ Then, using this time difference ✓, they use a chart or calculation to work out how far away from that city the earthquake is located ✓.

iii While the seismograph station can determine how far away the earthquake occurred ✓, they can't separately determine direction ✓. With two seismic stations, the two circles intersect twice, but it takes three stations where the circles are concurrent (intersecting at the same point) for seismologists to be certain where the epicentre is. ✓ This is an example of triangulation.

GEOLOGICAL HISTORY OF AUSTRALIA
Plate tectonics Pages 137–138

1 **a** ✓✓✓ *(take 1 mark off for each incorrect line)*

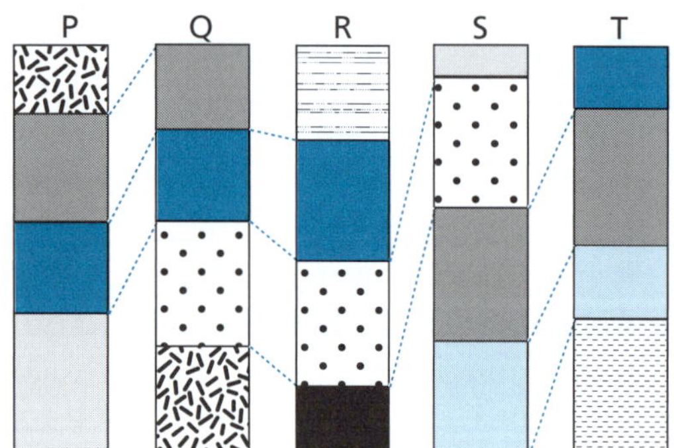

b i column T ✓ The oldest are at the bottom.

ii column P ✓ The youngest are at the top.

2 B; E; J; C; A; G; D; I; F; H ✓✓✓ *(take 1 mark off for each incorrect)*

3 D, A, E, B, C ✓✓ *(take 1 mark off for each incorrect)*

4 **a** I ✓

b A ✓

c H ✓

d J ✓ (Notice the movement on either side of it.)

e the fault ✓ (The dyke came later as it cuts through the fault.)

f **i** before ✓; **ii** before ✓; **iii** before ✓

g H ✓ The dyke is an igneous (molten rock) intrusion. Fossils can't form in igneous rock as the extreme temperatures of the magma would just destroy them.

h erosion ✓ (to get the landscape flat before more deposition)

i a river or stream ✓

5 **a** Animals and plants have been evolving in isolation ✓ from other organisms, especially large predators that existed on other continents, for approximately 45 million years. The evolution of life was also influenced to a large extent by the increasingly arid climate and the poor soils. Such environments lead to different biological solutions for survival. ✓ When other continents came together, their flora and fauna were challenged with substantial invasions. For example, when North and South America joined there was an exchange of animals. Many species had to quickly adapt to cope with the new competitors or predators. Those which couldn't became extinct. In Australia, the driving force to adapt was mostly climate change as Australia drifted north from the Antarctic Circle towards the Equator. As the climate became drier over much of the continent, organisms had to cope with a more arid environment.

b Not being at the margins of a tectonic plate, Australia has no active volcanoes and few earthquakes. ✓✓

6 **a** Radioactive decay is a natural process that occurs when an atomic nucleus becomes unstable and releases radioactive particles. In a rock a radioactive 'parent' element decays at a constant rate into a stable 'daughter' element. ✓ Scientists know how fast different isotopes decay and so the relative concentrations of these isotopes within a rock or mineral are used to measure the age. ✓ For an element to be practical for geochronology (measuring geological time), the isotope must be reasonably plentiful and produce daughter isotopes at a reasonable rate.

b The rock must contain carbon, and most rocks don't. ✓ In general, radiocarbon dating is only suitable for organic materials less than 50 000 years old because beyond that point the amount of carbon-14 becomes too little to be reliably measured. ✓

7 In order to determine the age of sedimentary rock layers, researchers locate neighbouring layers that include igneous rock, such as volcanic ash. These layers are like bookends; they give a start and an end to the time period when the sedimentary rock formed. ✓ Radioactive dating determines the age of the igneous rocks on either side, and so geologists can accurately establish the age of the sedimentary deposits between them. ✓

Strand: Physical sciences

HEAT ENERGY
Energy on the move Pages 144–145

1 **a** infra-red radiation ✓

b Radiant energy is converted to heat energy. ✓

c Heat is transferred to the sore muscles from the hot water via conduction through the rubber wall of the hot water bottle and skin. ✓

2 **a** PRQ ✓

b The warmer land surface transfers heat to the air near the ground. ✓ The heated air expands and rises. The rising air starts to cool and becomes denser and starts to sink. ✓ This denser, cooler air replaces the air that has moved inland off the ocean. ✓

c As the cool breeze comes in from the sea, the warm air over the land is pushed up. This upward movement carries the kite upwards. ✓

3 See the following table.

	Conduction	Convection	Radiation
How heat travels	particles vibrate and bump into their neighbours ✓	particles move from one place to another ✓	travels as invisible rays ✓
Can heat travel through empty space?	no ✓	no ✓	yes ✓

4 **a** B, C, D, A ✓

b A ✓

c The heated atoms of copper vibrate and collide with neighbouring atoms and pass on this energy. ✓ Copper also has mobile electrons. As they are heated they gain kinetic energy and the electrons conduct the heat more rapidly than the vibrating atoms. ✓

5 White, shiny clothes are selected for summer sports to help to reduce absorption of radiant energy from the Sun. ✓ Dull, dark clothes should be worn in winter, to increase the absorption of radiant energy. ✓

6 Not only is glass a poor conductor of heat, but the vacuum between the glass walls reduces heat losses or gains by conduction or convection because there is no matter to transfer the heat energy. ✓ The silvered surfaces reduce heat transfer by radiant energy. ✓

7 **a** meniscus ✓; **b** 43.4 °C ✓

SOUND ENERGY
Energy on the move Pages 152–154

1 **a** The explosive charge produces a flash of light. Light rays travel extremely fast (300 000 km/s) and therefore the observer can start the watch at the same time as the bell is rung. ✓

b compression wave ✓

c Greater, because the water particles are closer together than the particles in air, where it takes a longer time for compressions to form and transfer energy. ✓

d $v = \frac{d}{t} = 13\,500 \text{ m} \div 9 \text{ s}$ ✓ $= 1500 \text{ m/s}$ ✓

2 **a** See the following table. ✓

Singing voice	Frequency range (Hz)
alto	80–400
tenor	170–500
alto	200–600
soprano	250–1350

b Bass, as thick vocal cords produce lower pitched sounds. ✓

3 **a** false ✓; **b** true ✓; **c** true ✓; **d** false ✓; **e** true ✓

4 $v = f\lambda \therefore \lambda = \frac{v}{f}$ ✓ $= \frac{350}{256} = 1.37 \text{ m}$ ✓

5 **a** Sound energy is transmitted by the motion of air particles. With fewer air particles there is less transmission of sound energy leading to reduced sound levels. ✓

b Mechanical waves require a medium in which to travel. The absence of air prevents the sound from being propagated as sound is a mechanical wave. ✓

6 The shortest time will occur when the distance travelled is the shortest (i.e. straight down and up).
distance = 8000 m ✓
time = 1.14 s

$$\text{speed} = \frac{\text{distance}}{\text{time}} = \frac{8000}{1.14} = 7018 \text{ m/s}$$ ✓

Therefore, the rock type is most likely to be gabbro. ✓

7 **a** Refer to the following line graph. *(one mark for correct plotting of data points and one mark for line of best fit)*

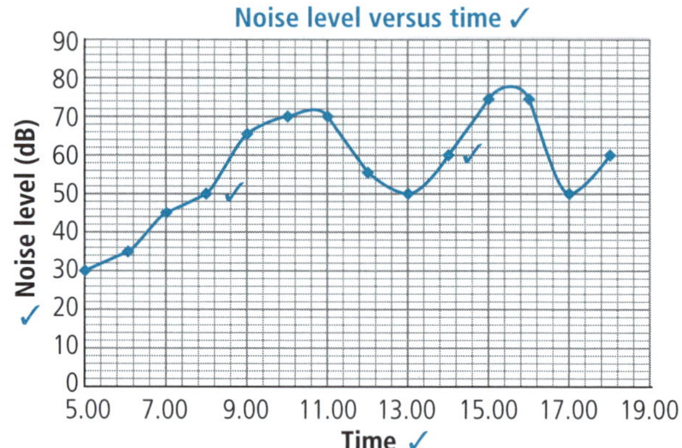

b **i** 5.00 ✓; **ii** 15.00 – 16.00 ✓; **iii** increases in road traffic ✓

c No, they are below 90 dB. ✓

LIGHT ENERGY AND COMMUNICATIONS
Energy on the move Pages 161–163

1 gamma rays; X-ray, ultraviolet; visible light; infrared; microwaves; radio waves ✓✓ *(one mark off for each out of sequence)*

2 **a** false ✓

b true ✓ That is where the word *electromagnetic* comes from: electric field + magnetic field.

c true ✓ The more energy, the shorter the wavelength.

d true ✓

e false ✓ We can perceive only a small part of the electromagnetic spectrum with our eyes.
f false ✓ All electromagnetic radiation travels at the speed of light (3×10^8 m/s) in a vacuum.
g true ✓
h true ✓ Too much exposure to sunlight can burn your skin. It is ultraviolet (UV) rays that can cause skin damage and cancers.
i true ✓
j false ✓ Light rays are bent (refracted) as they pass between materials of different densities unless, of course, they are incident at right angles to the surface.

3 AM radio signals are transmitted by changing (modulating) the amplitude of the radio waves. ✓ The transmitted frequency remains unaffected. ✓ FM radio signals are transmitted by changing (modulating) the frequency of the radio waves. ✓ The transmitted amplitude remains unaffected. ✓

4 a towards the normal ✓ As waves move from a medium that is less dense to one that is more dense, they are slowed down. This causes them to bend. ✓
b White light is composed of all the colours of the rainbow. ✓ Refracting them like this separates out the different frequencies (colours), just as chromatography can be used to separate out the different colours present in felt tip pen ink.
c The shorter the wavelength (higher frequency) the more the light is bent. ✓ The order of colours in the rainbow is the same as the order in the electromagnetic spectrum, so violet light bends most and red light bends least. ✓ (When white light passes through glass, the two sides are usually parallel to each other. So while each colour bends slightly differently as it enters the glass, the same is true when it exits. This brings the colours back into line again and so only white light is seen. In a prism the sides are not parallel and the bending is accentuated.)
d As light enters water droplets in the sky or in a mist at an angle, the different colours in white light slightly separate due to their different wavelengths. ✓ These reflect off the back of the droplet before exiting. ✓ As they exit they are further refracted. ✓ Wavelengths of light associated with a specific colour arrive at your eye from the collection of droplets. ✓ The net effect of the enormous collection of droplets is that a circular arc of ROYGBIV colours is observed across the sky. ✓

5 6 ✓; 4 ✓; 12 ✓; 1 ✓; 5 ✓; 9 ✓; 2 ✓; 10 ✓; 8 ✓; 7 ✓; 3 ✓; 11 ✓ *(½ mark for each)*

6 a iris ✓; b sclera ✓; c cones ✓;
d aqueous humor ✓; e optic nerve ✓;
f cornea ✓; g lens ✓; h vitreous humor ✓;
i retina ✓

ELECTRICAL ENERGY
Energy on the move Pages 169–171

1 C ✓ Only C is connected correctly to make a complete circuit.

2 a true ✓
b false ✓ They are called conductors.
c false ✓ A switch controls the flow of current; it does not (and should not) affect the size of the current.
d false ✓ All three globes are in series.
e true ✓ The circuit will no longer be complete.
f false ✓ Adding another globe increases the resistance in the circuit, and the voltage remains the same, so the globes get dimmer as less current flows.

g true ✓ There are other pathways around to complete the circuit.

h false ✓ Change A and V around. Ammeters are connected in series; voltmeters in parallel.

i true ✓ There is greater voltage so more current flows. Of course, that is assuming the increase in voltage is not enough to blow the globe.

j true ✓

3 $R = \frac{18}{3} = 6\ \Omega$ ✓✓ (6 ohms; the symbol Ω is the Greek capital letter omega)

4 **a** A = ammeter ✓; V = voltmeter ✓

b resistor ✓

c From $R = \frac{V}{I}$, rearranging gives $I = \frac{V}{R}$.

Hence $I = \frac{12}{30} = 0.4$ A ✓✓ (amps)

d From $R = \frac{V}{I}$, rearranging gives $V = IR$.

Hence $V = 1.2 \times 5 = 6$ V ✓✓ (volts)

e Now $R = \frac{V}{I}$, so any values of voltage and current where voltage divided by current = 8. For example: $V = 8$ V, $I = 1$ A; or $V = 16$ V, $I = 2$ A; or $V = 4$ V, $I = 0.5$ A ✓✓

5 **a** At present, the voltage is $V = 2 \times 10 = 20$ V. ✓ If the resistance doubles, $V = 2 \times 20 = 40$ V. ✓ The voltage doubles. ✓

b At present, the current is $A = \frac{20}{10} = 2$ A. ✓ If the resistance doubles, $A = \frac{20}{20} = 1$ A. ✓ The current halves.

6 **a** When electrons flow through a wire, globe or another load, there is some obstruction to the current. ✓ This obstruction, or hindrance, is called electrical resistance. For the electron, the journey around a circuit is not the shortest route. Rather, it is a zig-zag trail resulting from numerous collisions with fixed atoms within the conducting substance. ✓

b The conductor, or resistor, gives out heat when the current passes through it. ✓ (After running an electrical heater, which draws a lot of power, unplug it and feel the metal prongs. Hot, aren't they?)

c **i** Different materials have different resistances. ✓ Some metals are better conductors than others. Copper is a good conductor, that's why it is used in electrical wiring. (Silver is an even better conductor. Why isn't it used in household wiring?)

ii The longer the conductor the higher is the resistance. ✓ In every length of wire there is a certain resistance. Increasing that length increases the total resistance experienced.

iii The smaller its cross-sectional area the higher its resistance. ✓ Similarly, there is less resistance for a wire with a large cross-sectional area. Think of a bottleneck on a road. As several lanes merge down to one, there is a build up of traffic as cars are impeded from travelling. This is why wires carrying heavy loads are reasonably thick, and why the wires in your computer or calculator are fairly thin.

7 **a** water pump ✓; **b** top of the fountain ✓; **c** water ✓; **d** water pressure ✓

8 gains ✓; gains ✓; positive ✓; loses ✓; loses ✓; positive ✓; should ✓; should not ✓

Skills

RESEARCH AND FIRST-HAND INVESTIGATIONS
Investigations and problem solving Pages 177–179

1 **a** Sedimentary rocks are able to be weathered more readily than igneous or metamorphic rocks. A river that is flowing through a sedimentary landscape will erode more material than in an equivalent river flowing through an igneous landscape where the rocks do not erode as readily so there will be a greater amount of suspended particles in this river. ✓

b Yes, his hypothesis can be tested. ✓ He could collect at least five water samples from different rivers that flow through shale areas and a similar number through granite areas. By comparing the amount of sediment in each he should be able to determine whether his hypothesis is supported. ✓

c **i** 1 Measure out 100 mL of the well-stirred shale suspension using a measuring cylinder. ✓

2 Weigh a filter paper and then fold it into a cone. Set up a filter funnel on a stand and place a clean beaker underneath. Place the folded filter paper in the funnel. ✓

3 Pour the shale suspension through the filter paper and collect all the sediment. ✓

4 Open the filter paper and place it on a watch glass in an oven to allow it to dry. Weigh the filter paper and sediment when dry. ✓

5 Repeat steps 1 to 4 with the second suspension and compare the masses of sediment collected. ✓

ii His results will not be reliable as he did not repeat each experiment a minimum of five times. ✓

2 **a** false ✓; **b** false ✓; **c** true ✓; **d** true ✓; **e** true ✓

3 Lawrence has a very small sample size of five people. ✓ In human populations very large sample sizes (e.g. thousands of people) would need to be studied to ensure the reliability of the conclusions. ✓

4 A/H ✓; B/J ✓; C/G ✓; D/F ✓; E/I ✓

5 **a** Solubility of potassium nitrate in water versus temperature. ✓

b independent = temperature ✓; dependent = solubility of potassium nitrate ✓

c mass of water ✓

d Method:

1 Place 100 g of water in a beaker and lower the temperature to 10 °C using an ice-water bath. ✓

2 Add weighed amounts of potassium nitrate to the water and stir to dissolve the crystals. ✓

3 Continue to add known weights of potassium nitrate until no more crystals dissolve. Record the total mass added. ✓

4 Repeats steps 1–3 with 100 g of water held at:
i 20 °C; **ii** 30 °C; **iii** 40 °C; **iv** 50 °C; **v** 60 °C; **vi** 70 °C.
Use a warm water bath to change the temperature. ✓

e As the temperature of the water increases, the solubility of the potassium nitrate increases. ✓

f No ✓ Her experiment only involves potassium nitrate salt. Other salts may not produce the same results. ✓

6 **a** The second nail is a control. ✓

b *(any three)* volume of salt water; concentration of salt water; mass of nails; voltage supplied by the dry cell; temperature ✓✓✓

c Felicity's experiment needs to be repeated at least five times to ensure reliability. ✓

7 Teflon has many uses related to its properties. These uses include:

- coatings for wires, cables and bearings as it is a good insulator ✓

- braided hoses as it is flexible even under high pressures ✓
- containers and surface as Teflon does not absorb water ✓
- food containers as it does not absorb odours or colours ✓
- coatings for pots and pans as food will not stick to it. ✓

MODELS, SIMULATIONS AND THE MEDIA
Investigations and problem solving Pages 185–187

1 **a** C_4H_{10} ✓
b chemical bonds ✓
c The model shows the three-dimensional shape of the molecule and the angles of its chemical bonds. ✓

2 **a** both models are cubic in shape ✓; both models have atoms at the corners of the cube ✓
b Iron crystals have an atom in the centre of each face of the cube whereas copper crystals do not. ✓ Copper crystals have an atom at the centre of the cube whereas iron crystals do not. ✓

3 **a** No ✓
b No ✓ Most people associate fish oil with omega-3 fatty acids and so the journalist used the simpler term. ✓
c No ✓
d A placebo is a medication in which there are no active ingredients. It is used as a control. ✓
e No ✓ The groups were too small for reliable results. When testing medications, test groups should be in the hundreds or thousands so as to take account of genetic variability in a population. ✓
f Null results rarely make it into newspapers as they are not considered interesting or newsworthy. Journalists tend to report on research that is controversial with large negative impact or research that is life-saving. ✓
g Newspaper stories that link improved learning to fish oil will boost sales as people, particularly parents, want their child's learning to improve so they do well at school. They tend to believe what they read and do not query the accuracy of the report. ✓

4 **a** past climate data ✓
b The predictions have been conservative. This means they have tended to under-estimate the rise in sea levels. ✓
c Scientists will incorporate the new data into the computer model so as to improve its predictions of future changes in sea levels. ✓

5 **a** Education experts and science experts ✓ are consulted to ensure that the content is presented at an appropriate age level. They need to ensure that a mix of visual content and fun activities is included. ✓
b **i** chemistry ✓; **ii** physics ✓; **iii** biology ✓; **iv** astronomy ✓

6 The media can have a positive and negative effect. ✓
The media can promote the benefits of vaccination by showing videos and science programs that demonstrate the health benefits of the vaccination program. They can provide historical context and information about the massive numbers of deaths in past pandemics. ✓
Social media can also have a negative effect as non-experts can pass on anecdotal stories and false data from one person to another. This is demonstrated by the need to counter the effects of anti-vaccination groups who spread false information. ✓

7 **a** A computer program provides the simulation. ✓
b It removes the danger of flying and lets trainees practise their skills safely. ✓
c This is a mechanical model rather than a computer model. ✓

REVISION TESTS

Answers

CHECK YOUR ANSWERS

QUANTITATIVE ANALYSIS OF RESULTS

Investigations and problem solving Pages 193–195

1 **a** around 20 kPa (kilopascals) ✓
b around 30 kPa ✓; interpolation ✓
c around 70 kPa ✓; extrapolation ✓
d 100 °C ✓; the trend shown so far continues ✓

2 **a** **i** time ✓ as this is the independent variable ✓
ii temperature ✓ as this is the dependent variable ✓
b See the graph below (plotting points correctly ✓; curve joining points ✓; appropriately labelling each axis ✓✓; heading ✓).

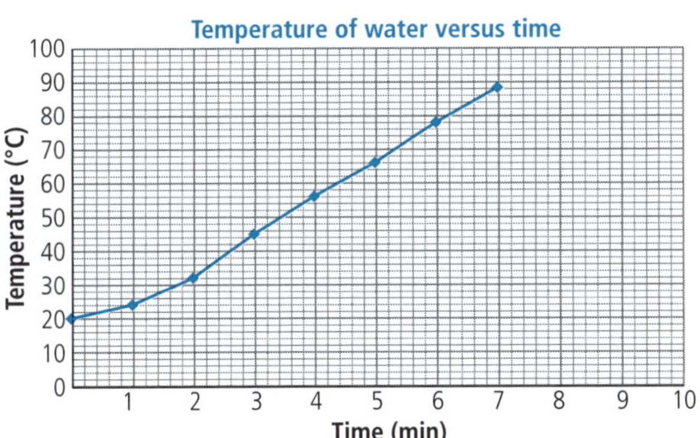

c 20 °C ✓
d 61 °C to 62 °C ✓
e 100 °C ✓; that the temperature is increasing at the same rate ✓
f No, he is assuming that the temperature keeps rising ✓ whereas, in fact, once the water reaches its boiling point it will stay there until all the water has boiled off. ✓

3 **a** 69 ✓ is very different to the other values. This is an outlier. She can remove it from the set and replace it with another measure she makes. ✓ If measurements are made correctly, outliers should not be encountered very often.
b **i** range = 33 – 27 = 6 clicks per minute ✓
ii mean = (32 + 27 + 33 + 29 + 31) ÷ 5 = 30.4 clicks per minute ✓
c Experimentation often generates multiple measurements of the same thing. These measurements are subject to error. ✓ By calculating the average, which provides an estimate of the true mean, it reduces the random error in the measurement that occurs. ✓ The reduction in the uncertainty in the estimate of the mean results from having a larger number of observations in the sample. The more measurements, the less the error. For typical school experiments, five repetitions of the measurement should suffice.

4 **a** Average = (8.8 + 3.1 + 4.2 + 6.2 + 7.6 +3.6) ÷ 6 = 5.58333. Given all calculations are only to one decimal place, his mean cannot be more accurate than that. He should report 5.6 mg/L. ✓
b **i** range = 8.8 – 1.8 = 7.0 mg/L ✓
ii mean = (8.8 + 3.1 + 4.2 + 6.2 + 7.6 + 3.6 + 5.2 + 8.6 + 6.3 + 1.8 + 6.8 + 3.9) ÷ 12 = 5.5 mg/L ✓
c There is very little difference between the two means. ✓ This provides some evidence that the original six observations are representative.

5 **a** column graph ✓
b Each state/territory is separate and the data is not continuous. ✓
c sector graph or pie chart ✓
d It shows parts of a whole. ✓ Another suitable form is a divided bar graph.
e A ✓ It is the fourth largest state.
f percentage = $\frac{7.3}{22.6} \times \frac{100}{1} = 32\%$ ✓✓
The 7.3 value is obtained from the column graph. Alternatively, you could measure the sector angle in the pie

chart. You would obtain a value about 116°. And then calculate $\frac{116}{360} \times \frac{100}{1}$ = 32%.

6 **a** Refer to the following graph. *(Take off 1 mark for each incorrect element.)*

Composition of cliche metal

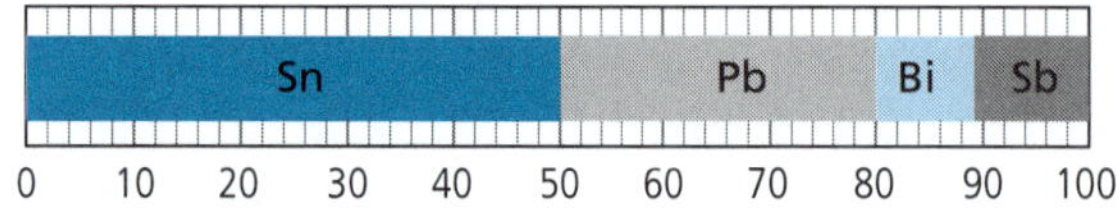

b Sb ✓

c 10.5 ÷ 100 × 60 = 6.3 kg ✓

PROBLEM SOLVING
Investigations and problem solving Pages 201–203

1 **a** Measurements are not exact. ✓ These measurements were made to the nearest millimetre. The true measure of each of these could be give or take 0.5 mm. ✓ So, in this instance, the true length lies anywhere between 296.5 mm and 297.5 mm.

b 297 + 0.5 = 297.5 mm ✓

c 210 – 0.5 = 209.5 mm ✓

d The area of a rectangular piece of paper is length × width. ✓ Using the largest values for both of these produces the maximum area of the paper. ✓

e Using the two extreme smallest values 296.5 × 209.5 = 62 116.75 mm^2. ✓ (Often the area would be simply given as 297 × 210 = 62 370 mm^2, which is near the middle of the two extreme areas and closely approximates the true area.)

2 There is no way his acceleration is accurate to 5 decimal places ✓, even though this (and more places) were provided by his calculator. Given the numbers in the question are provided to the nearest whole (integer), he should round his answer to 24 m/s. ✓

3 **a** 1800 ✓; 6300 ✓

b 9 460 000 000 000 ✓

c 300 000 000 ✓

d 0.000 02 ✓; 0.000 002 ✓

e 0.0103 ✓

f 0.000 000 000 025 ✓
(From this you should realise that writing very large or very small numbers in scientific notation is useful.)

4 **a** 2.01×10^7 ✓

b 1.496×10^8 ✓

c 1.073×10^5 ✓

d 5×10^{-2} ✓; 1×10^{-1} ✓

e 3×10^{-9} ✓

f 5.5×10^{-7} ✓

5 **a** for acetaldehyde:
MW = 2 × 12.01 + 1 × 16.00 + 4 × 1.008
= 44.052 u ✓
for ethylene oxide:
MW = 2 × 12.01 + 1 × 16.00 + 4 × 1.008
= 44.052 u ✓

b The molecular weights of both compounds are the same. ✓ These two compounds contain the same atoms. It is the arrangement of the atoms within the molecules that gives distinctive properties.

6 Number of beats in a year
= 72 × 60 × 24 × 365.25
= 37 869 120 ✓
Total number of beats in 80 years
= 37 869 120 × 80 = 3 029 529 600 ✓
= 3.03×10^9 ✓
This is around 3 billion!

7 First, notice that the Earth–Sun distance is given in kilometres, while the speed of light is in metres per second. Units need to be compatible. So 3.0×10^8 m/s = 300 000 000 m/s which is 300 000 000 ÷ 1000 = 300 000 km/s ✓
Using the formula $v = \frac{d}{t}$ and rearranging to get $t = \frac{d}{v}$ ✓ = 150 000 000 ÷ 300 000 = 500 seconds ✓
Converting this to minutes and seconds: 500 ÷ 60 = 8⅓ min = 8 min 20 s ✓

8 a There are 6 carbon atoms, 12 hydrogen atoms, and 6 oxygen molecules so $x = 6$ ✓; $y = 12$ ✓; $z = 6$ ✓.

b $\text{MW} = 6 \times 12.01 + 12 \times 1.008 + 6 \times 16.00$
$= 180.156$ u ✓✓

c Carbon constitutes $6 \times 12.01 = 72.06$ ✓ of the total molecular weight of glucose.
$\frac{72.06}{180.156} \times \frac{100}{1} = 40\%$ ✓

9 $f = \frac{v}{\lambda} = \frac{3.0 \times 10^8}{6.15 \times 10^{-7}} = 4.88 \times 10^{14}$ Hz ✓✓

10 a $r = 20\,200 + 6400$
$= 26\,600$ km ✓

b $C = 2 \times \pi \times 26\,600$
$= 167\,100$ km ✓

c speed $= \frac{\text{distance}}{\text{time}}$
$= 167\,100 \div 12$
$= 13\,900$ km/h ✓

11 a Each of the three divisions represent 1 minute ✓, so each division represents 20 seconds. ✓

b 50 seconds ✓

c speed $= \frac{50}{10} = 5$ m/s ✓✓

d between Q and R, S and T, U and V ✓✓

e at rest $= 90 + 40 + 60 = 190$ s ✓✓ (approximately) or 3 min 10 s

f distance $= 70 + 70 = 140$ m ✓✓ (The particle travelled 70 m out, then 70 m back.)

12 If the guillotine has a frequency of 25 Hz, then it can cut 25 times each second. This means the time between cuts is $\frac{1}{25}$ second. ✓ In that time you want the roll to travel 297 mm.

Using the formula speed $= \frac{\text{distance}}{\text{time}}$, then:

speed $= 297 \div \frac{1}{25}$
$= 297 \times 25$
$= 7425$ mm/s. ✓✓

This is 7.425 m/s or around 7.43 m/s. ✓

SAMPLE EXAM PAPER 1

Pages 205–210

Part A: Multiple-choice questions

1 **D** ✓ The other responses are incorrect as the lungs and diaphragm are not parts of these systems.

2 **A** ✓ B is incorrect as this is the role of connector neurons. C is wrong as this is the role of sensory neurons. D is wrong as the transmission is away from the central nervous system.

3 **B** ✓ All the remaining answers are infectious diseases.

4 **C** ✓ X rays, not gamma rays, are used to detect tumours, so A is incorrect. IR rays, not X-rays, are used to treat sore muscles and stiff joints, so B is incorrect. X-rays, not IR rays, can be used to kill and treat certain types of cancers, so D is incorrect.

5 **D** ✓ All the remainder are examples of abiotic components.

6 **A** ✓ B is incorrect because carnivores eat the carnivores that eat the herbivores. Decomposers break down dead remains, so C is incorrect. Plants and algae are producers, so D is incorrect.

7 **C** ✓ Birds migrate as well as some mammals, but amphibians do not so A is wrong. Some mammals hibernate but fish cannot, so B is wrong. D is wrong as evergreen trees do not lose their leaves.

8 **D** ✓ Thomson proposed answer A. Rutherford proposed answer B. Bohr proposed answer C.

9 **B** ✓ A is wrong as 11 electrons cannot fit in the L shell. C is wrong as aluminium only has 13 electrons. D is wrong as 3 electrons cannot fit in the K shell.

10 **D** ✓ There are 97 protons are in the nucleus so A is wrong. B is wrong as 249 is the total of protons plus neutrons. C

is wrong as protons are not found in the shells, only electrons.

11 **A** ✓ Becquerel discovered nuclear radiation and showed that uranium is active by using a photographic plate, so B and C are wrong. D is wrong as Villard discovered gamma rays.

12 **B** ✓ A is wrong as dinosaurs lived over 60 million years ago and C-14 dating cannot work on such old fossils. C is wrong as 98.9% of natural carbon is C-12 not C-14. D is wrong as the minerals in rocks usually do not contain C-14.

13 **A** ✓ B is a decomposition reaction. C is a combustion reaction. D is wrong as two new compounds form from a compound reaction with an element (oxygen).

14 **D** ✓ All the others are wrong as sulfuric acid producers form sulfate salts.

15 **B** ✓ Hydrocarbons are composed only of two elements, and the complete combustion of a hydrocarbon forms two products so A is wrong. Only two products form (a salt and water) when an oxide is neutralised so C is wrong. D is wrong as only one product will form from two elements. The reactants modelled are both compounds.

Part B: Restricted-response questions

16 10 ✓

17 safety ✓

18 nerves ✓

19 radiation ✓

20 ionises ✓

21 chemical ✓

22 precipitate ✓

23 oxygen ✓

24 extinction ✓

25 filter ✓

Part C: Written knowledge and skill questions

26 **a** combustion ✓

b exothermic ✓

c No ✓ It is not all absorbed as large amounts of heat are lost to the surroundings. ✓

d mass of water = 80 × 1 = 80 g ✓ so
$E = 4.2 \times 80 \times 34.0$
$= 11\,424$ J ✓

e Energy content = 24 000 × 1.2
= 28 800 J ✓

f No ✓ Only about 40% of the chemical energy of the almond has been transferred to the water during the combustion. Apart from heat losses to the environment the almond does not burn completely and the burnt nut retains some of its energy. ✓

27 **a** five ✓

b prawns ✓

c Humans are second order consumers as they eat prawns which are first order consumers. ✓ They are also third order consumers as they eat large fish which are second order consumers. ✓

d If the semaphore crabs die then the egrets need to eat more of their other food sources. ✓ Sea snails and small fish are eaten by egrets and so the population of the sea snails and small fish will decrease. ✓

28 **a** acetic acid ✓

b crimson ✓

c The crimson colour fades to pink and then the mixture becomes colourless. ✓

d The pH will decrease from around 10 to 7 and then when excess acid is added it will drop to below 7. ✓

e Neutralisation is an exothermic reaction and so heat is liberated. ✓

29 *(any three)* very dry and living things will dry out and plant roots find little fresh water; high winds make it difficult for plants to remain in one spot; low shelter for animals; poor or no soils and so few nutrients for plants to absorb; salt spray kills plants that are not adapted to these areas ✓✓✓

30 **a** stimulates glucose uptake into cells ✓

b stimulates heart rate and increases blood pressure ✓

SAMPLE EXAM PAPER 2

Pages 211–218

Part A: Multiple-choice questions

1 **B** ✓ A voltage is required to drive the electric current around the circuit. A switch can regulate the flow, but does not initiate it, hence A is wrong. Both C and D are incorrect as these are measuring instruments for voltage and current respectively.

2 **D** ✓ As light passes from a less dense to a more dense medium it bends towards the normal. C is incorrect as it shows no refraction, while B is incorrect as bending is away from the normal. A is wrong as the light ray is reflected from the surface and does not enter the plastic.

3 **A** ✓ As the opposite sides of the transparent plastic are parallel, then the light ray which comes out of the plastic will be parallel to the ray which entered the plastic. All other answers are incorrect.

4 **A** ✓ The angle shown in the diagram is between the mirror and the incident ray. As the normal is at right angles to the mirror's surface, the incident angle is $90° - 65° = 25°$. The incident angle is the angle between the light ray and the normal. In reflection angle of incidence = angle of reflection, so the angle of reflection is also 25°. The other answers are incorrect.

5 **C** ✓ Erosion is the process where the breakdown products by the action of water, glaciers, winds and waves at the surface of the Earth are carried away. A, B and D are examples of weathering.

6 **B** ✓ Energy provided through the student's muscles is used to generate electrical energy that makes the globe glow. The other answers are incorrect.

7 **A** ✓ Conduction is the process by which heat or electricity is directly transmitted through a substance; convection is the movement caused within a fluid (such as air in this example) where hotter (less dense) materials rise, and colder (more dense) materials fall; radiation is the energy that originates from a source and travels through some material (air) or through space as electromagnetic waves. The other answers are out of sequence and so incorrect.

8 **C** ✓ Notice that 12.0 m represents 1½ wavelengths, so $\lambda = 8.0$ m. Using the formula $v = f\lambda$, $v = 4.0 \times 8.0 = 32.0$ m/s. The other answers are incorrect calculations.

9 **C** ✓ P waves arrive when the first disturbance is noted. S waves arrive when the first sudden large deflection occurs. These are around 20 seconds apart. The other answers are incorrect.

10 **D** ✓ This is the correct sequence. All other answers are not in the correct sequence.

11 **C** ✓ Rearranging the equation $V = IR$ gives $R = V \div I$. From the graph, a voltage of 4.0 V generates a current of 2.0 A. Hence $R = 4.0 \div 2.0 = 2.0\ \Omega$ (ohms). (Taking any other voltage and current point along the line yields the same results.) The other solutions are incorrect.

12 **B** ✓ Convection currents are created by the heat rising from the core into the mantle. The convection currents are in the soft upper section of the mantle, the asthenosphere, and are kept there by the hard, rigid layer above it, the lithosphere (the outer solid part of the Earth which includes the crust and is about 100 km thick). These lithosphere plates 'float'

on the ductile asthenosphere. The plates behave as rigid bodies with some ability to flex and the built-up heat allows them to move. Other answers are incorrect.

13 **C** ✓ The vitreous humor lies between the lens and the retina. The pupil is the opening between the coloured part of the eye (iris) and appears black. The other answers are incorrect.

14 **A** ✓ Different forms of electromagnetic radiation differ in their wavelengths, frequencies and periods, but their speed in a vacuum is constant at 300 000 km/s.

15 **A** ✓ Only the two globes closest to the battery (cell) will continue to glow as there is a continuous circuit. With globe X burnt out, there is no complete circuit for the remaining three globes, so they will not glow.

Part B: Restricted-response questions

16 380 000 000 ✓

17 fault ✓

18 earthquakes ✓

19 seismograph ✓

20 rarefactions ✓

21 quantitative ✓

22 retina ✓

23 reflection ✓

24 insulator ✓

25 Richter ✓

Part C: Written knowledge and skill questions

26 $1342 \div 1225 = 1.10$ ✓

27 **a** six ✓

b Originally there were 24 grey (and undecayed) circles. After one half-life there were 12 grey circles; after a second half-life there were six; after a third half-life only three remained. ✓

Time elapsed: $3 \times 15\,000 = 45\,000$ years. ✓

28 **a** As S is recent (now) ✓ and Q is 6 million years old, then R is 3 million years old. ✓ P is about as far from Q as R is, so it must be another 3 million years old, that is 9 million years. ✓

b That the plate is moving throughout this time at a constant rate (speed). ✓

29 They are all the same age ✓ as they occur in the same stratum (layer).

30 **a** Potassium-40. The 40 represents the atomic weight of the element. ✓ Another way of writing this is ^{40}K.

b argon and calcium ✓

c With 25% K-40 left, it is 2.6 billion years old. ✓

d 10% ✓

31 Earthquake P is closer as there is less time difference between the arrival of the P wave and S wave. ✓ Both P and S waves are generated at the focus of the earthquake and travel outwards. Since P waves travel faster, the longer they travel, the greater is the time difference between the arrival of the two waves. ✓

32 **a** subduction zone ✓

b The older, heavier plate curves and plunges steeply through the athenosphere, descending into the Earth. ✓ As it does so, it forms a trench that can be several kilometres deep. Frictional heating leads to melting of the rock and magma formation. ✓ The magma rises to create a volcano. Most volcanoes, and many earthquakes, occur on land parallel to and inland from the boundary between the two plates. ✓

33 **a** Extension is on the vertical axis because it is a dependent variable. ✓ Force is the independent variable and is placed along the horizontal axis. The extension

depends on the force applied; the applied force does not depend on the extension produced.

b extrapolation ✓

c Perhaps, that is assuming the spring extends the same amount with each additional weight (force). In other words, the trend continues until at least 5.0 N. But what if the force on the spring is too large and causes it to snap? ✓

34 Layers of sedimentary rocks were laid down in the order conglomerate, shale, sandstone, siltstone, limestone and breccia. ✓ An igneous intrusion of basalt then made its way though these layers. ✓ Finally faulting occurred where the right-hand block was moved upwards relative to the left-hand block. ✓ Erosion of the surface features of breccia is currently occurring.

TEST & EXAM RESULTS

Transfer your **Percentage Score** that you calculated in the **Your Feedback** box at the end of each **test** and **exam** to the table below. This will help you work out your areas of strength and weakness.

Topic	Percentage Score
Review of major body systems	%
Nervous and endocrine systems	%
Microbes and disease	%
Human health	%
Studying an ecosystem	%
Food chains and food webs	%
Ecosystems and natural change	%
Impacts on ecosystems	%
Historical development of atomic theory	%
Atomic structure	%
Radioactivity	%
Applications of radioactivity	%
Observing and modelling chemical reactions	%
Chemical equations	%
Neutralisation	%
Combustion	%
Tectonic plates	%
Volcanoes	%
Earthquakes	%
Geological history of Australia	%
Heat energy	%
Sound energy	%
Light energy and communications	%
Electrical energy	%
Research and first-hand investigations	%
Models, simulations and the media	%
Quantitative analysis of results	%
Problem solving	%
Sample Exam Papers	
Paper 1	%
Paper 2	%

FEEDBACK CHECKLIST TO IMPROVE YOUR TEST & EXAM RESULTS

Do you want to improve your scores in the Revision Tests and the Sample Exam Papers? Check that:

You are ready to do a test or exam.

- You need to revise your class notes.
- You need to revise the relevant sections of your school textbook.
- Do further revision using the ***Excel*** *Science Study Guide* (available for Years 7, 8, 9 and 10).
- Use the Quick Revision and Revision Summaries in this book to prepare yourself for each Revision Test and Sample Exam Paper.

Your standards are realistic.

- Use these Revision Tests and Sample Exam Papers to determine your strengths and weaknesses.
- If you make mistakes then you need to learn from them. This is the ultimate key to success.
- Once you identify any weaknesses, spend more time trying to improve your understanding in these areas by doing more revision using your notes, textbook and science study guide.

You have planned your answers.

- Do not rush your responses to each question. Spend a little time thinking about the best way to construct a good reply.
- For mathematical questions, show all your working. Once you have completed your calculation, double-check to make sure that you have not made any errors.
- For written questions, present your well-planned responses in a logical order. Sub-headings may be useful for longer responses.
- Present all relevant information, even if you think some of this is common sense.

Your answer correctly responds to the verbs used in the question.

- In science, the verbs used in the question require different types of responses.
- The verb *identify* requires you to provide a name for a thing or process.
- The verb *explain* requires you to relate cause and effect and show the relationships between things.
- The verb *describe* requires you to provide characteristics and features of an object or living thing.
- The verb *discuss* requires you to identify issues and provide points for and/or against.
- The verb *predict* requires you to suggest what may happen based on available information.
- The verb *account* requires you to state reasons for an event or process.
- The verb *evaluate*, in a mathematical sense, means 'calculate or find the value of'.

INDEX

INDEX

INDEX

INDEX

INDEX